Color Constancy

Wiley-IS&T Series in Imaging Science and Technology

Series Editor:
Michael A. Kriss

Consultant Editors:
Anthony C. Lowe
Lindsay W. MacDonald
Yoichi Miyake

Reproduction of Colour (6th Edition)
R. W. G. Hunt

Colour Appearance Models (2nd Edition)
Mark D. Fairchild

Colorimetry: Fundamentals and Applications
Noburu Ohta and Alan R. Robertson

Color Constancy
Marc Ebner

Published in Association with the Society for Imaging Science and Technology

Color Constancy

Marc Ebner
Julius-Maximilians-Universität
Würzburg, Germany

Copyright © 2007   John Wiley & Sons Ltd, The Atrium, Southern Gate, Chichester,
West Sussex PO19 8SQ, England

Telephone (+44) 1243 779777

Email (for orders and customer service enquiries): cs-books@wiley.co.uk
Visit our Home Page on www.wileyeurope.com or www.wiley.com

All Rights Reserved. No part of this publication may be reproduced, stored in a retrieval system or transmitted in any form or by any means, electronic, mechanical, photocopying, recording, scanning or otherwise, except under the terms of the Copyright, Designs and Patents Act 1988 or under the terms of a licence issued by the Copyright Licensing Agency Ltd, 90 Tottenham Court Road, London W1T 4LP, UK, without the permission in writing of the Publisher. Requests to the Publisher should be addressed to the Permissions Department, John Wiley & Sons Ltd, The Atrium, Southern Gate, Chichester, West Sussex PO19 8SQ, England, or emailed to permreq@wiley.co.uk, or faxed to (+44) 1243 770620.

Designations used by companies to distinguish their products are often claimed as trademarks. All brand names and product names used in this book are trade names, service marks, trademarks or registered trademarks of their respective owners. The Publisher is not associated with any product or vendor mentioned in this book.

This publication is designed to provide accurate and authoritative information in regard to the subject matter covered. It is sold on the understanding that the Publisher is not engaged in rendering professional services. If professional advice or other expert assistance is required, the services of a competent professional should be sought.

Other Wiley Editorial Offices

John Wiley & Sons Inc., 111 River Street, Hoboken, NJ 07030, USA

Jossey-Bass, 989 Market Street, San Francisco, CA 94103-1741, USA

Wiley-VCH Verlag GmbH, Boschstr. 12, D-69469 Weinheim, Germany

John Wiley & Sons Australia Ltd, 42 McDougall Street, Milton, Queensland 4064, Australia

John Wiley & Sons (Asia) Pte Ltd, 2 Clementi Loop #02-01, Jin Xing Distripark, Singapore 129809

John Wiley & Sons Canada Ltd, 6045 Freemont Blvd, Mississauga, Ontario, L5R 4J3, Canada

Wiley also publishes its books in a variety of electronic formats. Some content that appears in print may not be available in electronic books.

British Library Cataloguing in Publication Data

A catalogue record for this book is available from the British Library

ISBN   978-0-470-05829-9 (HB)

Typeset in 10/12pt Times by Laserwords Private Limited, Chennai, India
Printed and bound in Singapore by Markono Print Media Pte Ltd
This book is printed on acid-free paper responsibly manufactured from sustainable forestry in which at least two trees are planted for each one used for paper production.

Dedicated to my wife Monika and our daughter Sophie.

Contents

Series Preface   xi

Preface   xiii

1 Introduction   1
    1.1 What is Color Constancy? 1
    1.2 Classic Experiments 3
    1.3 Overview 7

2 The Visual System   9
    2.1 Eye and Retina 9
    2.2 Visual Cortex 16
    2.3 On the Function of the Color Opponent Cells 30
    2.4 Lightness 31
    2.5 Color Perception Correlates with Integrated Reflectances 32
    2.6 Involvement of the Visual Cortex in Color Constancy 35

3 Theory of Color Image Formation   39
    3.1 Analog Photography 41
    3.2 Digital Photography 46
    3.3 Theory of Radiometry 47
    3.4 Reflectance Models 52
    3.5 Illuminants 56
    3.6 Sensor Response 60
    3.7 Finite Set of Basis Functions 63

4 Color Reproduction   67
    4.1 Additive and Subtractive Color Generation 68
    4.2 Color Gamut 69
    4.3 Computing Primary Intensities 69
    4.4 CIE XYZ Color Space 70
    4.5 Gamma Correction 79
    4.6 Von Kries Coefficients and Sensor Sharpening 83

5 Color Spaces — 87
- 5.1 RGB Color Space 87
- 5.2 sRGB 87
- 5.3 CIE $L^*u^*v^*$ Color Space 89
- 5.4 CIE $L^*a^*b^*$ Color Space 92
- 5.5 CMY Color Space 93
- 5.6 HSI Color Space 93
- 5.7 HSV Color Space 96
- 5.8 Analog and Digital Video Color Spaces 99

6 Algorithms for Color Constancy under Uniform Illumination — 103
- 6.1 White Patch Retinex 104
- 6.2 The Gray World Assumption 106
- 6.3 Variant of Horn's Algorithm 113
- 6.4 Gamut-constraint Methods 115
- 6.5 Color in Perspective 121
- 6.6 Color Cluster Rotation 128
- 6.7 Comprehensive Color Normalization 129
- 6.8 Color Constancy Using a Dichromatic Reflection Model 134

7 Algorithms for Color Constancy under Nonuniform Illumination — 143
- 7.1 The Retinex Theory of Color Vision 143
- 7.2 Computation of Lightness and Color 154
- 7.3 Hardware Implementation of Land's Retinex Theory 166
- 7.4 Color Correction on Multiple Scales 169
- 7.5 Homomorphic Filtering 170
- 7.6 Intrinsic Images 175
- 7.7 Reflectance Images from Image Sequences 188
- 7.8 Additional Algorithms 190

8 Learning Color Constancy — 193
- 8.1 Learning a Linear Filter 193
- 8.2 Learning Color Constancy Using Neural Networks 194
- 8.3 Evolving Color Constancy 198
- 8.4 Analysis of Chromatic Signals 204
- 8.5 Neural Architecture based on Double Opponent Cells 205
- 8.6 Neural Architecture Using Energy Minimization 209

9 Shadow Removal and Brightening — 213
- 9.1 Shadow Removal Using Intrinsic Images 213
- 9.2 Shadow Brightening 215

10 Estimating the Illuminant Locally — 219
- 10.1 Local Space Average Color 219
- 10.2 Computing Local Space Average Color on a Grid of Processing Elements . 221

	10.3 Implementation Using a Resistive Grid	230
	10.4 Experimental Results	237

11 Using Local Space Average Color for Color Constancy   239
    11.1 Scaling Input Values 239
    11.2 Color Shifts 241
    11.3 Normalized Color Shifts 246
    11.4 Adjusting Saturation 249
    11.5 Combining White Patch Retinex and the Gray World Assumption 251

12 Computing Anisotropic Local Space Average Color   255
    12.1 Nonlinear Change of the Illuminant 255
    12.2 The Line of Constant Illumination 257
    12.3 Interpolation Methods 259
    12.4 Evaluation of Interpolation Methods 262
    12.5 Curved Line of Constant Illumination 265
    12.6 Experimental Results 267

13 Evaluation of Algorithms   275
    13.1 Histogram-based Object Recognition 275
    13.2 Object Recognition under Changing Illumination 279
    13.3 Evaluation on Object Recognition Tasks 282
    13.4 Computation of Color Constant Descriptors 290
    13.5 Comparison to Ground Truth Data 299

14 Agreement with Data from Experimental Psychology   303
    14.1 Perceived Color of Gray Samples When Viewed under Colored Light . . . 303
    14.2 Theoretical Analysis of Color Constancy Algorithms 305
    14.3 Theoretical Analysis of Algorithms Based on Local Space Average Color . 312
    14.4 Performance of Algorithms on Simulated Stimuli 316
    14.5 Detailed Analysis of Color Shifts 319
    14.6 Theoretical Models for Color Conversion 320
    14.7 Human Color Constancy 324

15 Conclusion   327

Appendix A Dirac Delta Function   329

Appendix B Units of Radiometry and Photometry   331

Appendix C Sample Output from Algorithms   333

Appendix D Image Sets   339

Appendix E Program Code	**349**
Appendix F Parameter Settings	**363**
Bibliography	**369**
List of Symbols	**381**
Index	**385**
Permissions	**391**

Series Preface

It may not be obvious to you, but when you enter your home during mid day and go down to the lower level where there are no windows and only tungsten or fluorescent lights illuminate your path, the white walls appear white and all furniture colors look '*right*' just as they did on the upper level. This is color constancy. The human visual system has evolved in such a manner as to maintain a dynamic '*white balance.*' In the evening, while a tungsten light bulb might appear a bit yellow, the surroundings maintain their respective color balance, just as they did when daylight was pouring through the windows. This curiosity was first demonstrated by Edwin H. Land when he illuminated a '*Mondrian*' consisting of a random set of color patches with three illuminants: red, green and blue. The intensity of the three illuminants are adjusted so that the white patch appears to be neutral. Then, say, the red patch is illuminated with the three sources in such a way that the reflected light from the red patch has the same spectrum as the white patch in the initial experiment (increased blue and green intensity). One would assume that the red patch would move to a neutral color. However, to the observer, the red patch still appears red, the white patch appears neutral, the green patch appears green, and so forth. Now contrast this observation with that of a man purchasing a suit in a men's store. He picks out a suit that appears to be dark brown under the ambient tungsten illumination. Much to his surprise and consternation, in daylight the suit looks to be a very ugly shade of green. In this case the human visual system is able to detect the impact of the illuminant (the tungsten light source has very little green light). How does one reconcile these different experiences? The answer can only be found in experimental and theoretical studies of the human color visual system. The study of color constancy is an important, growing aspect of the research directed toward understanding the human color vision system, which has important practical and commercial implications. Perhaps the most common potential application is in the reproduction of conventional or digital photographic images. Conventional silver halide-based films and digital cameras using CCD- or CMOS-based imaging sensors have fixed spectral sensitivities, which means that when they are balanced for a given illuminant they produce a neutral color from a neutral object, and any change in the illuminant will skew the color reproduction in the direction of the illuminant: toward the red in tungsten lighting, toward the green in some fluorescent lighting, and toward the blue in daylight shadows. Films solved this problem by using different spectral sensitivities for daylight and tungsten illuminants, and color-correcting filters for other illuminants. Digital cameras have preset and dynamic means of estimating the illuminant, which allows for a shift in the relative gain of the red, green, and blue channels to retain a '*white balance.*' These corrections allow the '*imaging system*' to produce a print or display that '*looks*' like the original to the observer. However, assume that there are mixed illuminants such as tungsten and daylight, tungsten and fluorescent, or direct daylight and blue shadows. The human color visual system is able to adjust as it scans the scene and create a '*memory*' of the image. However, when an image is captured by film or digital cameras with fixed spectral sensitivities and only a single '*white balance*' adjustment, the resulting image will not be what the observer remembered. The study of color constancy focuses on such problems and the results of this ongoing research will result in the development of software to process an image in order to bring it back to what the observer '*saw and remembered*' or, for digital cameras, in-camera processing, which does the processing before the image is stored and subsequently printing is done as normal color image file based on the data generated by the sensors. Another application would be advanced object recognition where the capture image content is obscured by shadows and multiple illuminants.

Color Constancy by Marc Ebner is the fourth offering from the Wiley-IS&T Series in Imaging Science and Technology. Dr Ebner provides a systematic introduction to the human visual system, color image formation, color reproduction, and color spaces, which are needed to form a basis to understand the complexity of color constancy. The bulk of the text is devoted to the experimental and theoretical understanding of color constancy and the associated methods to estimate the spectrum of '*local*' illuminants in complex scenes. The results of current theories and algorithms are compared with the relevant experimental psychological data in published papers. This text is a '*must*' for all scientists and engineers working in any aspect of complex color scene recognition and reproduction.

Dr Ebner has been conducting research in computer vision since 1996, and after receiving his Doctorate in 1999, he has been teaching computer graphics, virtual reality, and evolutionary algorithms. His research focuses mainly on computer vision and evolutionary algorithms. He is the author of over 30 peer-reviewed publications and is a frequent speaker at international conferences, particularly in areas such as machine intelligence, computer vision, biologically inspired systems, and genetic programming. Over the last few years, he has had a growing interest in applying evolutionary methods in the area of computer vision. In 2000, he started working on the problem of color constancy. He realized that when developing visual algorithms it is very important that the algorithms are sufficiently robust to allow automated tasks inside as well as outside. However, he understood that the lighting conditions inside are very different from the lighting conditions outside, thus complicating the development of robust algorithms. These color vision problems have defined the research done by Dr Ebner, as he focuses on the development of parallel algorithms for color constancy, which could also be mapped to what is known about the human visual system. With his outstanding contributions to color constancy research, Dr Ebner joins an elite group of researchers, including Graham Finlayson, Mark Fairchild, and Brian Funt, in the forefront of research in this field. Dr Ebner brings his expertise, knowledge, and passion for color vision research to this fourth offering in the Wiley-IS&T Series in Imaging Science and Technology.

MICHAEL A. KRISS
Formerly of the Eastman Kodak Research Laboratories and the University of Rochester

Preface

A human observer is able to recognize the color of objects irrespective of the light used to illuminate the objects. This ability is called *color constancy*. In photography, color constancy is known under the name *white balance*. Most amateur photographers have probably experienced the following problem at one time or another when a photograph is taken. Light from a light source is reflected from the objects. The sensors measure the reflected light. The measurements depend on the type of light source used. For instance, if yellow light falls on a white wall, the sensor measures the yellow light that is reflected from the wall. Thus, the resulting colors may not be the same as the colors that were perceived by the observer. The wall will nevertheless appear to be white to a human observer. Digital cameras use postprocessing to achieve an approximately color constant or white-balanced image.

Obtaining a color constant descriptor from the image pixels is not only important for digital photography but is also very important for computer vision. Many algorithms work only under one set of lighting conditions but not under another. For instance, an algorithm may work very well under natural lighting but the same algorithm may not work as well when used under artificial illumination. Color constant descriptors are also very important for color-based object recognition. At present, it is not known how color constant descriptors are computed by the human visual system. However, a number of algorithms have been proposed to address the problem of color constancy. The book describes all of the major color constancy algorithms that are known from the literature together with recent research done by the author.

Human color perception is only approximately constant as you probably have noticed when buying clothes in a store. If you select a set of seemingly black trousers you may very well find out at home that the selected set of trousers is actually kind of bluish. Color perception is also influenced by the colors which are present in the surround of an object. In this case, color perception could have been influenced by the lack of a sufficiently complex surround. You can find out the color of a set of trousers by putting the trousers next to another set of trousers. If you place a seemingly black set of trousers next to another one you may find out that one is actually kind of dark bluish whereas the other one is indeed black. If you take the black trousers and place them next to black velvet you will think that the trousers are kind of dark grey and that the velvet is actually black. Why color perception sometimes behaves as just described will become clearer after reading this book.

This book is intended as a general introduction into the field of color constancy. It may be used as a textbook by students at the graduate level. It is assumed that you have at least some background knowledge in image processing or computer vision, and that you have read at least one introductory textbook to image processing or a general computer vision textbook. The book is intended to be read from front-to-back. Readers with advanced knowledge in the field may of course skip directly to the chapters of interest. Chapters 3, 4, 5, 6 and 7 could be used as an introductory course to computational color constancy.

The book is addressed to professionals working with images in computer science where images are processed solely in software. Students as well as professionals in electrical engineering may consult this book when implementing color constancy algorithms into scanners, digital cameras, or display devices. Researchers studying the human visual system may also find it helpful in understanding how image processing is done in a technical field. Given the algorithmic solutions that are presented in this book, it may be possible to arrive at a better understanding on how the human visual system actually processes the available data.

When reading the chapter on the human visual system, you should keep in mind that I do not have a medical or biological background. My background is in computer science. Since the field of color constancy is deeply intertwined with human color perception, I have tried my best to give you an accurate as well as an extensive overview about how the human visual system works. Thus, bear with me when I sometimes take a very algorithmic view of human image processing.

This book is based on my Habilitationsschrift, '*Color Constancy*,' which was submitted for review to the faculty of Mathematics and Computer Science at the Universität Würzburg, Germany. A number of changes have been made to the original version. In particular, I have tried to further unify the terminology and symbols. The chapter about color spaces has been added. Several figures were updated or improved. Mistakes, which I have become aware of, were corrected. All algorithms are now described in the appendix using pseudocode. This should give a better understanding of what the algorithms do and how they work. Note that accurate color reproduction could not be ensured during printing. Why color reproduction is very difficult is explained in depth in Chapter 4. Thus, the color images shown in this book should not be taken as an absolute reference. The printed images should only be taken as being indicative about the general performance of the algorithms. Hopefully, there are no remaining mistakes in the printed final version. However, should you come across a mistake or have any suggestions, I would be grateful to hear about it (m.ebner.1@alumni.nyu.edu).

I am particularly grateful to Prof. Albert for supporting my research. I am also very grateful to the two reviewers of the Habilitationsschrift, Prof. Dietrich Paulus from the Universität Koblenz-Landau, Germany, and Prof. Brian Funt from the Simon Fraser University, Canada, who have helped to a great extent by providing their detailed review of the original version. Vinh Hong from the Universität Koblenz-Landau also carefully read the earlier version. His corrections were also very helpful and thorough and I am thankful for the same. I also thank Kelly Blauvelt for proofreading parts of this book. My parents have always supported me. I thank them very much for everything. But most of all, I thank my wife Monika and my daughter Sophie for their love.

Würzburg, Germany
September, 2006

Marc Ebner

1

Introduction

The human visual system is a remarkable apparatus. It provides us with a three-dimensional perception of the world. Light is reflected from the objects around us. When the reflected light enters the eye, it is measured by the retinal cells. The information processing starts inside the eye. However, most information processing is done inside the brain. Currently, how this information is actually processed is still largely unknown. In computer science, we try to imitate many of the feats the human brain is capable of. Much work is still ahead if our algorithms are to match the capabilities of the human visual system. Among the many problems addressed by the computer vision community are structure from motion, object recognition, and visual navigation.

Color processing performs a very important role in computer vision. Many tasks become much simpler if the accurate color of objects is known. Accurate color measurement is required for color-based object recognition. Many objects can be distinguished on the basis of their color. Suppose that we have a yellow book and a red folder. We can distinguish the two easily because one is yellow and the other is red. But color can also be used in other areas of computer vision such as the computation of optical flow or depth from stereo based on color and shading. In this book, we will have an in-depth look at color perception and color processing.

1.1 What is Color Constancy?

Color is actually not an attribute that can be attached to the objects around us. It is basically a result of the processing done by the brain and the retina. The human visual system is able to determine the colors of objects irrespective of the illuminant. This ability is called *color constancy* (Zeki 1993). Mechanisms for color constancy also exist in other species, ranging from goldfish to honeybees (Tovée 1996). Color is an important biological signaling mechanism. Without color constancy, objects could no longer be reliably identified by their color. The visual system is somehow able to compute descriptors that stay constant even if the illuminant changes. Why is this remarkable? The receptors in the human eye only measure the amount of light reflected by an object. The light reflected by the object varies with the illuminant.

Let us follow the path of a ray of light to understand what actually happens. The ray of light leaves the light source and reaches the object at some particular point. The reflectance at that particular point, as a result of the material properties of the object, determines how much of the incident light is reflected. This reflected light then enters the eye, where it is measured. Three types of receptors exist inside the human retina, which measure the incident light. They absorb light with wavelengths in the long, middle, and short part of the spectrum. In order to accurately describe how much light is reflected by the object, we would have to measure the reflectance for the entire visible spectrum. The reflectance could be measured by illuminating the objects with white light or some other illuminant whose power distribution is known. However, a measuring device such as a camera does not know the power distribution of the illuminant. It does not measure the reflectance. Instead, it measures the product of the reflectance of the object and the amount of light hitting the object. Here we have two unknowns. We do not know the reflectance nor do we know the type of illuminant we have.

If we only take a measuring device with no additional processing, then the output will vary depending on the illuminant. Every amateur photographer has probably experienced this effect at one time or another. A photograph may look very different depending on the type of light source used. Sunlight or candle light produces much warmer colors compared to the colors obtained when a flash is used. It does not matter whether an analog or a digital camera is used to capture the image. Suppose that we have a yellowish light source. Light from the light source falls onto the objects of the scene. A white surface will reflect the incident light equally for all wavelengths. Thus, the white surface will appear to be yellow in the photograph. However, if we have a digitized image, we can process this image to obtain a better color reproduction. Many popular programs, such as Adobe's Photoshop, Google's Picasa, the GIMP, or Digital Arts Xe847 Photoshop plug-in, can be used to process images to obtain a color-corrected image.

In photography, this process is known as automatic white balance. Digital cameras usually have several options to handle white balance. If the type of illuminant that illuminates the scene is known, i.e. the illuminant is either a light bulb, a neon light, sun light, or a cloudy sky, then the white balance of the camera can be set to the appropriate option. The camera knows the color of these illuminants and is able to compute an image of the same scene as it would appear under a white illuminant. For some cameras, it is also possible to take an image of a white patch. The camera then uses the contents of this image to compute the color of the illuminant. Some cameras also have extra sensors to measure the ambient light in order to determine the color of the illuminant. The option automatic white balance is supposed to automatically select the best option in order to obtain an image as it would appear under a white illuminant. The ultimate goal is, of course, to just use the data that is available from the image, in order to obtain an image of a scene that looks exactly as it did to a human observer who took the photograph.

Human color perception correlates with integrated reflectance (McCann et al. 1976). Other experiments have shown that the human visual system does not actually estimate the reflectance of objects (Helson 1938). What is known about the visual system is that color processing is done in an area denoted as V4 (visual area no. 4). In V4, cells have been found that respond to different colors irrespective of the type of illuminant (Zeki

1993; Zeki and Marini 1998). Even though the visual system somehow computes a color constant descriptor, this color constant descriptor is not equal to the reflectance of objects. From a machine vision point of view, we are also interested in determining the reflectance of objects. If the reflectance was known, we could use the reflectance for color-based object recognition. The reflectance could also be used to segment scenes or for computation of optical flow based on color. In general, the reflectance information is very important for all aspects of color-based computer vision. For instance, an autonomous service robot should continue to work whether the environment of the robot is illuminated using artificial light or sunlight. If the computer vision algorithm uses the reflectance, which by definition is independent of the illuminant, then the environment of the robot (artificial versus natural lighting) would not make a difference.

Thus, there are two main roads to follow in developing color constancy algorithms. One goal is to determine the reflectance of objects. The second goal is to perform a color correction that closely mimics the performance of the visual system. The first goal is important from a machine vision point of view, whereas the second goal is very important for consumer photography. Since we are trying to determine the color of objects from just three measurements but do not know anything about the illuminant, or about object geometry, some assumptions have to be made in order to solve the problem. In the course of this book, we review many of the known algorithms that can be used to achieve better color reproduction. We also present new algorithms. All algorithms are evaluated in detail on different image sets in order to determine how accurate they are.

The problem of color constancy has fascinated scientists for a long time. Edwin H. Land, founder of the Polaroid Corporation, is one of the most famous researchers in the area of color constancy. In 1959, he performed a series of experiments with quite startling results (Land 1959a,b,c, 1964). Following this, he developed one of the first computational algorithms of color constancy (Land 1974, 1986a,b). Land had a major influence on the field of color constancy. His algorithms inspired many researchers, who then developed variants and improvements over the original algorithm.

1.2 Classic Experiments

Land (1962, 1964) realized very early on that the perceived color of an object depends on the rank order of the amount of light reflected for a given wavelength compared to the rank order of the amount of light reflected for another wavelength of the spectrum. It does not depend on the absolute values of reflected light. Land assumed that independent sets of receptors exist, i.e. one set for red, one for green, and one for blue, that operate as a unit to produce the perceived color. He named this system, which computes color constant descriptors, the "retinex." The name retinex is a mixture of the words retina and cortex because at the time, Land did not know whether the color constant descriptors are computed inside the retina or whether the visual cortex of the brain is also involved. According to the retinex theory, the visual information processing starts with the receptors of the retina. Inside the retina, three types of sensors measure the light in the red, green, and blue parts of the spectrum. This visual information is then independently processed for the three color bands.

Land and McCann (1971) then developed a computational theory for color constancy, the retinex theory. In their experiments, they used a stimulus similar to the famous paintings

Figure 1.1 A Mondrian image similar to the one used by Land to develop his retinex theory. Land used colored sheets of paper and arranged them randomly. The resulting image reminded him of the abstract paintings that were drawn by the Dutch artist Piet Mondrian. This is why this stimulus is called a *Mondrian image*.

by Dutch artist Piet Mondrian. Later, Land (1983) noted that this stimulus actually looks more like a van Doesburg painting. Land and McCann used rectangular colored papers and arranged them randomly as shown in Figure 1.1. The colored papers were matted to reduce the influence of specular reflectance. Three projectors with sharp-cut bandpass filters were used to illuminate this Mondrian-like pattern. The first filter allowed only short wavelengths, the second allowed middle-length wavelengths, and the third allowed long wavelengths. The transmittance characteristics of the filters are shown in Figure 1.2. The amount of light emitted by each projector could be varied by a separate transformer.

All projectors were turned on and the transformers were set such that the papers of the Mondrian pattern appeared deeply colored. Also, the whites of the pattern had to be "good whites" (Land and McCann 1971). Next, a telescopic photometer was used to measure the light reflected from a particular area on the pattern. Land and McCann selected a white rectangle. They measured the reflected light, the luminance (see Table B.2 for a definition), using the telescopic photometer with only one projector turned on at a time. In other words, they measured the luminance reflected by the rectangle when the first projector was turned on. Next, they measured the luminance when only the second projector was turned on, and finally they measured the luminance with only the third projector turned on. In their experiment, Land and McCann obtained 60 short-wave units, 35 middle-wave units and 6 long-wave units for the white rectangle. Next, the telescopic photometer was turned to measure the luminance reflected from another rectangle. Land and McCann chose a rectangle that looked dark brown. The transformers were set such that the luminance reflected from the dark brown rectangle was equivalent to the luminance that was earlier measured for the white rectangle. In other words, the luminances were again 60, 35, and 6 for the short, middle, and long wavelengths, respectively. Even though the measured luminance was now

INTRODUCTION

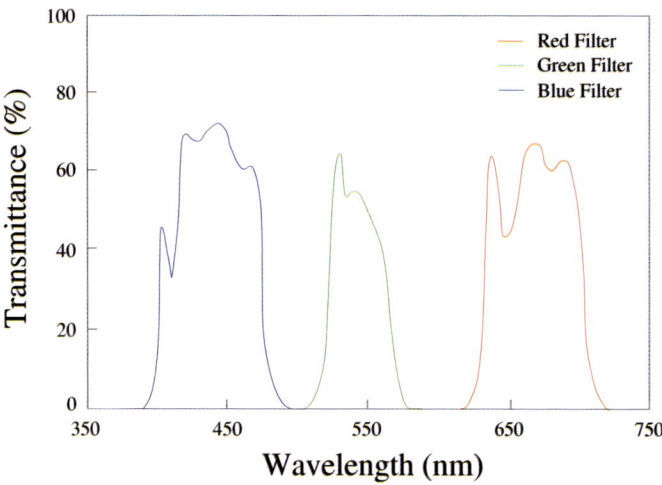

Figure 1.2 Spectral transmittances of the filters. The filters were placed in front of the projectors which were used to illuminate the Mondrian image. (Reproduced from Land EH and McCann JJ 1971 Lightness and retinex theory. Journal of the Optical Society of America 61(1), 1-11, by permission from The Optical Society of America.)

equivalent to that of the white rectangle, the rectangle was still perceived as dark brown. This experiment was repeated with different colored rectangles: bright yellow, blue, gray, lime, red, and green. For each rectangle, the transformers were set such that the measured luminances were 60, 35, and 6 for the short, middle, and long wavelengths, respectively.

In each case, the perceived color remained the same. In other words, to a human observer, the bright yellow patch remained bright yellow even though the measured luminance was equivalent to that of a white rectangle. Also, all the colors of the other rectangles of the Mondrian image also remained almost unchanged (in a few cases, there were some changes) when the transformers were adjusted to match the luminance from a specific rectangle to the luminance originally measured for the white rectangle. Land attributed the few cases where there were some changes to the shapes of the receptor responses of the retina. These receptors do not have the shape of a sharp-cut bandpass filter. Instead, they are Gaussian shaped with considerable overlap, especially between the red and the green receptors. Land later repeated the same experiment using a more standardized setting and obtained the same results (Land 1983).

With this experiment, Land and McCann vividly demonstrated that the perceived color of an object does not depend on the light reflected by the object. The perceived color depends on the reflectance, which specifies how much of the incident light is reflected. The reflectance determines the color of the object. The reflected light is essentially proportional to the product of the irradiance (see Table B.1 for a definition) and the reflectance of the object. A human observer is somehow able to derive the reflectances for the objects in view regardless of the illuminant used. In contrast, a digital or an analog camera can do no more than the telescopic photometer can. It only measures the reflected light. In order

Figure 1.3 Experiment with two color Mondrians. Each Mondrian is illuminated by a set of three projectors.

to arrive at an image that is similar to the one observed by a human, we need to do some postprocessing of the measured data.

Land (1986a) and Land and McCann (1971) also experimented with small exposure time in order to see if the perceived image somehow depends on the amount of time the image is viewed. Observers also viewed the Mondrian through photographic shutters. This allowed Land and McCann to limit the exposure time to a fraction of a second. By limiting the length of time for which the Mondrian was seen, they were able to rule out that any adaptation mechanisms or eye motion could be involved in the mechanism for color constancy. The experimental setup was exactly the same as for the first experiment. The transformers were set such that the luminance reflected from a rectangle was equivalent to the luminance reflected from a white rectangle that was previously measured. The results were equivalent to the first experiment even if observers were only able to view the Mondrian image for one-hundredth of a second.

To see if the environment has an impact on the perceived color, Land and McCann also moved the rectangular patches around on the Mondrian. They report that the color sensation does not change significantly if the rectangle is moved to a new neighborhood where it is surrounded by different colored rectangles.

Land and McCann (1971) came to the conclusion that color perception involves structures of the retina as well as the visual cortex. On the basis of these experiments, Land and McCann developed a computational theory of color constancy, the retinex theory. Land later developed additional versions of his retinex theory. The retinex theory of color vision is discussed in detail in Chapter 7.

Land (1974) also performed an experiment using two color Mondrians. Each Mondrian was illuminated by three projectors (Figure 1.3): one projector with a short-wave filter, the second with a middle-wave filter, and the third with a long-wave filter. A white patch was selected from one of the Mondrians. The three projectors that illuminated this Mondrian were turned on and the amount of light reflected for each of the three color bands was measured. Next, a colored patch was selected from each Mondrian. The patch selected from the first Mondrian was green and the patch selected from the second Mondrian was yellow. The projectors were adjusted such that the amount of light given off by the two patches was equivalent to the amount of light measured for the white patch earlier. When all six projectors were turned on, observers reported that the color of the first patch was green and the color of the second patch was yellow. The reflected light was equivalent to the amount of light reflected from the white patch for both of them.

INTRODUCTION

Figure 1.4 A long tube with internal baffles and an adjustable aperture. When observers view colored patches through such a tube, the patches look grayish-white. When the entire Mondrian is viewed, the patches appear to have a color that corresponds to the reflectance properties of the patch.

However, when both patches were viewed through viewing tubes as shown in Figure 1.4, the colored patches appeared to be grayish-white. The tubes were quite long with internal baffles and an adjustable aperture. The tubes were positioned such that the center of the green patch could be viewed with the left eye and the center of the yellow patch could be viewed with the right eye. The design of the tube ensured that only light from the center of the patches and none from the surrounding area entered the eye. When the whole Mondrian could be viewed, the patches appeared to be green and yellow. However, if the tubes were used to view the same two patches, they appeared to have the same color, i.e. grayish-white. From this, we see that the type of the environment of an object does have an influence on the perceived color of the object.

1.3 Overview

We now briefly describe how this book is organized. First, we have summarized Edwin H. Land's classic experiments on color perception. In Chapter 2, we have an in-depth look at the human visual system. Except for the wiring and response type of some cells, little is known about the actual algorithmic processing that is done by the human visual system. In Chapter 3, we have a look at the theory of color image formation. We will see how analog and digital color images are created. The course of light will be followed from the light source to the object and into the lens of the camera. At the end of the chapter, we arrive at an image that can be processed to achieve color constancy. Once an image is obtained, it also has to be reproduced, i.e. shown on a computer screen or printed on a piece of paper. Accurate color reproduction is discussed in Chapter 4. Color spaces are described in Chapter 5. Algorithms for color constancy are discussed in Chapter 6 and Chapter 7. Algorithms that assume a uniform illumination are covered in Chapter 6, whereas algorithms that do not assume a single uniform illuminant are covered in Chapter 7. Algorithms that assume a single uniform illuminant often try to estimate the illuminant. The estimate is then used to compute the reflectances for the object points. Such algorithms cannot be used for scenes with varying illumination. In practice, we usually have a scene with multiple light sources, for instance, a combination of natural and artificial light or simply several lamps placed at different locations of the room. Chapter 8 attempts to describe color constancy algorithms from a set of samples. Shadows are treated in Chapter 9, where we discuss methods for shadow removal and shadow attenuation. Chapter 10 explains how local space average color may be used to estimate the illuminant locally for each image pixel. In Chapter 11, we show how local space average color can be used to calculate a color-corrected image. Chapter 12 focuses on the computation of local space average color for an illuminant that varies nonlinearly over the image. In Chapter 13, the performance of all algorithms is evaluated on a set of images from a variety of sources. Chapter 14 compares the results obtained with the different algorithms to data obtained from experimental psychology. The most important points of this work are summarized in Chapter 15.

2

The Visual System

The human brain contains on the order of 10^{11} neurons and more than 10^{15} synapses (Tovée 1996). Different areas of the brain can be distinguished by grouping the neurons with similar response properties and similar patterns of connections. Some areas of the brain seem to be devoted to specialized tasks, e.g. to process visual motion or color. In the following text, we will have a detailed look at the visual system of primates with a focus on color perception. This review is based on the monographs by Dowling (1987), Tovée (1996), and Zeki (1993, 1999). A succinct summary and general introduction can be found in (Colourware Ltd 2001; Mallot 1998; Wyszecki and Stiles 2000). An in-depth review about color vision in particular is given by Lennie and D'Zmura (1988).

2.1 Eye and Retina

Information processing by the human visual system starts with the eye. Light falls through the lens and onto the retina, which is located at the back of the eyeball. A schematic drawing of the human eye is shown in Figure 2.1. Six extraocular muscles are attached to the eyeball. These muscles can be used to focus the eye on any desired target. Muscles are also used to alter the shape of the lens. The shape of the lens determines the focal length. By altering the shape of the lens, the image can be focused on the back of the retina. In front of the lens lies a ring of two muscles called the *iris*. The central opening of this ring is the pupil. The diameter of the pupil determines the amount of light that enters the eye. One muscle is used to enlarge the diameter of the pupil, while the other is used to reduce it.

The retina basically consists of a set of receptors that measure the incoming light. It is structured as three main layers. The receptors, as shown in Figure 2.2, are located at the bottom of the retina. Retinal ganglion cells are located at the top of the retina. Bipolar cells are located in between these two layers. The incoming light has to pass through the top two layers to reach the receptors, which are located at the back of the retina. A light-absorbing material is located below the receptor layer. This prevents any diffuse illumination of the retina caused by stray light. Two types of photoreceptors exist, namely, rods and cones.

Color Constancy M. Ebner
© 2007 John Wiley & Sons, Ltd

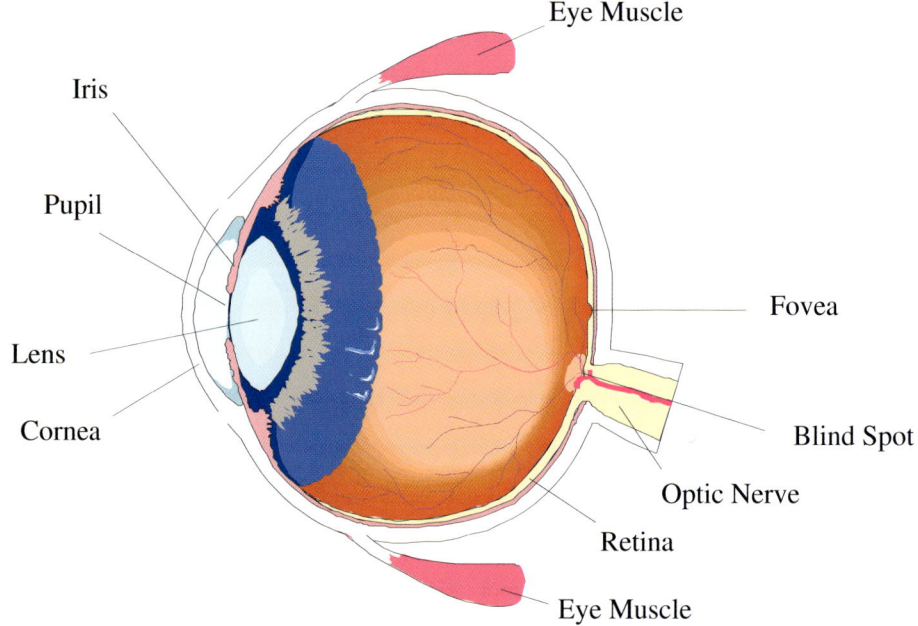

Figure 2.1 A cross section of the human eye. Illustration from LifeART Collection Images © 1989–2001 by Lippincott Williams & Wilkins used by permission from SmartDraw.com.

The cones are used for color vision in bright light conditions (*photopic vision*), whereas the rods are used when very little light is available (*scotopic vision*).

The sensitivities of rods and cones are shown in Figure 2.3. Rods are approximately 25 times more sensitive than cones (Fain and Dowling 1973). Three types of cones can be distinguished (Brown and Wald 1964; Marks et al. 1964). The first type responds to light in the red part of the visible spectrum, the second mainly responds to light in the green part, and the third mainly responds to light in the blue part. Photopigments are contained in the outer segment of the photoreceptors (Tovée 1996). The photopigment molecules are light sensitive, in that the molecule breaks into two parts when a photon of light is absorbed. The absorbance characteristics for the photopigments of the rods and cones are shown in Figure 2.4. The receptors and retinal neurons are maintained in an on-state in the dark. Light entering the retina in fact turns them off (Dowling 1987).

Not all mammals have three types of receptors (Tovée 1996). For instance, nonprimate mammals have only two pigments. Some birds, i.e. pigeons, even have five types of pigments. Prosimians, the most primitive extant primates, have only one pigment. A minimum of two pigments is required for color perception. With only a single pigment, a change of wavelength cannot be distinguished from a change of intensity. In humans, some forms of color blindness are caused by a genetic defect that results in a missing photopigment. In some humans, a red pigment could be missing. This causes the person to confuse wavelengths in the range 520–700 nm. In some, a green pigment could be missing. This causes

THE VISUAL SYSTEM

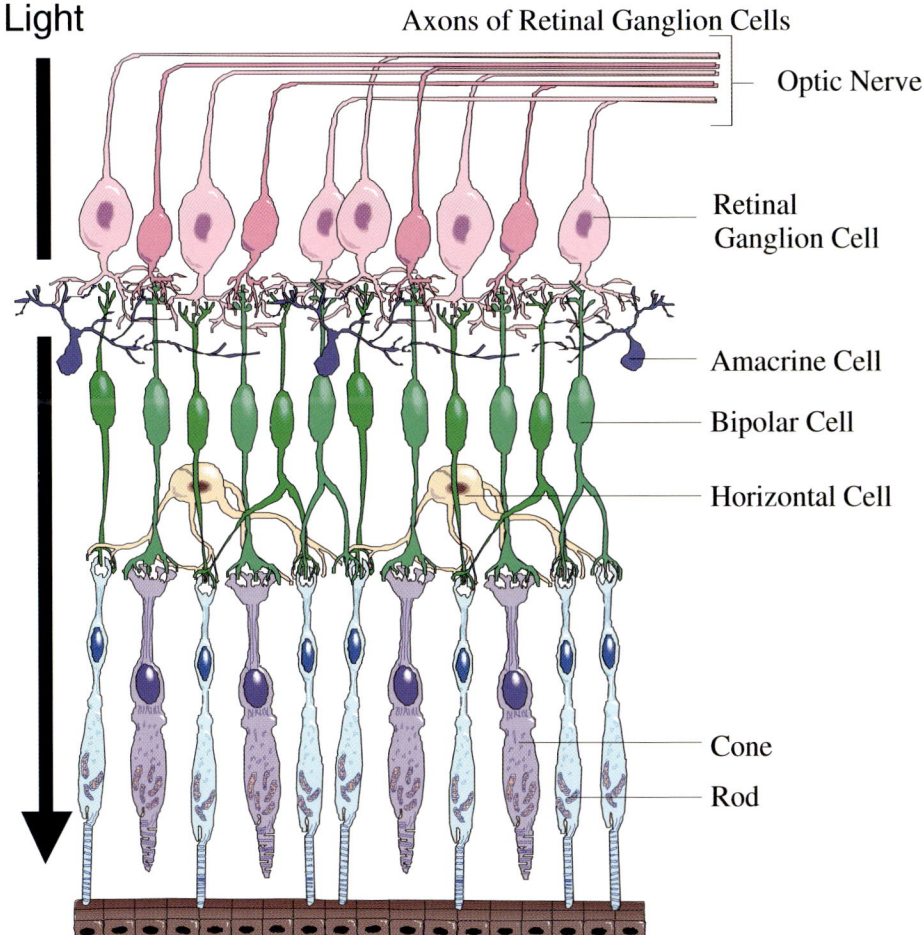

Figure 2.2 The retina consists of three layers. Retinal ganglion cells are located at the top followed by a layer of bipolar cells and receptors at the bottom. Light has to pass through the top two layers to reach the light-sensitive sensors. Information then travels upward from the receptors to the bipolar cells and on to the retinal ganglion cells. Information is also exchanged laterally through amacrine and horizontal cells. (Retina illustration from LifeART Collection Images © 1989–2001 by Lippincott Williams & Wilkins, used by permission from SmartDraw.com.

the person to confuse wavelengths in the range 530–700 nm. In some, a blue pigment could be missing, which causes them to confuse wavelengths in the range 445–480 nm. However, the last case is very rare. In some persons, the spectral absorbance characteristics are altered to some extent compared to a normal observer. The spectral peaks may be shifted by a few nanometers. The exact position is genetically determined. Color blindness may also be caused by high light intensities, which may damage the cones. Another cause

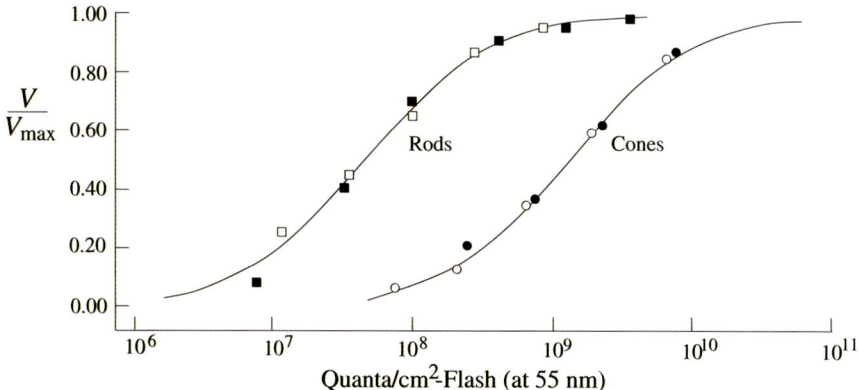

Figure 2.3 Sensitivities of rods and cones of the mudpuppy (a large aquatic salamander) retina. The graph shows the intensity-response curves, which were measured intracellularly from the retina of a dissected eye. The response is shown as a fraction of the response at saturation. A fit to the data is also shown. Rods are approximately 25 times more sensitive than cones. (Reprinted with permission from Gordon L. Fain and John E. Dowling. Intracellular recordings from single rods and cones in the mudpuppy retina. Science, Vol. 180, pp. 1178–1181, June, Copyright 1973 AAAS).

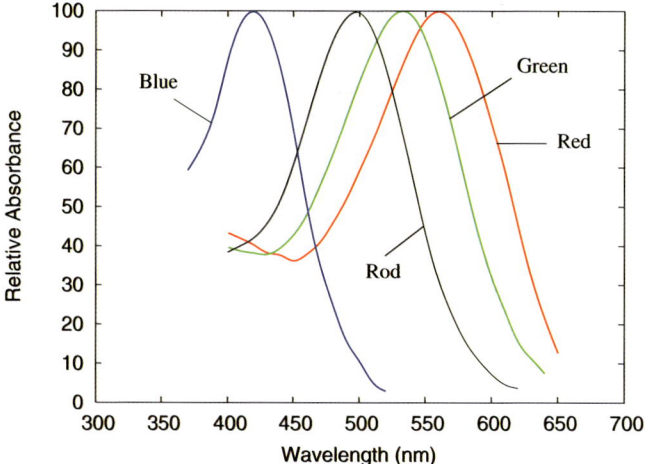

Figure 2.4 Absorbance characteristics for the photopigments found in the rods and cones. The maximum absorbance of the red, green, and blue cones is located at 559 nm, 531 nm, and 419 nm, respectively. The maximum absorbance of the rod is located at 496 nm (data from Dartnall et al. 1983).

THE VISUAL SYSTEM

of color blindness is oxygen deprivation. In particular, the blue cones are very sensitive to either high light intensities or oxygen deprivation.

A photopigment consists of a protein, called *opsin*, that is connected to a lipid called *retinal*. On absorbing a photon of light, the photopigment separates into two parts. This changes the potential of the cell membrane, which in turn signals the presence of light. The maximum response of the rod pigment is located at 496 nm. The red, green, and blue cone pigments peak at 559, 531, and 419 nm, respectively (Dartnall et al. 1983). Owing to these response characteristics, visible light covers the spectrum from 380 to 700 nm (Tovée 1996). Note that Figure 2.4 only shows relative sensitivity. Rods have a much higher sensitivity than cones. This is due to the fact that rods have a larger diameter and are longer. Also, it takes longer for the rod pigments to regenerate. Therefore, if a photon is absorbed by a rod, it is quite likely that it will absorb a second photon while still being excited. This may be enough to reach a certain level of excitation where information is transmitted on to the bipolar and ganglion cells.

The receptors measure the amount of light entering the retina. We now briefly have a look at what exactly the receptors measure. An introduction to the field of radiometry is given in Section 3.3. The total energy measured by a single receptor can be calculated by integrating over all wavelengths, λ. Let us consider the three types of cones that are sensitive to the red, green, and blue part of the spectrum. Let $S_i(\lambda)$ be the response curves of these three receptors with $i \in \{r, g, b\}$. Let $E(\lambda)$ be the irradiance, i.e. the flux per area, falling on a receptor for a given wavelength λ. Then, the energy Q_i measured by a single receptor is essentially given by

$$Q_i = \int S_i(\lambda) E(\lambda) \, d\lambda. \tag{2.1}$$

The irradiance E falling onto the receptor when viewing a matte surface can be considered to be the product between the reflectance of the object R (the percentage of incident light that is reflected by the object) and the irradiance L falling onto the object.

$$E(\lambda) = R(\lambda) L(\lambda) \tag{2.2}$$

The details (such as the dependence of the reflected light on the orientation of the surface) are deferred until Section 3.3. Therefore, the energy Q_i can be modeled by

$$Q_i = \int S_i(\lambda) R(\lambda) L(\lambda) \, d\lambda. \tag{2.3}$$

The receptors transmit the information to the bipolar cells located in the second layer. The bipolar cells in turn form synaptic connections with the retinal ganglion cells. There are on the order of 126 million photoreceptors inside the retina. Approximately 120 million receptors are rods and the remaining 6 million are cones. Most of the cones are located inside a small area called the *fovea*, which is located at the center of the retina. The distribution of rods and cones is shown in Figure 2.5. The resolution of the image at the center of the retina is much higher compared to that at the periphery. Even though there are approximately 126 million receptors inside the retina, there are only 1 million retinal ganglion cells that transport this information to the lateral geniculate nucleus (see Figure 2.9) and from there to the visual cortex. The information must be compressed by some means before it is transmitted through the axons of the ganglion cells. Image

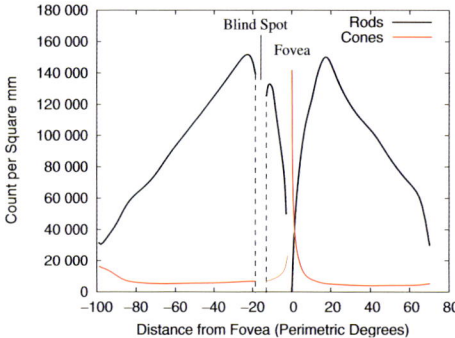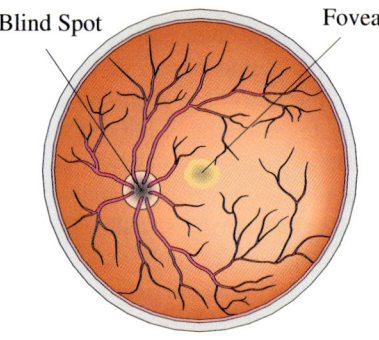

Figure 2.5 Distribution of rods and cones. Cones are mostly located in the center of the retina, the fovea. Spatial resolution is much lower in the periphery compared to the fovea (data from Østerberg 1935). A frontal view of the retina is shown in (b). (Frontal view of retina from LifeART Collection Images © 1989–2001 by Lippincott Williams & Wilkins, used by permission from SmartDraw.com.

processing already starts inside the retina, where discontinuities are detected. Horizontal and amacrine cells are used for a lateral exchange of information. Horizontal cells are coupled to each other through gap junctions (Herault 1996). The gap junctions behave as resistors, which results in spatiotemporal smoothing of the input signal. The bipolar cells compute the difference between the signal and its low-pass version, which results in high-pass filtering.

Each ganglion cell is connected to several receptors via bipolar cells. Ganglion cells receive direct input from bipolar and amacrine cells (Dowling 1987). Bipolar cells show a sustained response, whereas amacrine cells show a transient response. Some ganglion cells are mainly connected to bipolar cells and others are mainly connected to amacrine cells. The receptors that may influence the output of a given retinal ganglion cell are called the *ganglion cell's receptive field*. There is considerable overlap between the receptive fields of adjacent retinal ganglion cells. Kuffler has measured the response of ganglion cells from the cat's retina (Kuffler 1952, 1953). He stimulated the receptive fields with small spots of light. Some ganglion cells respond when light falls on the center of the receptive field. They are inhibited by the light falling on the surround of the receptive field. Such cells are said to have a center-surround characteristic and are called *on-center cells*. The intermediate area produces an on–off response. The size of the receptive fields varies with the intensity of the spot. Other cells, which show the opposite behavior, also exist. They are activated by the light falling on the surround of the receptive field and are inhibited by the light falling on the center. These are called *off-center cells*. The two different response characteristics are shown in Figure 2.6. According to De Valois and Pease (1971), cells with a center-surround organization are mainly used for contour-enhancing. Edges can be located by combining the output of several ganglion cells. This is done in the first cortical visual area.

The so-called on–off ganglion cells are sensitive to motion and many of these cells also show direction-sensitive responses (Dowling 1987). They receive most of their input from amacrine cells. On–off ganglion cells respond to spots of light appearing or disappearing inside the receptive field. They are also activated by spots of light moving in a particular

THE VISUAL SYSTEM

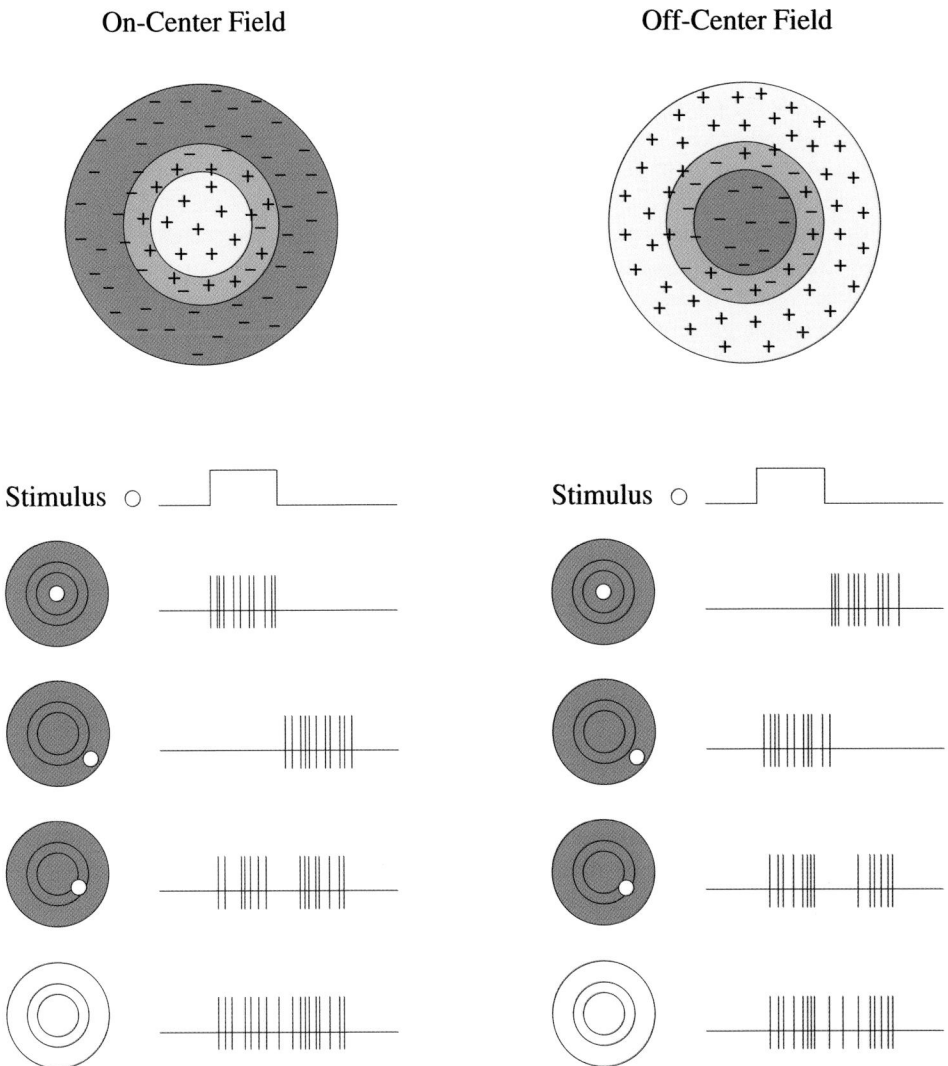

Figure 2.6 Idealized responses from the ganglion cells of the cat when stimulated with small spots of light. Some cells show an on-center surround response while others show an off-center surround response.

direction. Some of these cells are also inhibited by the light moving in the opposite direction. Usually these cells respond equally well to a bright spot moving in front of a dark background as they do to a dark spot moving in front of a bright background. Retinas of cats and monkeys have many more on- and off-center ganglion cells than on–off ganglion cells. Cold-blooded vertebrates in contrast have more on–off ganglion cells.

Some colors mix well while others do not (Tovée 1996). For instance, one cannot speak of a greenish-red. If we mix red and green, we get yellow. Yellow seems to mix well with red and green but not with blue. If we mix yellow with red we get orange and if we mix it with green we get a yellowish-green. The reason for this is the presence of three color opponent mechanisms, which are used for color discrimination. The first compares the output of the red and green cones. The second compares the output of the blue cones to the sum of the red and green cones, that is, yellow. The third is an achromatic channel that is used to detect differences in luminance. The three color channels are shown in Figure 2.7. The output is again compared using a center-surround organization of the retina. McCann et al. (1976) argue that color opponent cells may be used to guarantee accurate transmission of spectral or color information. They suggest that it is reasonable to transmit a small difference between two large numbers as opposed to transmitting both the large numbers.

The information gathered by the receptors leaves the eye through the optic nerve. The optic nerve is formed by all the axons of retinal ganglion cells. Since these axons must leave the eye at some point, this causes a blind spot. There are no receptors at this blind spot. Even though a small area near the center of the image does not contain any sensors, we do not consciously perceive this lack of information. The human brain fills in the missing information.

The human visual system is able to see over an astonishingly large range of illumination levels. At noon, the sun has an intensity on the order of 10^{10} candelas/m^2, and a white paper placed under sunlight has an intensity of approximately 10^4 candelas/m^2. Comfortable reading is possible when the intensity of light is 10 candelas/m^2. The intensity of the weakest visible light is 10^{-6} candelas/m^2 (Tovée 1996). Color vision covers a range of approximately $10^7 : 1$, while scotopic vision covers a range of $1 : 10^{-6}$. However, the response range of the optic nerve fibers is only 100:1. A mechanism called *light adaptation* ensures that small changes in lightness (see following text for a definition) can still be discriminated. The response of a retinal ganglion cell depends on the average illumination of the surround. The illumination of the surround shifts the response function of the retinal ganglion cell. Figure 2.8 shows the shift of the response recorded intracellularly from a single cone of a turtle. The response is shown as a function of ambient intensity. The resting membrane potential in the dark is 0 mV. The shift of the response curve enables the visual system to discriminate between changes of illumination of less than 1%. Because of this mechanism, a surface will appear equally light relative to the other surfaces of the scene, independent of the illumination. This phenomenon is known as lightness constancy.

2.2 Visual Cortex

Processing of visual information is done in at least 32 separate cortical areas. The function of 25 of these areas is primarily visual. The remaining areas are used for visual tasks in combination with other actions such as visually guided motor control. Each aspect of the visual information, such as the shape of an object, its color, and direction of motion, is processed in separate visual areas. Each area both receives and sends signals (Zeki 1999). There is no master or terminal area where the data is finally analyzed. The information is processed in a distributed fashion. The analysis takes place in parallel using separate pathways. According to Wilson et al. (1993), the pathways can be divided into two broad categories: what and where. Information such as shape and color are processed using the what pathway (Tovée 1996). Spatial information is processed using the where pathway, which mainly deals with motion information.

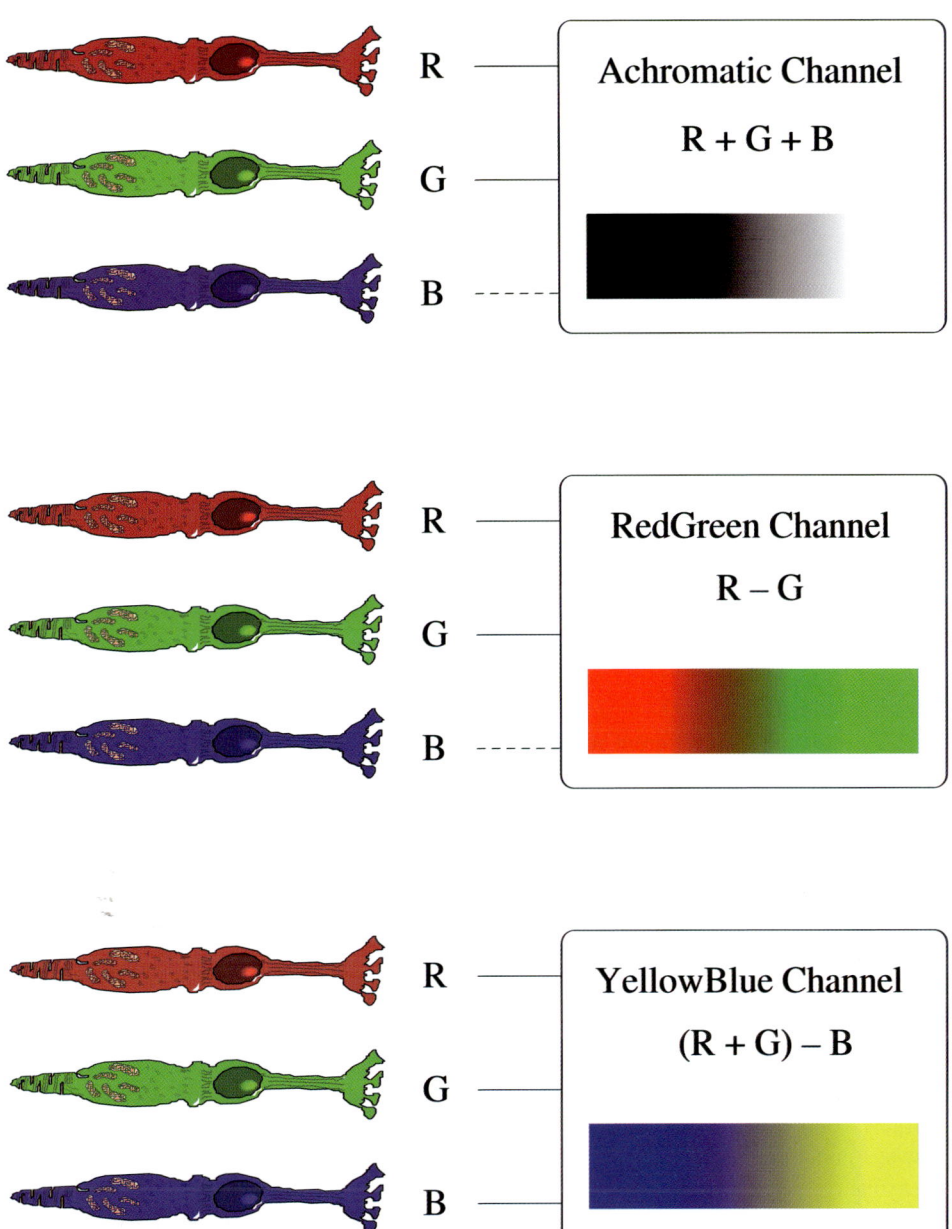

Figure 2.7 Three color opponent mechanisms. The first is an achromatic channel. The second compares the red and green channels. The third compares yellow and blue. The connection to the blue cones in the achromatic, as well as the red–green channel, is drawn with a dotted line as the role of the blue cones is unclear. (Cone illustration from from LifeART Collection Images © 1989–2001 by Lippincott Williams & Wilkins, used by permission from SmartDraw.com.)

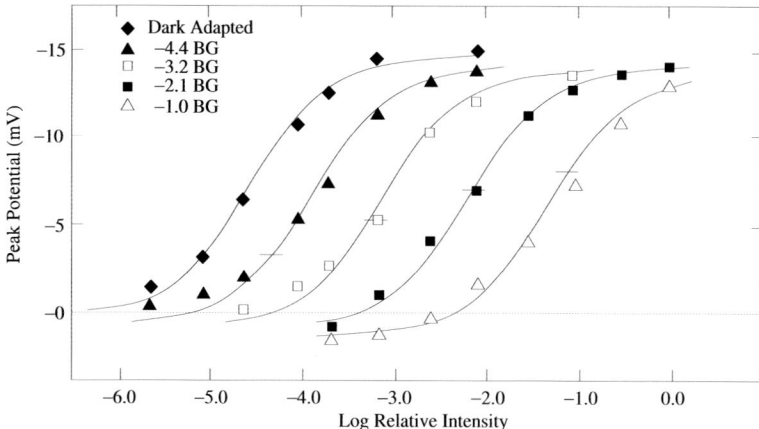

Figure 2.8 Light- and dark-adapted intensity-response curves recorded intracellularly from a red cone of turtle retina. The dark-adapted potential is 0 mV. This shift of the response curve enables the visual system to discriminate small increments or decrements about the background level. (Reproduced from R. A. Normann and I. Perlman. The effects of background illumination on the photoresponses of red and green cones. Journal of Physiology, Vol. 286, pp. 491–507, 1979, by permission of Blackwell Publishing, UK.)

This separation into different pathways already starts inside the retina. Approximately 80% of the retinal ganglion cells are the so-called parvocellular or P cells. The P cells are selective for different wavelengths. They also respond to high spatial frequencies. P cells have a slow sustained response. Another group of cells account for approximately 10% of the retinal ganglion cells. These are called *magnocellular* or *M cells*. They respond to low spatial frequencies and are not wavelength sensitive. They have a transient response and the conduction velocity of the signal is higher. Also, the dendritic field of M cells is three times larger than the dendritic field of a nearby P cell. At least eight different types of retinal ganglion cells can be distinguished among the remaining 10% of cells.

Starting inside the retina, the signals travel along the axons of the retinal ganglion cells. Because of the distinction into M cells and P cells, the visual system can be divided into separate pathways, the M or P pathways. Motion is processed using the M pathway. It is largely color blind. Color is processed through the P pathway. The separation of the pathways is not complete as there exists some communication between the two pathways. The axons of the retinal ganglion cells end at the lateral geniculate nucleus. Some retinal ganglion cells also connect to the superior colliculus. On its way to the lateral geniculate nucleus, some axons cross to the other half of the brain at the optic chiasm as shown in Figure 2.9. The axons are connected in such a way that the right half of the visual field is processed inside the left hemisphere of the brain and the right half of the visual field is processed inside the right hemisphere of the brain. The lateral geniculate nucleus consists of six layers of neurons. Layers 2, 3, and 5 are connected to the eye that is located inside the same hemisphere as the lateral geniculate nucleus. Layer 1, 4, and 6 are connected to the opposite eye. Layers 1 and 2 are connected to the M cells of the retina, while layers 3, 4, 5, and 6 are connected to the P cells. The M cells of the first two layers are a bit

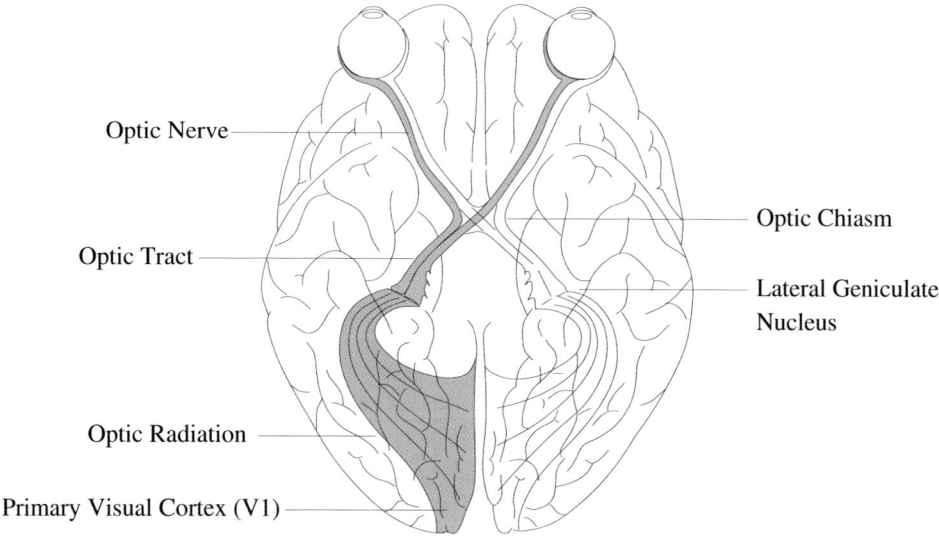

Figure 2.9 Path from the retina to the visual cortex. On its way to the lateral geniculate nucleus, some axons cross at the optic chiasm. Information from the right half of the visual field is processed inside the left hemisphere of the brain. Information from the left half of the visual field is processed inside the right hemisphere. (Reproduced from Semir Zeki. A Vision of the Brain. Blackwell Science, Oxford, 1993, by permission of Blackwell Science, UK.)

larger than the P cells of the remaining layers, and hence the name magnocellular cells. The connection to the lateral geniculate nucleus is made in such a way that the topography of the retina is maintained. If two retinal ganglion cells are located close together inside the retina, they connect to the nearby cells inside the lateral geniculate nucleus.

The neurons of the lateral geniculate nucleus are mainly connected to the primary visual cortex, which is also known as the *striate cortex*. It is located at the back of the brain. Since the primary visual cortex is the first visual area, it is also called *V1*. The connections between the retina and the primary visual cortex are genetically determined (Zeki 1999). Again, the topography is maintained in the mapping from the retina to area V1. The primary visual cortex is highly structured (Tovée 1996). It consists of a number of layers as shown in Figure 2.10. The larger M neurons of the lateral geniculate nucleus connect to sublayer $4C\alpha$ and also to the lower part of layer 6. The smaller P neurons of the lateral geniculate nucleus connect to sublayers 4A, $4C\beta$, and the upper part of layer 6. Sublayer $4C\alpha$ is connected to layer 4B, where orientation-selective cells are found. Most of the cells of layer 4B also respond to different directions of motion. Some cells also require the stimulation of both eyes and are selective for retinal disparity. Sublayer $4C\beta$ is connected to layers 2 and 3. The primary visual cortex is organized in vertical columns. The cells form alternating stripes with one stripe being connected to one eye and the next being connected to the other eye. Cells above layer 4C receive input from both eyes; however, one eye is usually dominant. Hence, these blocks of cells are called *ocular dominance columns*. Cells located at the center of a column are connected to only one eye. Inside each

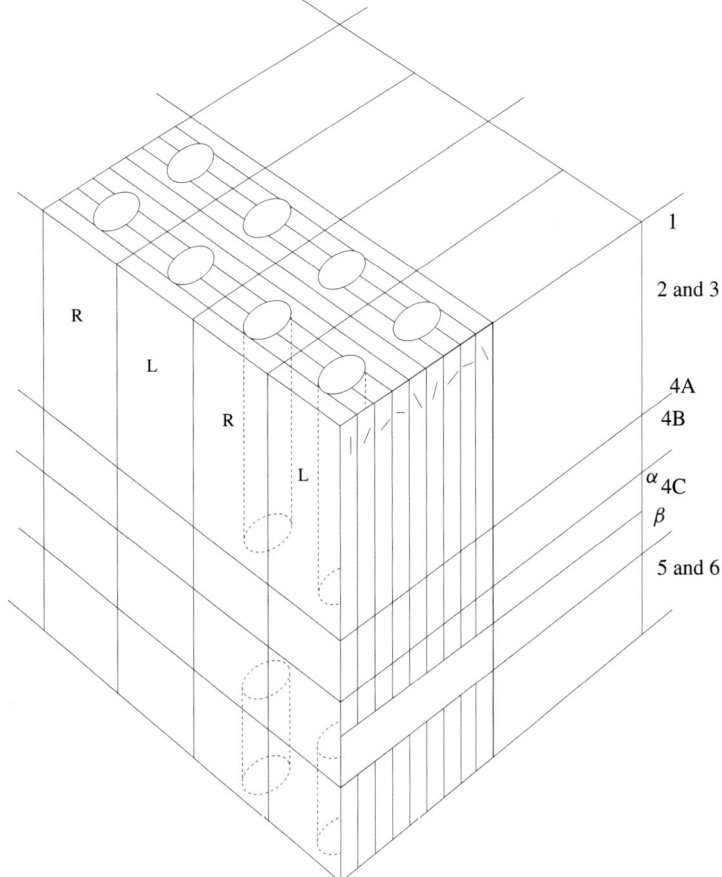

Figure 2.10 Structure of the primary visual cortex. The primary visual cortex consists of six layers. Vertical columns (blobs) are found in layer 2, 3, 5, and 6. (Copyright 1984 by the Society for Neuroscience. Reproduced by permission of Society for Neuroscience. M. S. Livingstone and David H. Hubel. Anatomy and physiology of a color system in the primate visual cortex. The Journal of Neuroscience, Vol. 4, No. 1, pp. 309–356, Jan., 1984.)

ocular dominance column, another vertical column is found in layers 2 and 3. Columns are also found in layers 5 and 6. These columns are called *blobs*. The area that surrounds the blobs is called the *interblob region*. The blobs receive their information from both the P and the M pathways.

The width of an ocular dominance column is approximately 0.4 mm (Hubel and Wiesel 1977). Two such columns containing the data for both eyes amount to roughly 1 mm. The orientation-selective cells are not arranged randomly. If a penetration is made horizontally to the cortical surface, recording the response of the cells, it is found that the optimal or preferred orientation to which the cell responds changes continuously. Roughly, a 1-mm displacement corresponds to a 180° change in orientation. In other words, a 2×2 mm^2

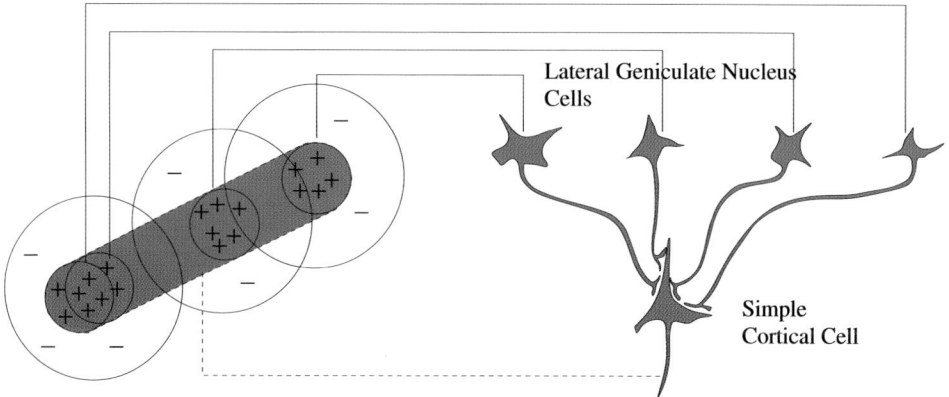

Figure 2.11 A simple cell responds to lines or edges oriented at a particular orientation. It may be constructed by receiving input from several of the center-surround neurons found in the lateral geniculate nucleus. (Reproduced from D. H. Hubel and T. N. Wiesel. Receptive fields, binocular interaction and functional architecture in the cat's visual cortex. Journal of Physiology, Vol. 160, pp. 106–154, 1962, by permission of Blackwell Publishing, UK.)

block contains cells that analyze a particular region of the visual field for all possible orientations with both eyes. Ocular dominance columns and orientation columns do not necessarily have to be orthogonal to each other.

Several types of neurons can be distinguished in V1. Cells in layer 4C have a center-surround response characteristic like the cells found in the lateral geniculate nucleus. The receptive fields of the neurons found in the visual cortex are more or less square or rectangular. The size of the receptive fields vary from one visual area to another. A receptive field has three main features – a position, a shape, and a specificity (Zeki 1999). The so-called simple and complex neurons respond to oriented bars or differences in illumination (Tovée 1996). A simple cell is probably constructed using input from several center-surround cells as shown in Figure 2.11. It responds to a bar of light oriented at a certain angle. It has an on-center region and an off-center surround or vice versa. It is also possible that the receptive field consists of only two rectangular regions. In this case, one region is excitatory and the other is inhibitory. Some cells are sensitive to the length of the bar. They have an additional antagonistic area at one end of the bar. These cells are called *end-stopped* cells. Simple cells are found mostly in layers 4 and 6. Complex cells (Figure 2.12) respond to bars of fixed width and a certain orientation. Some of the complex cells respond to a bar of a certain size while others respond best to end-stopped stimuli. Approximately 10–20% of the complex cells respond to moving edges or bars. In contrast to the simple cells, the exact position of the stimulus within the receptive field is not important. Complex cells are found mostly in layers 2, 3, and 5. Some of the cells found in V1 also respond to sine-wave gratings of a particular orientation and frequency. Some neurons also seem to respond to some kinds of texture.

The structure of layer 2 and 3 creates two new pathways: the P-B (parvocellular-blob pathway) and the P-I (parvocellular-interblob pathway). The cells located inside the blobs of layer 2 and 3 are either color or brightness selective. They are not orientation selective.

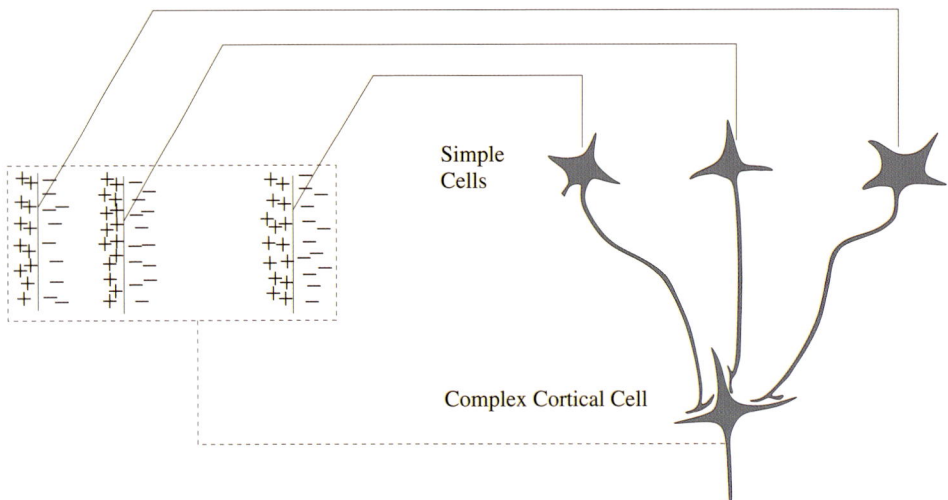

Figure 2.12 A complex cell responds to lines or edges oriented at a particular orientation irrespective of the exact position of the stimulus inside its receptive field. It may be constructed by receiving input from several simple cells. (Reproduced from D. H. Hubel and T. N. Wiesel. Receptive fields, binocular interaction and functional architecture in the cat's visual cortex. Journal of Physiology, Vol. 160, pp. 106–154, 1962, by permission of Blackwell Publishing, UK.)

They respond to light of a particular wavelength and either show a color opponent or a double opponent behavior. These cells receive their input from color opponent cells of the lateral geniculate nucleus. They have a larger receptive field compared with the cells found inside the lateral geniculate nucleus. Double opponent cells show a particular behavior in the center, i.e. some cells give an on-response to red and an off-response to green, and show exactly the opposite behavior in the surround, i.e. an off-response to red and an on-response to green. The behavior of single opponent cells as well as double opponent cells is illustrated in Figure 2.13. Double opponent cells may be created by combining the output of color opponent cells. The neurons of a single blob show either a red/green opponent behavior or a blue/yellow opponent behavior. These two forms of opponency are not found together inside a single blob. The distribution of the red/green opponent retinal ganglion cells and the blue/yellow retinal ganglion cells is reflected in the distribution in the different blob types. The retina contains more red and green cones, and hence the proportion of red/green blobs is larger than that of the blue/yellow blobs. Two blobs are often paired with a ladderlike connection between them. The rungs of this ladder connect the two blobs. These connections also contain color-selective cells. They show a mixed spectral selectivity, i.e. neither red/green nor blue/yellow.

Most cells of the interblob region respond to lines or bars of a particular orientation. They do not respond to color or show any color opponency. In contrast to the cells found inside the blobs, the receptive field of the cells found in the interblob region is very small. The response characteristic of the neurons of the interblob regions is arranged

THE VISUAL SYSTEM

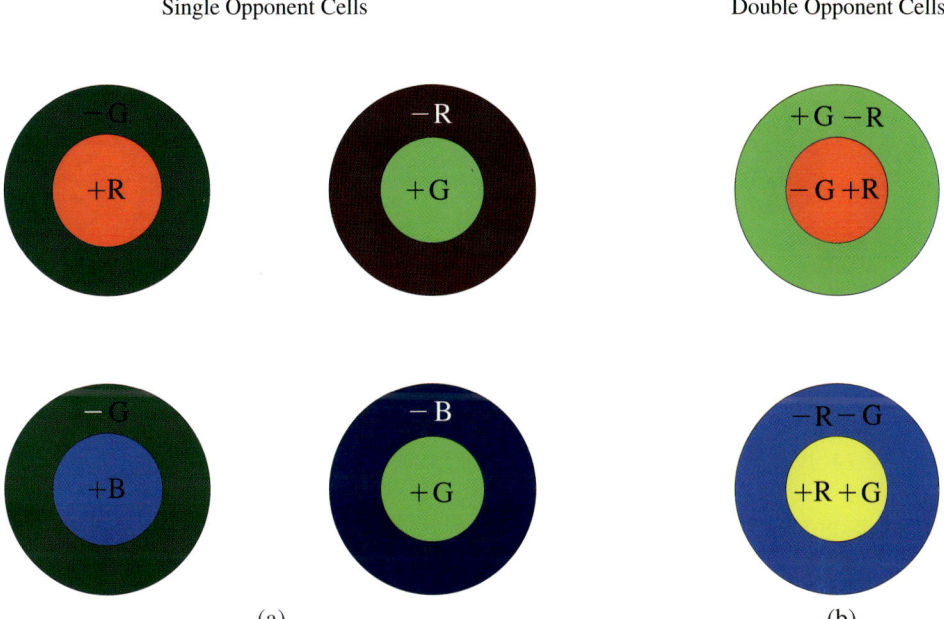

Figure 2.13 Four different single opponent cells are shown in (a). Two double opponent cells are shown in (b).

systematically. If one cuts horizontally through the upper part of V1, one finds an orderly progression of response characteristics. Suppose that the first neuron located at the beginning of a particular module responds to lines of a particular orientation. Then a neuron located a distance 25 µm away will have a response characteristic rotated by approximately 10°. From start to end, the response characteristics cover the entire 360° range. The response characteristics along a vertical cut, however, stay constant.

If V1 is damaged, visual perception is seriously impaired. All main visual pathways pass through V1 before being distributed to other visual areas. These areas are called *V2, V3, V4*, and *V5*. Damage to V1 leads to the so-called scotomas. These are holes in the visual field where no information is consciously perceived. However, some patients are able to act on the basis of information perceived in their scotomas. This type of perception without consciously knowing it is called *blindsight*. It is likely to be mediated by connections from the retina to the superior colliculus (Tovée 1996). There also exists a direct connection between the retina and area V5, the motion center. This connection does not pass through V1 (Ffytche et al. 1995; Zeki 1999). It allows V5 to process fast motion and detect its direction. Each visual area is specialized to process a particular aspect of the visual information. The location of the visual areas inside the brain is shown in Figure 2.14. Hubel and Wiesel (1977) give a summary of the functional architecture of the visual cortex of the macaque monkey. Hubel and Wiesel used extracellular microelectrodes to record the responses, as they were interested in the all-or-none impulses of the cells. A summary on the functional

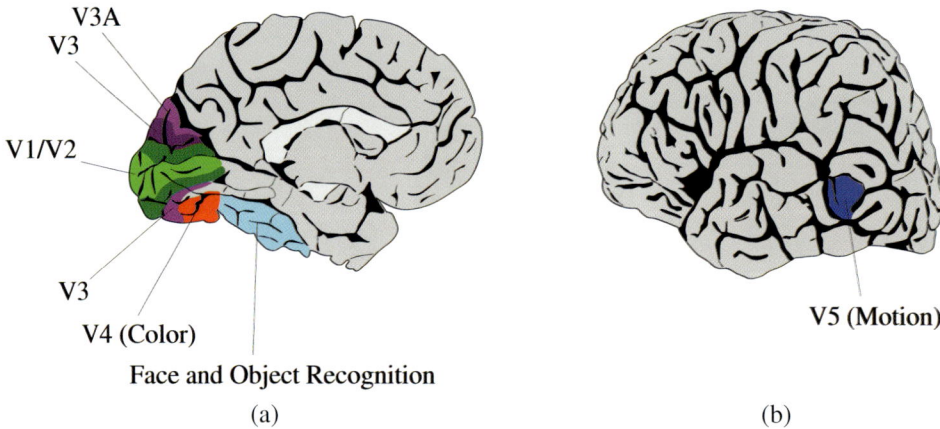

Figure 2.14 Position of the visual areas V1, V2, V3, V4 and V5. The image in (a) shows the medial side of the left hemisphere of the brain. Color processing is done in V4, and motion is processed in V5. (Reproduced from S. Zeki. An Exploration of Art and the Brain. Oxford University Press, Oxford, 1999 by permission of S. Zeki, University College London, UK.)

specialization in the visual cortex of the rhesus monkey is given by Zeki (1978). Similar functional specialization is also found in the human visual cortex (Zeki et al. 1991).

V1 mainly connects to area V2, which surrounds V1 (Tovée 1996). Area V2 seems to be organized into three types of stripes, the so-called thick, thin, and interstripes. The stripes seem to be used to process visual orientation (thick stripes), color (thin stripes), and retinal disparity (interstripes). Adjacent stripes respond to the same region of the visual field. Neurons of layer 4B of V1 connect to the thick stripes. Cells found inside the thick stripes are selective for orientation and movement. Many of the cells also respond to retinal disparity. The neurons of the blobs are connected to the thin stripes. These cells are not orientation selective. More than half of these cells respond to color. Most show a double opponent characteristic. The cells of the interblob region connect to the interstripes. Neurons of the interstripe region respond to different orientations but neither to color nor to motion. A condition know as chromatopsia is caused by damage to certain parts of V1 and V2. Individuals who suffer from chromatopsia are not able to see shape or form. However, they are still able to see colors.

Motion and color are processed by different cortical areas. A difference in the luminance of a moving stimulus is required for coherent motion perception (Ramachandran and Gregory 1978). Cells found in V3 respond to lines of different orientation and also to motion (Tovée 1996). They do not respond to color. V3 is believed to process dynamic form. Some cells of V3 are able to discount the movement of the eye. They only respond to a stimulus that moves relative to the eye. These cells also receive information about the eye position (Zeki 1999). Cells in V3 are connected to layer 4B of V1. V3 is also connected to the thick stripes of V2.

Cells located in the thick stripes (color) of V2 as well as cells found in the interstripe region (form) of V2 are connected to V4 (Tovée 1996). The subdivision of the parvocellular

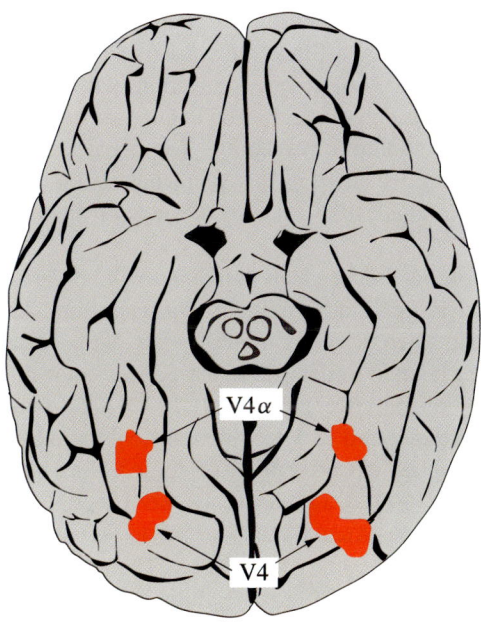

Figure 2.15 Position of the V4 complex. It consists of V4 at the back and V4α in front. (Redrawn from Figure 1 (page 1373) S. Zeki and A. Bartels. The clinical and functional measurement of cortical (in)activity in the visual brain, with special reference to the two subdivisions V4 and V4α) of the human colour centre. Proceedings of the Royal Society of London. Series B, 354, pp. 1371–1382, The Royal Society, 1999, used by permission.)

pathway starting with the blob and interblob regions seems to continue in V4. The location of V4 is shown in Figure 2.15. It actually consists of two subareas V4 and V4α (Zeki and Bartels 1999). V4 also has a retinotopic organization, whereas the retinotopic mapping is absent in V4α. Some cells found in V4 respond to the actual color of objects instead of to the light reflected from the objects.

The three cortical stages of color processing in the human brain are described by Zeki and Marini (1998). The cells found in V1 or V2 only respond to light of a particular wavelength. However, the cells found in V4 respond to the color of objects independent of the light that illuminates the objects. It is believed that V4 compares the intensity of the light from a patch with the intensity of the surrounding patches by some as-yet-unknown mechanism. The next stage of color processing includes the classification of color constant signals. Finally, objects are associated with colors. Some cells found in V4 of the macaque monkey are also orientation selective. Some of these cells also have variable degrees of color preference. In contrast to the orientation cells found in V2, V3, and V3A, the orientation-selective cells found in V4 have greater orientational tolerances. Like other visual areas, V4 seems to be compartmentalized. Regions of heavy concentration of orientation-selective cells seem to be separated from one another by cells with strong color or wavelength selectivities.

The receptive field of the cells found in V4 is quite large. The corpus callosum connects both hemispheres of the brain. Since the corpus callosum also plays a role in color perception (see Section 2.6) and V4 also has connections with the corpus callosum, it is the first possible area where color constant cells can occur. As we will see later on, the entire scene plays an important role in color perception. The perceived color of a point is dependent not only on the light reflected from that point but also on the light that is reflected from the surrounding area (Judd 1960). It also depends on the colors seen previously by the same observer.

If the entire V4 complex is damaged, an individual will not be able to see or understand color (Zeki 1999). Such a condition is called *achromatopsia*. Another condition called *dyschromatopsia* is caused when the V4 complex is only partially damaged. In this case, an individual may no longer be able to determine the correct color of objects. The perceived color varies with the illuminant (Kennard et al. 1995). The perceptual capabilities are probably reduced to the cells found in V1, which are only wavelength selective. In other conditions of dyschromatopsia, the perception of some colors is affected more than that of others (Zeki and Bartels 1999). Cases have also been reported where achromatopsia is restricted to one quadrant alone. Area V4 also seems to be used for object recognition as the cells of V4 connect to the temporal visual cortex. The temporal visual cortex integrates form and color and is in charge of object recognition. The neurons of the temporal visual cortex respond to faces, objects, or other complex patterns. Some cells show responses that are independent of the size and position of the stimulus that activates them.

Motion information and stereoscopic depth are believed to be processed in V5. This area is also called the *middle temporal visual area* or MT. If area V5 is damaged, a person may no longer be able to see objects that are in motion (Zeki 1999). This condition is known as *akinetopsia*. Movshon et al. (1985) give a description of the behavior of neurons found in V5 compared to the behavior of neurons found in V1. Neurons found in V5 respond to a particular speed and direction of motion (Zeki 1999). Some also respond to motion irrespective of the direction of motion. But the large majority of cells have a preferred direction of motion. They do not respond to motion in the opposite direction. A motion selective cell and its receptive field are shown in Figure 2.16. Some cells are driven binocularly. They respond to motion toward or away from the observer. The cells of V5 are indifferent to color. Like other visual areas, V5 is highly structured. If the response of cells inside V5 is sampled in a direction parallel to the cortical surface, one finds that the directional selectivity of the cells changes in an orderly way.

Neurons found in V1 respond to component motion; however, neurons found in V5 respond to global motion (Movshon et al. 1985; Tovée 1996). This is illustrated in Figure 2.17. If a large object moves from left to right, neurons of V1 will respond to the component motion, i.e. in this case, the neurons respond to the local motion of the edges of the car, which appear to move diagonally. However, global motion is purely horizontal. Neurons found in V5 respond to the global motion of the car. Also, V5 seems to possess two subdivisions. One part seems to analyze the motion of independently moving objects in our environment. The other part analyzes motion that occurs as a result of our own movements. Some types of coherent motions, which may also have been caused by eye or head movements, do not appear to activate V5 strongly (McKeefry et al. 1997).

Information from V3 and V5 finally reaches the parietal cortex (Tovée 1996). The parietal cortex seems to encode a spatial representation of the environment. It also receives

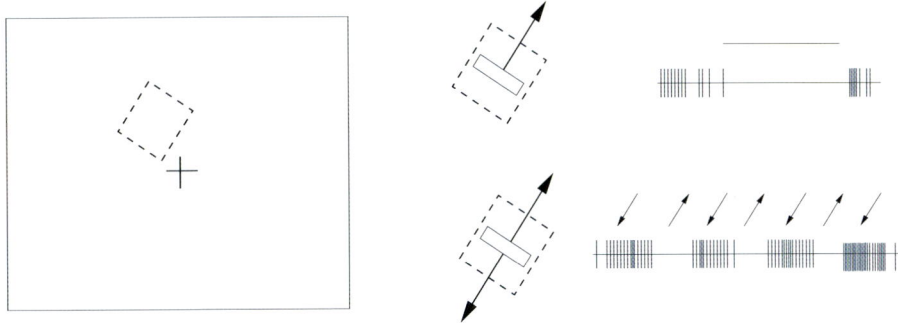

Figure 2.16 Motion selective cell. The receptive field of the cell is shown on the left. The cell only responds to motion in a preferred direction. (Reproduced from S. Zeki. An Exploration of Art and the Brain. Oxford University Press, Oxford, 1999 by permission of S. Zeki, University College London, UK.)

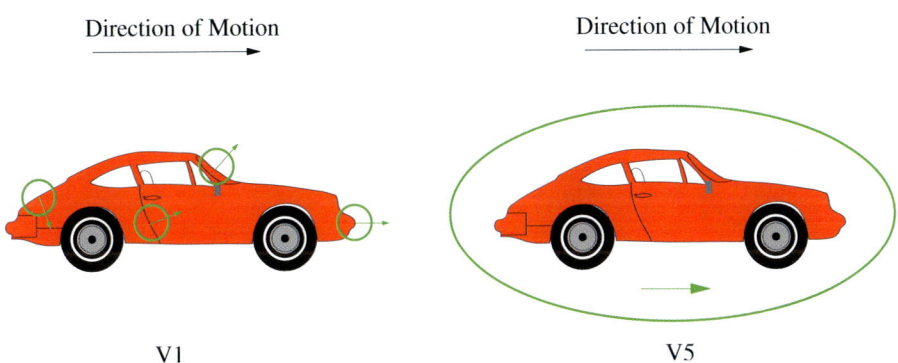

Figure 2.17 Difference between component motion and global motion. Cells found in V1 respond to component motion. Cells found in V5 respond to the global motion of objects. (Car image from SmartDraw used by permission from SmartDraw.com.)

auditory, somatosensory, and vestibular sensory inputs. The organization of the visual system is shown in Figure 2.18. Information processing starts with the retinal ganglion cells. Approximately 80% of the retinal ganglion cells are parvocellular cells and 10% are magnocellular cells. The retinal ganglion cells are connected to the P and M cells of the lateral geniculate nucleus. The W cells of the lateral geniculate nucleus are believed to be connected to one of the cells of the remaining 10%. The information then reaches V1. From there, information is sent to areas V2 through V5 and finally to the posterior parietal cortex and the inferior temporal cortex . Motion perception is done largely through the magnocellular pathway. The parvocellular pathway splits into a parvocellular-blob pathway and a parvocellular-interblob pathway. Color is processed largely through

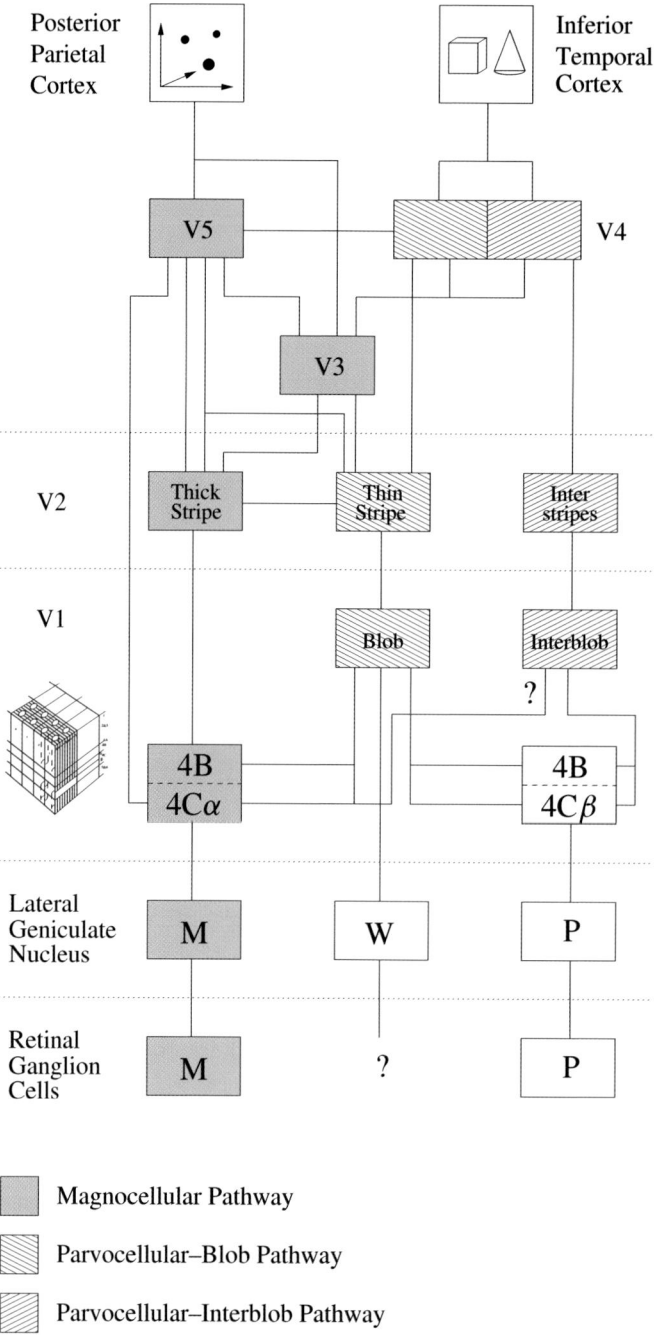

Figure 2.18 Organization of the visual system. Two main pathways, the magnocellular and the parvocellular pathways, can be distinguished. (Reproduced by permission from Massachusetts Institute of Technology. D. C. Van Essen and E. A Deyoe. Concurrent Processing in the Primate Visual Cortex. M. S. Grazzanida (ed.), The Cognitive Neurosciences, The MIT Press, pp. 383–400, 1995.)

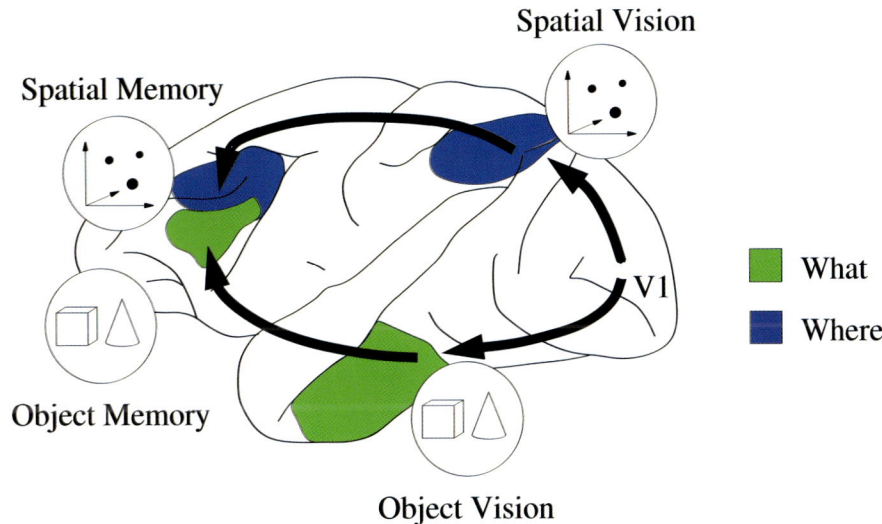

Figure 2.19 What and where pathways of the primate brain. The what pathway starts at V1 and passes through the inferior temporal cortex. The where pathway passes through the posterior parietal cortex. (Reprinted with permission from F. A. W. Wilson, S. P. Ó Scalaidhe, and P. S. Goldman-Rakic. Dissociation of object and spatial processing domains in primate prefrontal cortex. Science, Vol. 260, pp. 1955–1958, June, Copyright 1993 AAAS.)

the parvocellular-blob pathway. Spatial information is processed in the posterior parietal cortex. Object recognition is done in the inferior temporal cortex. Therefore, the two pathways are also called the *what* and *where* pathways The locations of the inferior temporal cortex and the posterior parietal cortex are shown in Figure 2.19. Instead of making a distinction between what and where, it may sometimes be more appropriate to make a distinction between what and how, that is, an action-oriented processing of visual information.

The different aspects of vision, color, form, and motion appear to be precisely registered in time. However, this is only correct if one looks at timescales larger than 1 s (Zeki 1999). Recent experiments have shown that these attributes are not perceived at the same time. Color is perceived before form and form is perceived before motion. Color is perceived earlier than motion and the difference in time is 70–80 ms (Moutoussis and Zeki 1997).

Color perception starts with the receptors of the retina. Wavelength-selective cells are found inside the primary visual cortex, V1. The response of these cells is similar to a measuring device. Cells found in V4 respond to the color of objects irrespective of the illuminant. The output of these cells are likely to be used for object recognition. But there are also additional factors such as memory, judgment, and learning in color processing (Zeki 1999; Zeki and Marini 1998). Color information processing also depends on the type of objects viewed. If the viewed scene contains familiar objects, other areas like the inferior temporal cortex, the hippocampus, and the frontal cortex will also be involved.

2.3 On the Function of the Color Opponent Cells

Land (1986a) notes that the transformation performed by the double opponent cells may be a simple rotation of the coordinate system. Let R, G, and B be the energy measured by the red, green, and blue cones, respectively. Let BW be the output of the achromatic channel, i.e. the black–white axis. The output of the achromatic channel is computed by adding the energies from the red, green, and blue cones. Let RG be the output of the red–green channel. The output of the red–green channel is computed by subtracting the energy measured by the green cones from the energy measured by the red cones. Finally, let YB be the output of the yellow–blue channel. The output of this channel is computed by subtracting the energy measured by the blue cones from the energies measured by the red and the green cones. With suitable coefficients, the operation performed by the three channels can be written as

$$\text{BW} = \frac{1}{\sqrt{3}}R + \frac{1}{\sqrt{3}}G + \frac{1}{\sqrt{3}}B \qquad (2.4)$$

$$\text{RG} = \frac{1}{\sqrt{2}}R - \frac{1}{\sqrt{2}}G \qquad (2.5)$$

$$\text{YB} = \frac{1}{\sqrt{6}}R + \frac{1}{\sqrt{6}}G - \sqrt{\frac{2}{3}}B. \qquad (2.6)$$

The coefficients do not represent the actual coefficients but are simply meant to illustrate the case.

This is just a rotation of the coordinate system, which can also be written as

$$\begin{bmatrix} \text{BW} \\ \text{RG} \\ \text{YB} \end{bmatrix} = \begin{pmatrix} \frac{1}{\sqrt{3}} & \frac{1}{\sqrt{3}} & \frac{1}{\sqrt{3}} \\ \frac{1}{\sqrt{2}} & -\frac{1}{\sqrt{2}} & 0 \\ \frac{1}{\sqrt{6}} & \frac{1}{\sqrt{6}} & -\sqrt{\frac{2}{3}} \end{pmatrix} \begin{bmatrix} R \\ G \\ B \end{bmatrix}. \qquad (2.7)$$

The inverse transformation is given by

$$\begin{bmatrix} R \\ G \\ B \end{bmatrix} = \begin{pmatrix} \frac{1}{\sqrt{3}} & \frac{1}{\sqrt{2}} & \frac{1}{\sqrt{6}} \\ \frac{1}{\sqrt{3}} & -\frac{1}{\sqrt{2}} & \frac{1}{\sqrt{6}} \\ \frac{1}{\sqrt{3}} & 0 & -\sqrt{\frac{2}{3}} \end{pmatrix} \begin{bmatrix} \text{BW} \\ \text{RG} \\ \text{YB} \end{bmatrix}. \qquad (2.8)$$

In principle, the algorithms for color constancy, which are described in the following chapters, can also be applied to the rotated coordinates (Figure 2.20). Land notes that, in practice, the transformation may not be that simple if the algorithm is nonlinear, i.e. contains a thresholding operation. We see later, e.g. in Section 6.6, Section 7.5, or Section 11.2, that some of the algorithms for color constancy are also based on a rotated coordinate system where the gray vector, i.e. the achromatic axis, plays a central role.

THE VISUAL SYSTEM

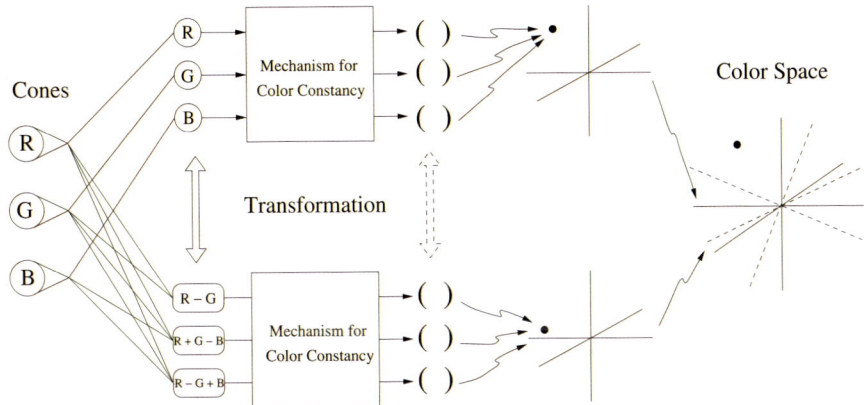

Figure 2.20 The transformation performed by the color opponency mechanism may be a simple rotation of the coordinate system. In this case, the mechanism for color constancy may operate on the rotated coordinates. (Reprinted from Vision Res., Vol. 26, No. 1, Edwin H. Land, Recent advances in retinex theory. pp. 7–21, Copyright 1986, with permission from Elsevier.)

2.4 Lightness

Reflectance, the amount of incident light reflected by an object for a given wavelength, defines the color of an object. The biological analog of reflectance is called *lightness* (McCann et al. 1976). Suppose a human observer is looking at several achromatic pieces of paper ranging from white to black. For each paper, there exists a unique sensation in the observer's brain. This sensation is called *lightness* (Land 1974). Lightness cannot be measured by a photometer but can only be experienced with the human visual system. The relationship between reflectance and lightness is nonlinear, as shown in Figure 2.21.

This is also shown by the following experiment. Human observers are given several achromatic papers and asked to pick the brightest paper and the darkest paper. Next, they are asked to pick the piece of paper lying intermediate between the two extremes. After the middle one is picked, the observers are asked to pick the one lying halfway between the darkest one and the middle one. The observers are also asked to pick the one lying halfway between the middle one and the brightest one. This is continued recursively. As a result, one obtains the scale shown in Figure 2.21 with the pieces of papers shown at the bottom of the graph. The paper chosen as having a lightness of 3 actually has a reflectance of 0.08, the paper chosen as having a lightness of 5 actually has a reflectance of 0.25, and the paper chosen as having a lightness of 7 actually has a reflectance of 0.55. Lightness and reflectance are not linearly related.

As early as 1920, it was suggested that the relationship between the reflectance R of gray samples and its position Y along a scale of equal visual steps from black to white is closely approximated by the following relation:

$$Y = kR^{\frac{1}{2}} \tag{2.9}$$

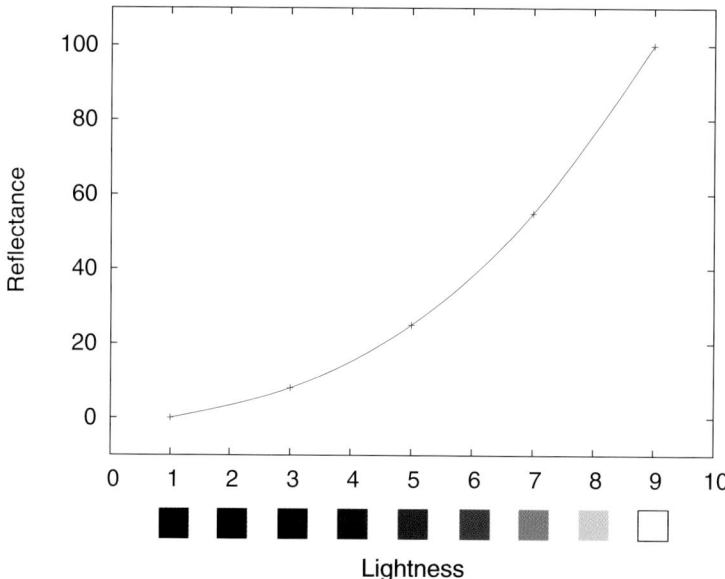

Figure 2.21 Relationship between lightness and reflectance. Achromatic papers are arranged on an equally spaced lightness scale by human observers. For each paper, the corresponding reflectance is shown in the graph (data from Land 1974).

for some constant k (Hunt 1957). Instead of a square root relationship, a cube root relationship has also been suggested. Figure 2.22 shows several different fits for the relationship between reflectance and lightness. The relationship is best approximated by the function

$$Y = 9.033 R^{0.4296} - 0.0151 \qquad (2.10)$$

with a root mean squared error (RMSE) of 0.028. The logarithmic (excluding zero), the cube root, and the square root fits have an RMSE of 0.323, 0.352 and 0.216, respectively. Here, the reader should take note of the fact that, even though a logarithmic relationship is not the best approximation to the data, the logarithmic fit is better than the cube root fit. We will come back to this point in Chapter 14.

2.5 Color Perception Correlates with Integrated Reflectances

McCann et al. (1976) demonstrated that human color perception correlates with integrated reflectance. We will give a definition for integrated reflectance in a moment. Basically, McCann et al. performed an experiment to show that human color perception correlates with the reflectance of objects as opposed to the amount of light reflected by the objects (see also Land (1974)). For their experiments, McCann et al. used a color Mondrian similar to the one used by Land. The Mondrian contained 17 colored areas. It was illuminated using three

THE VISUAL SYSTEM

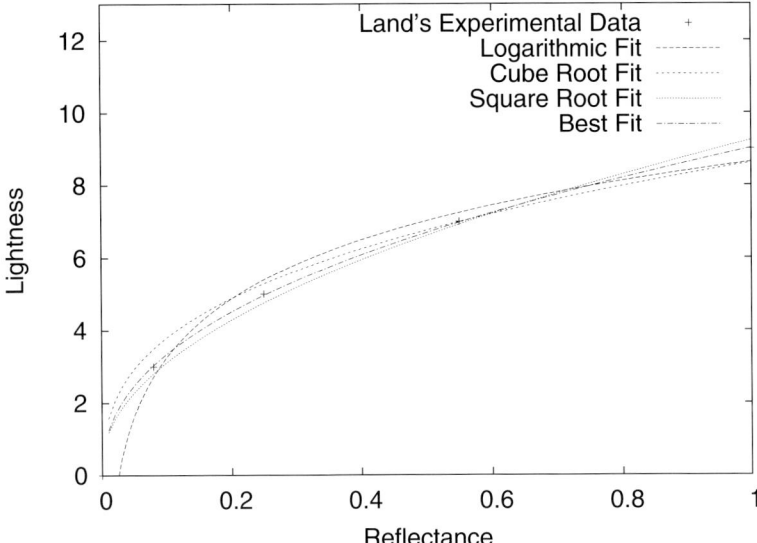

Figure 2.22 Different fits for the relationship between reflectance and lightness. Note that the axes have been switched, compared with Figure 2.21. Logarithmic (excluding zero), cube root and square root fits are shown for Land's experimental data. The data is best approximated by the function $Y = 9.033 R^{0.4296} - 0.0151$, which is also shown.

narrow-band projectors whose intensities could be varied. Subjects matched the color of the patches on the Mondrian to color chips in a book of colors (the Munsell book of colors).

The Munsell book of colors was placed inside a large five-sided box. A pair of goggles was placed in front of the open side of the box through which an observer could view the book. Black velvet was placed around the goggles to cover the remaining side of the box. The right eye of the goggles was covered such that an observer could view the book only with his left eye. The color Mondrian was also placed inside a box with attached goggles. The goggles that were used to view the color Mondrian had the left eye covered. Therefore, the color Mondrian could only be viewed with the right eye. The goggles were used to eliminate any binocular interactions between the two eyes. Three projectors with narrow-band filters were used to illuminate the Mondrian as well as the Munsell book. The narrow-band filters had a peak transmission at 450, 530, and 630 nm, respectively. The bandwidth was 10 nm at half-height. Both illuminants were set up to provide a uniform illumination. The powers of the three projectors were adjusted such that the white area appeared "the best white" (McCann et al. 1976). The same illumination was used to illuminate the color Mondrian.

Initially, the gray area of the Mondrian was chosen and the energy given off by the patch on the Mondrian was measured. The subject was asked to match the corresponding colors for all of the areas of the color Mondrian from the book of colors. Next, a red area of the Mondrian was selected. The illumination of the color Mondrian was adjusted such that the energy given off by the red area was equivalent to the gray area that had been previously matched. The illumination of the Munsell book of colors always remained the same. Again, the subject was asked to select the corresponding color from the book of colors. This was repeated for a green area, a blue area, and a yellow area of the Mondrian. For each patch of the Mondrian, a matched color was obtained from the book of colors. Since the color Mondrian contained 17 different colored patches, this procedure resulted in 17 matches for each of the five illuminants. The experiment was performed by three subjects. Each subject performed the experiment five times. The data was then averaged using the Munsell designation. Hue, value, and chroma were averaged separately and then the Munsell paper with the nearest matching was found.

Five areas of the Mondrian had been used to adjust the illuminant for the color Mondrian. The colors of these areas were gray, red, green, blue, and yellow. Note that even though the areas all had a different color, during the course of the experiment, the radiances given off by the four colored patches was equivalent to the radiance given off by the gray patch.

To quantify the color sensations, they used the following procedure. Human color perception was modeled using filters that approximate the response curve of the cones found in the retina. One set of filters was used to approximate the response curve of the receptors that mainly respond to red light, and another set of filters was used to approximate the response curve of the receptors that mainly respond to green light, and a third set of filters was used to approximate the response curve of the receptors that mainly respond to blue light. For each area of the Mondrian, the radiance was measured using a photometer equipped with one of the filters that approximate the response curve of the retina cones. This gave one measurement for the red cone, one measurement for the green cone, and one for the blue cone. Similarly, measurements of the color chips were also obtained from the Munsell books of colors.

The measurements were plotted into a graph with the measured radiance of the color Mondrian along one axis and the measured radiance of the matching chip from the Munsell book of colors along the other axis. From this graph, it could be seen that the correlation between the two measurements was not very good. Next, the integrated reflectance was compared. A large white paper was used as a reference standard. This paper was placed in front of the Mondrian and the radiance given off by this paper was measured using the same photometer–filter combination as before. Integrated reflectance was calculated as the radiance given off by an area divided by the radiance given off by the reference patch. Note that this value is not equal to the reflectance of the patch under consideration as there is considerable overlap between the response curves of the cones. Similarly, integrated reflectance was also measured for the chips of the Munsell book of colors.

In Section 2.1, we have seen that the energy Q_i measured by a receptor i with $i \in \{r, g, b\}$ is given by

$$Q_i = \int S_i(\lambda) R(\lambda) L(\lambda) \, d\lambda. \tag{2.11}$$

Since a narrow-band filter was used for each of the three projectors, we only need to consider the three wavelengths $\lambda_r = 630$ nm, $\lambda_g = 530$ nm, and $\lambda_b = 450$ nm. Let $R_p(\lambda)$ be the reflectance of the patch p for wavelength λ. The measured energy for the patch p by receptor i is given by

$$Q_i = \sum_{\lambda \in \{\lambda_r, \lambda_g, \lambda_b\}} S_i(\lambda) R_p(\lambda) L(\lambda). \tag{2.12}$$

THE VISUAL SYSTEM

Integrated reflectance is computed by dividing the measured energy for a given patch by the measured energy for the reference standard. Let $R_{\text{std}}(\lambda)$ be the reflectance of the reference standard for wavelength λ. Therefore, integrated reflectance R_i is given by

$$R_i = \frac{\sum_{\lambda \in \{\lambda_r, \lambda_g, \lambda_b\}} S_i(\lambda) R_p(\lambda) L(\lambda)}{\sum_{\lambda \in \{\lambda_r, \lambda_g, \lambda_b\}} S_i(\lambda) R_{\text{std}}(\lambda) L(\lambda)}. \tag{2.13}$$

Again, the integrated reflectances were plotted into the graph. The correlation between the integrated reflectances was much better compared to the correlation between the radiance measurements. Correlation improved even further when scaled integrated reflectance was compared. Scaled integrated reflectance takes into account that the fact that human perception of lightness is not linear as explained in Section 2.4. The measurements for several of these experiments line up along the diagonal, i.e. the scaled integrated reflectances are correlated. When the values of the measured energies are directly plotted into the graph, the measurements are no longer lined up along the diagonal. Therefore, scaled integrated reflectance is a much better indicator of the color of a patch compared to the reflected energy.

If we have an arbitrary illuminant that is defined by the power distribution $L(\lambda)$, then the integrated reflectance R_i would be given by

$$R_i = \frac{\int S_i(\lambda) R_p(\lambda) L(\lambda)}{\int S_i(\lambda) R_{\text{std}}(\lambda) L(\lambda)}. \tag{2.14}$$

Thus, the measured energy that is reflected from the current patch of interest with reflectance R_p is normalized by the energy reflected from a reference patch. This ratio is not independent of the illuminant (Brill and West 1981).

McCann et al. (1976) describe some small systematic deviations from a perfect correlation. This was attributed to differences in overall illumination between the color Mondrian and the Munsell book of colors. Note that scaled integrated reflectance is not entirely independent of the illuminant. This is because the response curves of the receptors overlap. The overlap is particularly large for the response curves of the red and the green cones. If narrow-band illumination is used, then the energy measured by a single cone is the sum of three products as described earlier. If any one of the illuminants is varied, then the corresponding product changes. Thus, increasing the intensity of one projector changes all the scaled integrated reflectances measured for the three cones. Nascimento and Foster (1997) also note that human color constancy is not perfect. That human color constancy cannot be perfect also follows from the existence of metamers (Brill and West 1981). Metamers are colors that appear to be the same under one illuminant and different under a second illuminant. Metamers cannot be distinguished by postprocessing the available data. Thus, color constancy can only hold approximately.

2.6 Involvement of the Visual Cortex in Color Constancy

Land (1983, 1986b); Land et al. (1983) also showed that the visual cortex is involved in color perception. Originally, it was not known whether the computation of color constant descriptors is carried out inside the retina or the visual cortex. Marr (1974) suggested in 1974 that lightness computations are carried out inside the retina. He gave a detailed

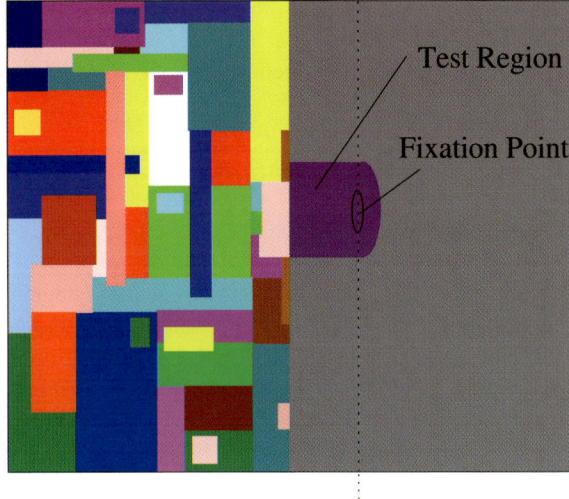

Figure 2.23 Stimuli used by Land to determine if the visual cortex is also involved in color perception. Note that the colors shown here are just meant to illustrate the experiment. They do not correspond to the colors used by Land et al. Land covered one half of a color Mondrian with black velvet. A purple test region was placed at the center of the Mondrian such that it was bordered on three sides by the black velvet. Human subjects were asked to fixate the point marked and report on the perceived color of the test region when asked. (Adapted by permission from Macmillan Publishers Ltd: Nature, E. H. Land, D. H. Hubel, M. S. Livingstone, S. Hollis Perry, M. M. Burns, Nature, Vol. 303, No. 5918, pp. 616–618, Copyright 1983.)

correspondence between Horn's retinex algorithm (Horn 1974) and cells found in the human retina. In 1983, Land was able to show that an intact corpus callosum is required for normal color perception. Therefore, color constant descriptors must be computed inside the visual cortex. Currently, it is known that most likely color perception occurs in an area called *V4* (Zeki 1993; Zeki and Marini 1998). V4 is the first area beyond V2 that contains wavelength-selective cells and also has callosal connections. It was found that visual area V4 contains cells that respond to the color of objects irrespective of the wavelength composition of the light reflected by the object.

Land again used a color Mondrian for his experiment. He covered one half of the Mondrian with black velvet. A purple piece of paper was placed in the center of the Mondrian as shown in Figure 2.23. The purple piece of paper was bounded by the velvet on three sides. A subject whose corpus callosum was cut off for medical reasons was asked to view the Mondrian. The subject was asked to fixate rotating bilaterally symmetrical letters that were shown at the center of the Mondrian. The subject's task was to report on the perceived color of the test region (the purple sheet of paper) when asked. The moving fixation target was used to reduce the risk of after-images. The fixation point was set at

THE VISUAL SYSTEM

3.7° to the right of the center of the Mondrian. This ensured that the colored part of the Mondrian was entirely in the subject's left visual field.

Since the subject's corpus callosum was cut off, the left half of the image was processed by the right half of the brain and vice versa. If computations required for color constancy actually occur in the retina, then events that modify only one half of the image should have no impact on color perception of objects viewed by the other half of the brain. Three slide projectors were used to illuminate the setup. One was equipped with a red filter, the second with a green filter, and the third with a blue filter. Neutral-density filters were used as slides in each of the projectors. The slides had a hole that exactly matched the shape of the purple piece of paper. The projectors were carefully registered such that the projected hole of the slide was in register with the test region, i.e. the purple piece of paper. By removing the slides from the projectors, the illumination could be varied outside of the test region whereas the illumination inside the test region remained constant. The filters used and the intensities of the projectors were chosen such that according to the retinex theory the color of the purple piece of paper was white. Most normal human subjects also reported the color of the test region to be white when the illuminant in the surrounding part of the image was attenuated. If the whole stimulus was uniformly illuminated, i.e. the neutral-density filters were removed from the projectors, then the test region appeared purple to a normal human observer.

However, the color perception of the subject with a cut corpus callosum was different. His speech center was in the left hemisphere of the brain. He was only able to describe things that he saw in his right visual field. He was not able to verbally describe things that he saw in his left visual field. The subject with a cut corpus callosum always reported perceiving the test region as white irrespective of the illumination, that is, he also perceived the test region to be white if the neutral filters, which attenuate the light outside of the test region, were removed from the projectors. However, when the subject with the cut corpus callosum viewed the setup in a mirror, his color perception matched those of a normal observer. When he viewed the setup through a mirror, he reported the test region to be purple if a uniform illumination was used and white whenever the illuminant was attenuated using the neutral-density filters.

Figure 2.24 shows the experimental setup. A subject with a cut corpus callosum views a color Mondrian. Half of the Mondrian is covered with black velvet. In the center of the Mondrian there is a purple test region. The right half of the stimulus is processed by the left hemisphere of the brain. The left half is processed by the right hemisphere of the brain. Since his verbal region is in the left hemisphere, the subject can only describe what he sees in the right visual field. The left hemisphere processes only the visual information obtained from the black velvet and the purple test region. The illuminant of the test region always stays constant. Since the color of the velvet is black, it does not matter if the illuminant is attenuated outside of the test region. In this case, the subject always reports the test region to be white. However, if the subject views the setup through a mirror, then the colored half of the Mondrian is processed by the left hemisphere. In this case, the illuminant outside of the test region does have an impact on the color perception of the test region. If the illuminant outside of the test region is attenuated, then the test region appears to be white; otherwise it appears to be purple. If the neutral-density filters are not used, then the relative color of the test region can be correctly determined. However, if the light outside of the

Direct View	Mirror Image

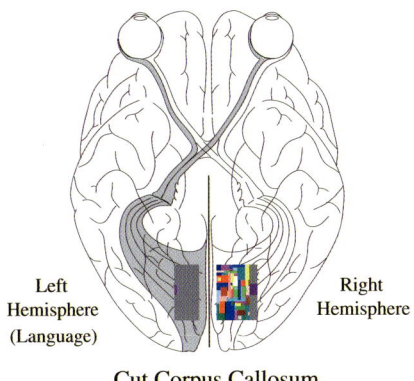	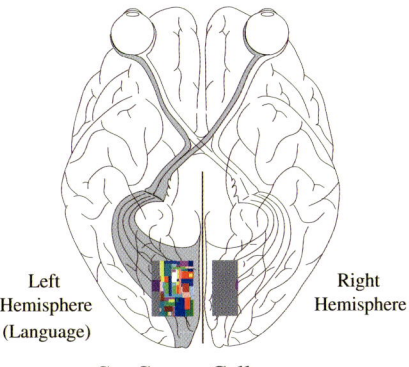

Figure 2.24 A human subject with a cut corpus callosum views a color Mondrian. Half of the color Mondrian is covered with black velvet. A purple test region is located at the center of the stimulus. The left half of the stimulus is processed by the right hemisphere of the brain and vice versa. The subject views the stimulus either directly or through a mirror. (Adapted by permission from Macmillan Publishers Ltd: Nature, E. H. Land, D. H. Hubel, M. S. Livingstone, S. Hollis Perry, M. M. Burns, Nature, Vol. 303, No. 5918, pp. 616–618, Copyright 1983. Figure 2.24 shows the visual path (Figure 2.9) which is reproduced from Semir Zeki. A Vision of the Brain. Blackwell Science, Oxford, 1993, by permission from Blackwell Science Ltd.)

test region is attenuated, then the test region is assumed to be white because the test region is the brightest patch in view.

A normal observer is able to exchange information between the left and the right hemisphere of the brain through the corpus callosum. A normal observer is therefore able to report on the correct color of the test region if the neutral-density filters are not used. In contrast, the subject with a cut corpus callosum can only report on the visual information processed by the left hemisphere of the brain. His color perception depends on whether a mirror setup is used. Therefore, the visual cortex is also involved in the process of color constancy.

3

Theory of Color Image Formation

We will now have a look at the theory of color image formation. How do we arrive at a digitized image of an everyday scene? We assume that we have placed some objects in front of our digital or analog camera. The objects have to be illuminated by a light source, otherwise they would not be seen. If we take our photographs outside, during daytime, the sun will provide the necessary illumination. At night, we may use artificial light sources such as neon lights or an ordinary light bulb. In the morning or evening hours, we may encounter a mixed illumination. Some illumination will be provided by the sun, while some will be provided by artificial light sources. We may also have indirect illumination. This can occur if we have a mirror or shiny object that reflects most of the incident light. Many indoor lamps illuminate the room by shining its light on the ceiling. From there, the light is reflected and illuminates the whole room. Because of indirect lighting we almost never have a single light source, but usually have many in an everyday scene.

If we trace the path of a light ray, starting from the light source, the ray will hit an object at some point. Part of the light will be absorbed, the remainder will be reflected. The reflected light may enter the lens of our camera. Most of the time, however, the ray of light will be reflected such that it hits another object or it may disappear into infinity. We can distinguish between matte and shiny objects. A matte object reflects light equally in all directions as shown in Figure 3.1(a). The amount of reflected light does not depend on the position of the viewer. A perfect mirror, in contrast, which is shown in Figure 3.1(b), will reflect all of the incident light.

Let $\mathbf{N}_{Obj}$ be the normal vector at the position where an incoming ray hits the object. Let $\mathbf{N}_L$ be a normalized vector that points to the direction of the light source. Let ϕ be the angle between the normal vector $\mathbf{N}_{Obj}$ and the vector $\mathbf{N}_L$ that points to the light source. Then the angle between the outgoing ray and the normal vector will also be ϕ. For a transparent object, part of the ray will be refracted and the remainder will be reflected.

A transparent surface is shown in Figure 3.2. Part of the incident light is absorbed by the surface. Some part of the remaining light is reflected back from the surface. The remaining part is refracted and passes through the surface. Let θ_L be the angle of incidence, and θ_T be the angle of transmission. The angle of incidence and the angle of transmission

THEORY OF COLOR IMAGE FORMATION

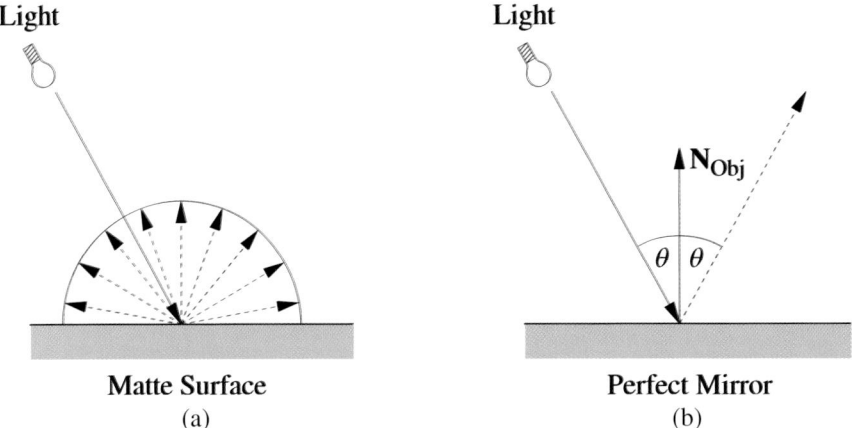

Matte Surface
(a)

Perfect Mirror
(b)

Figure 3.1 A matte surface is shown in (a) and a perfect mirror in (b). The matte surface absorbs part of the incoming light. The remaining light is reflected equally in all directions. The perfect mirror reflects all of the incoming light.

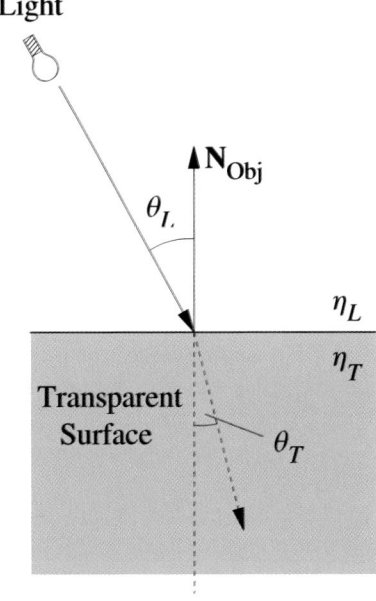

Figure 3.2 A transparent surface. Part of the incident light is absorbed by the surface, the remainder is refracted and reflected.

are related by Snell's law (Jähne 2002; Lengyel 2002).

$$\frac{\eta_L}{\eta_T} = \frac{\sin\theta_L}{\sin\theta_T} \tag{3.1}$$

where η_L is the refractive index of the medium above the object and η_T is the refractive index of the object.

THEORY OF COLOR IMAGE FORMATION 41

Eventually, the ray of light will enter the lens of the camera. In the back of the camera, we either have an analog film or a sensor array that measures the incident light. We will look at how the analog film responds to light in the next section. Digital photography is addressed in Section 3.2.

3.1 Analog Photography

Black and white film as well as color film consists of several layers. Figure 3.3 shows how a black and white image is created on film (Hedgecoe 2004). The top layer is a scratch resistant coating that protects the film. The second layer, the emulsion layer, contains silver halide crystals that are sensitive to light. The silver halide crystals are suspended in a binding agent that is called an *emulsion*. Gelatin is used as a binding agent. The third layer is the antihalation layer. This layer ensures that no light is reflected back through the emulsion. This would cause halos around bright areas, hence the name "antihalation layer". At the bottom, we have the plastic film base and an anticurl backing. The last layer ensures that the film does not curl up.

When the emulsion is exposed to light, silver ions are created. The silver ions are located at positions that were exposed to light. This is called a *latent image*, which is invisible to the eye. The latent image is then developed which produces a visible image. The silver ions create a black–silver image where light has struck. The development of the film has to be carried out in darkness in order to avoid additional exposure of the film. At this stage, the film still contains the silver halide crystals. These

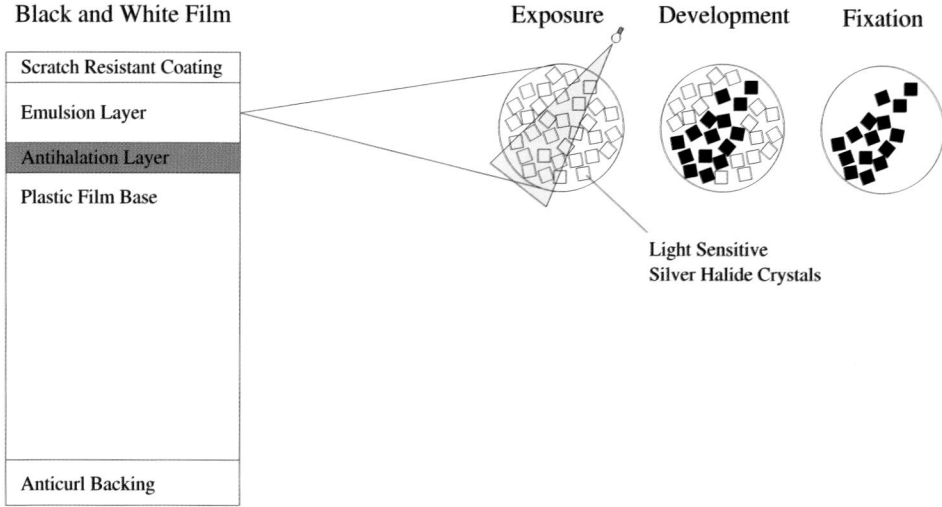

Figure 3.3 Exposure, development, and fixation for a black and white film. Black and white film consists of a single light sensitive layer (redrawn and modified from Hedgecoe 2004).

silver halide crystals have to be removed. Otherwise they would turn black when the film is again exposed to light. The silver halide crystals that were not exposed and developed are removed during a process called *fixation*. Fixation neutralizes the developer. Unused silver halide crystals are washed away. The result is a stable negative. The negative is used to create a black and white print by applying a similar process. The photographic paper also contains a light sensitive silver halide emulsion.

The size of the silver halide crystals determines the resolution, contrast, and speed of the film. The smaller the crystals or grains, the higher the resolution of the film. Small grains, however, require a lot of light, and such film is said to be slow. With larger grains, the resolution of the film decreases. Large grains require less light, such film is said to be fast. The grain size also determines the contrast or range of gray tones of the film.

Color film consists of three layers of emulsion. Since our visual system has three different cones that primarily respond to light in the red, green, and blue part of the spectrum, three light sensitive emulsions are used to capture the incident light in the red, green, and blue parts of the spectrum. Figure 3.4 shows the different layers of color negative film. The first layer is again a scratch resistant coating. Silver halides are naturally sensitive to light in the blue part of the spectrum. Blue skies turn out to be white on film, green grass or red flowers turn out to be darker in the picture. Thus, the first light sensitive emulsion is used to sense light in the blue part of the spectrum. A yellow filter is used as the next layer. The yellow filter blocks the light from the blue part of the spectrum. By adding diluted dyes to the emulsion one can make the emulsion sensitive to either red or green. A green sensitive emulsion is used as the second light sensitive layer. A magenta filter blocks out the green light. The third emulsion is used to sense light in the red part of the spectrum. An

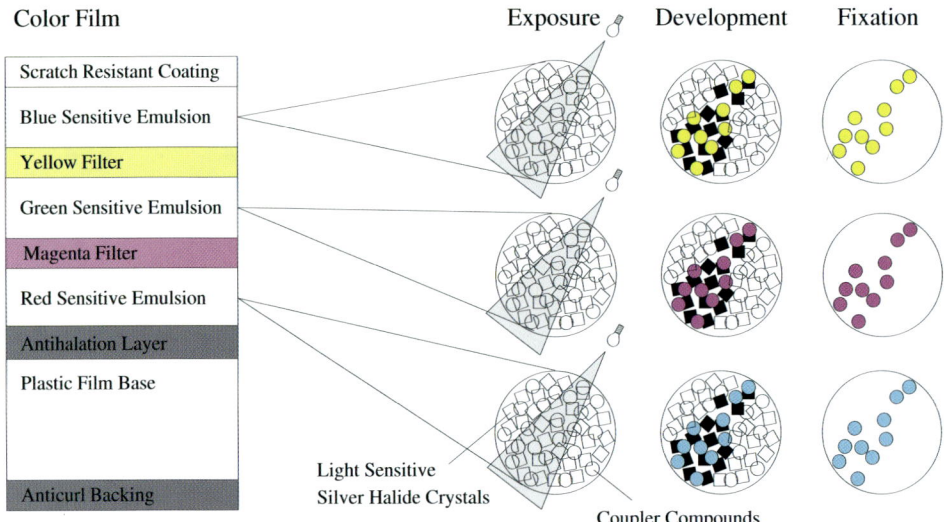

Figure 3.4 Exposure, development, and fixation for a color film. Color film consists of three light sensitive layers (redrawn and modified from Hedgecoe 2004).

THEORY OF COLOR IMAGE FORMATION

antihalation layer prevents light from being reflected back to the emulsion. The remaining layers are the plastic film base and the anticurl backing.

When the color film is exposed to light, silver ions are created inside the emulsions sensitive to the red, green, and blue part of the spectrum. The result is a latent image consisting of three separate black and white images. The latent images of all three layers are developed. A developer is added which becomes oxidized and combines with coupler compounds to produce dyes. Three types of dyes are used: yellow, magenta, and cyan. The result is a negative image that consists of three layers. The antihalation layer and the yellow and magenta filter become colorless and transparent after processing. The yellow dyes are located at positions where light struck the silver halides of the blue sensitive emulsion. The magenta dyes are located at positions where light struck the silver halides of the green sensitive emulsion. The cyan dyes are located at positions where light struck the silver halides of the red sensitive emulsion. Fixation neutralizes the developer. Unused silver halide crystals and coupler compounds are washed away. The result is a stable color negative.

Figure 3.5 shows the sensitivity curves of the red, green, and blue sensitive layers for two different types of slide film. Color slide film also consists of three light sensitive layers. The structure of the FUJICHROME 64T Type II Professional color reversal film is shown in Figure 3.6. A positive color image is created during development by first developing the latent image in all three layers. The remaining sensitive emulsion is exposed either chemically or with light. This is the color reversal step. A developer is added which combines with the coupler compounds to produce the three dyes yellow, magenta, and cyan. Silver and unused compounds are bleached out during fixation. The result is a positive color image. Figure 3.7 shows the spectral density curves of the three dyes yellow, magenta, and cyan.

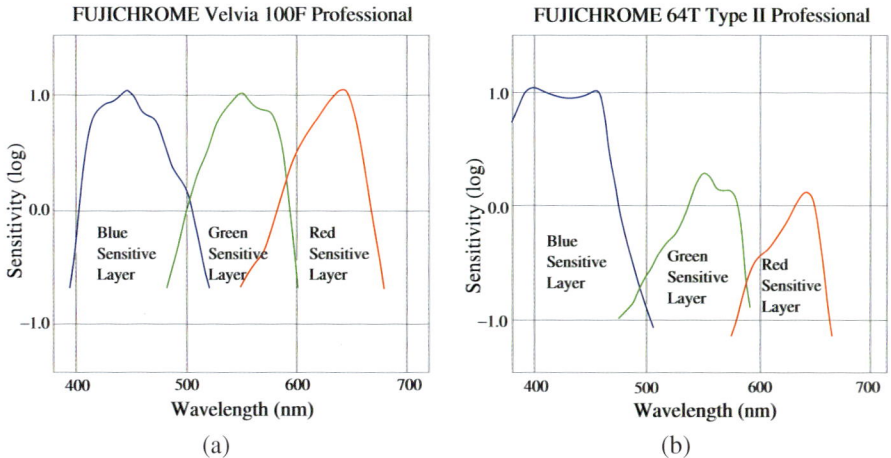

Figure 3.5 Sensitivity curves of the red, green, and blue sensitive layers for two types of film. Graph (a) shows the curves for the FUJICHROME Velvia 100F Professional film. Graph (b) shows the curves for the FUJICHROME 64T Type II Professional film (redrawn from Fuji Photo Film Co., LTD. 2005a,b). Reproduced by permission of FUJIFILM.

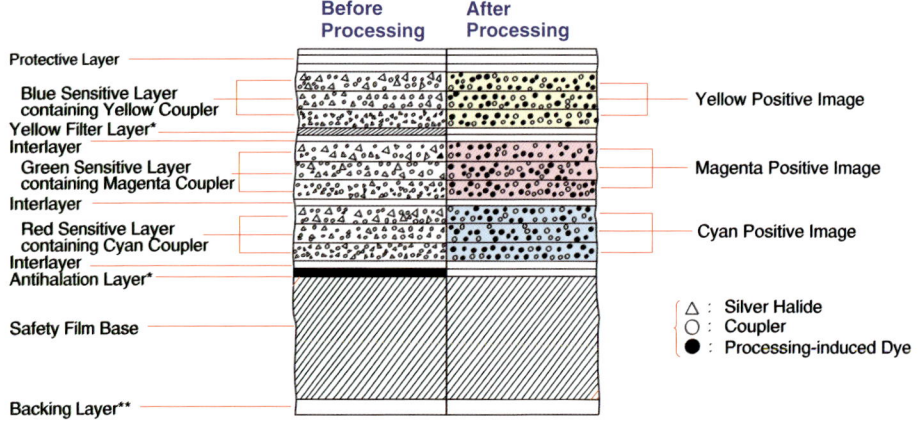

* These layers become colorless and transparent after processing.
** The backing layer is colorless and transparent both before and after processing.

Figure 3.6 Film structure of the FUJICHROME 64T Type II Professional color reversal film (from Fuji Photo Film Co., LTD. 2005a). Reproduced by permission of FUJIFILM.

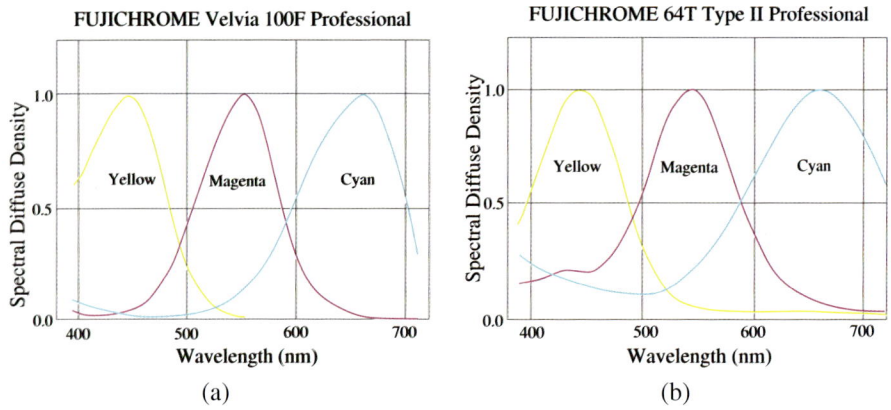

Figure 3.7 Spectral density curves of the three film dyes, yellow, magenta, and cyan for the FUJICHROME Velvia 100F Professional film shown in (a) and the FUJICHROME 64T Type II shown in (b) (redrawn from Fuji Photo Film Co., LTD. 2005a,b). Reproduced by permission of FUJIFILM.

Color prints are created in a similar way. The color negative is placed in front of an enlarger light. Photographic paper is exposed to light that shines through the color negative. The photographic paper also consists of three layers of emulsion. Again, a three-layered latent image is formed. The coupler compounds of the three layers combine with yellow, magenta, and cyan dyes to form a negative image. Since we have applied the same process twice, we have arrived at a color positive print. The color, which is seen in the print, is a function of the spectral sensitivity curves of the light sensitive emulsion and is also a function of the spectral dye density curves. A red ball will turn out to be red in the color print because the red light was absorbed by the red sensitive layer of the film. As a result, cyan dyes are deposited in the emulsion layer that is sensitive to red light. When the color print is created, white light illuminates the color negative. Cyan light illuminates the photographic paper. The green and blue sensitive layers respond to the cyan light. After development, yellow and magenta dyes are deposited in the green and blue layers. If the color print is now illuminated with white light, the yellow dye absorbs light in the blue part of the spectrum and the magenta dye absorbs light in the green part of the spectrum. Only the red light is reflected back to the viewer, making the ball red in the image.

When the film is placed inside the camera it measures the light that enters the lens. Since only reflected light is measured, the measured color will depend on the type of illuminant used. Special types of film exist, which can be used for daylight or tungsten light. When we look again at Figure 3.5, we see that the FUJICHROME Velvia 100F film is a film intended for daylight. The FUJICHROME 64T, in contrast, is intended for tungsten illumination. The blue sensitive layer is much more sensitive than the green and red sensitive layers and the green sensitive layer is slightly more sensitive than the red layer.

A color correction may also be achieved by using filters. Table 3.1 shows the type of filter used by professional photographers to achieve accurate color reproduction. The required filter depends on the type of illuminant and also on the type of film. The type of light source can be described using the temperature of a black-body radiator. A black-body radiator is a light source whose spectral power distribution depends only on its temperature (Jacobsen et al. 2000). The color temperature of a light source is the temperature of a black-body radiator, which essentially has the same spectral distribution in the visible region. The concept of a black-body radiator is formally introduced in Section 3.5.

Additional postprocessing may be done on a computer. The color negative can be either digitized using a film scanner or we can develop color prints that can then be digitized

Table 3.1: Filters that should be used for accurate color reproduction. The filter depends on the light source as well as the type of film used. The filter numbers are references to Kodak Wratten filters. The first digit designates the color and the letter the density of the filter (data from Hedgecoe (2004)).

Light Source	Temperature (K)	Filter for Tungsten Film	Daylight Film
Candle light	1930	82C + 82C	Not recommended
40 W light bulb	2600	80D	Not recommended
150 W light bulb	2800	82C	80A + 82B
Tungsten photographic lamps	3100	no filter required	80A
Halogen lamp	3400	82A	80B
Daylight	5800	85B	No filter required
Electronic flash	5800	85B	No filter required

with a flatbed scanner. In order to achieve accurate color reproduction, the scanner has to be calibrated.

3.2 Digital Photography

A digital camera works similar to an analog camera except that a sensor array replaces the analog film. Currently available digital cameras outperform film in image quality and sharpness (Castleman 2004). Professional or semiprofessional digital single-lens reflex (SLR) cameras with large sensors are particularly suitable when little light is available (Uschold 2002). Since our visual system has three different cones which primarily respond to light in the red, green and blue parts of the spectrum, we need three types of sensors to measure the incident light. A standard sensor array consists of a single type of light sensitive sensor. Filters are placed above the light sensitive sensors to make them respond to either the red, green, or blue part of the spectrum. Usually the red, green and blue filters are arranged in a GRGBpattern as shown in Figure 3.8. This type of sensor arrangement is called a *Bayer pattern* (Bayer 1976; Bockaert 2005). The red filter blocks all wavelengths except red light, the green filter blocks all wavelengths except green light, and the blue filter blocks all wavelengths except blue light. Other types of sensors that use red, green, blue, and cyan filters also exist. Such a sensor is used in the Sony DSC-F828 camera.

Since the red, green, and blue sensors are spatially separated, the camera does not measure the incident light at a single point in space. Instead, the measurements of the red, green, and blue intensities are interpolated to obtain the color at a given point. That is, we can only guess the correct color of an image point. If the image is slightly blurred the interpolated color will be correct. Otherwise, colors that are not actually in the image may be introduced during interpolation.

Figure 3.9 illustrates how an image is obtained from a standard Bayer sensor. Figure 3.9 (a) shows the image that is formed on the sensor array. Each sensor measures only the light from the red, green, or blue part of the spectrum. In a GRGB Bayer sensor, 25% of the pixels respond to red light, 50% respond to green light, and 25% respond to blue light. Figure 3.9 (b) shows the image as seen by the Bayer sensor. The individual channels are shown in (d), (e), and (f). For each pixel, we only have the exact information for one color channel. The data for the other color channels must be

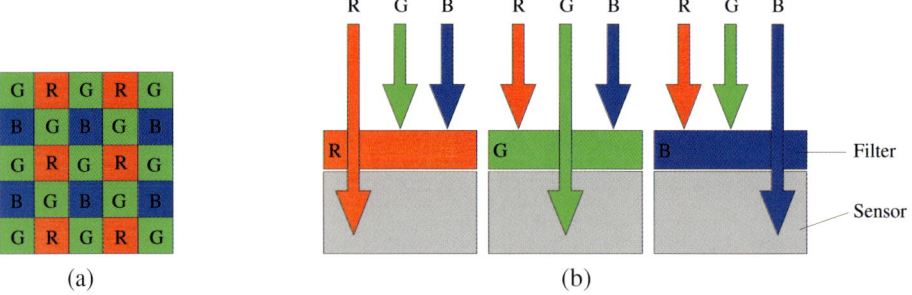

Figure 3.8 Standard arrangement of sensors (a). Three types of filters are used, which are placed above the sensors (b).

THEORY OF COLOR IMAGE FORMATION

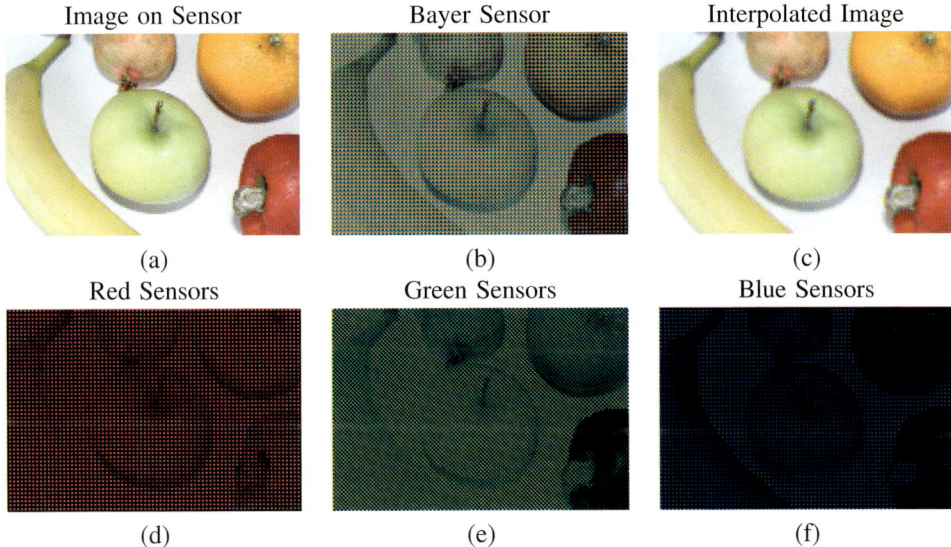

Figure 3.9 The actual image on the sensor is shown in (a). Image (b) shows what a standard Bayer sensor measures. The red, green, and blue channels are shown in (d), (e), and (f). Image (c) shows the interpolated image.

interpolated from surrounding pixels. The image shown in Figure 3.9(c) shows the interpolated image. Since we must interpolate to obtain the output image, the actual resolution of the output image is less than the number of pixels of the Bayer sensor.

An exception to this is the Foveon X3 sensor (Foveon Inc. 2002). Figure 3.10 shows how this sensor works. Here, the three sensors are placed above each other. The topmost sensor measures blue light, the one below measures green light, and the one at the bottom measures red light. This works because silicon, the material that the chips are made of, absorbs different wavelengths at different depths. Blue light is absorbed very early, while green light is able to penetrate a little deeper than blue, and red light is able to penetrate the deepest. Since the red, green, and blue sensors are stacked up behind each other, the Foveon chip is able to measure these three color components at every image point.

3.3 Theory of Radiometry

A single sensor measures the incident light at a particular point on the sensor array. We will now look at how the measured data depends on the light emitted by an object. See Horn (1986); Poynton (2003); Wyszecki and Stiles (2000) and Levi (1993) for a nice introduction to the theory of radiometry. The following account is mostly based on Horn (1986). In radiometry, the amount of light falling on a surface is called *irradiance* (Horn 1986; International Commission on Illumination 1983; Jain et al. 1995; Lengyel 2002; Mallot 1998; Radig 1993). Irradiance E is measured as the power or flux Φ per unit area

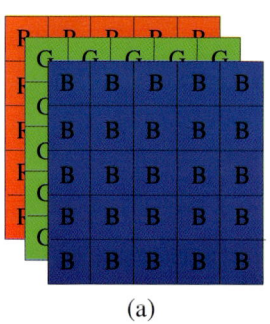
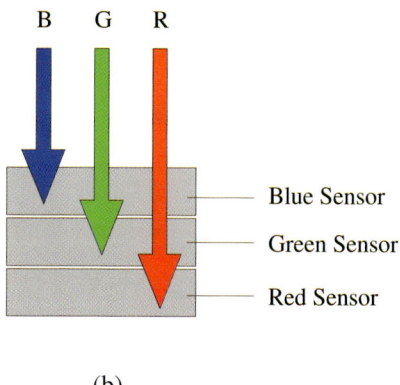

(a)   (b)

Figure 3.10 Instead of placing the sensors side by side, the three sensors are placed on top of each other (a). Silicon absorbs different wavelengths of light at different depths (b). Blue light is absorbed very early, green light is able to penetrate a little deeper, and red light penetrates even deeper. By placing the sensors at different depths, the Foveon chip is able to measure the three components at every image point.

A falling on a surface.

$$E = \frac{d\Phi}{dA}. \qquad (3.2)$$

It is measured in watts per square meter ($\frac{W}{m^2}$). If we focus our attention on a single point on the surface of an object, the incoming light is reflected into a hemisphere centered around this point. Let us consider the amount of light radiated per unit area per unit solid angle Ω of the surrounding hemisphere. The solid angle Ω is defined as the area of the projection of a planar area A_{Obj} onto the unit hemisphere as shown in Figure 3.11. The light radiated per unit area per unit solid angle is called the *radiance* or *radiosity L*

$$L = \frac{d^2\Phi}{d\Omega dA} = \frac{d^2\Phi}{d\Omega dA_0 \cos\theta} \qquad (3.3)$$

where $dA = dA_0 \cos\theta$ denotes the unit area in the specified direction θ. It is measured in watts per square meter per steradian ($\frac{W}{m^2 sr}$). Steradian, abbreviated sr, is the unit of solid angle measure. A matte surface has a uniform radiance distribution. In contrast, a perfect mirror reflects all of the irradiance in a single direction. Most actual surfaces radiate different amounts of energy in different directions.

The solid angle of a planar patch with area A_{Obj} at a distance r from the origin is calculated as

$$\Omega = \frac{A_{Obj} \cos\phi}{r^2} \qquad (3.4)$$

where ϕ is the angle between the normal of the patch $\mathbf{N}_A$ and a line between the patch and the given point (Figure 3.11). In other words, the size of the patch, when projected on to the hemisphere around the viewer is inversely proportional to the squared distance from the viewer and proportional to the angle between the normal of the patch and the direction

THEORY OF COLOR IMAGE FORMATION

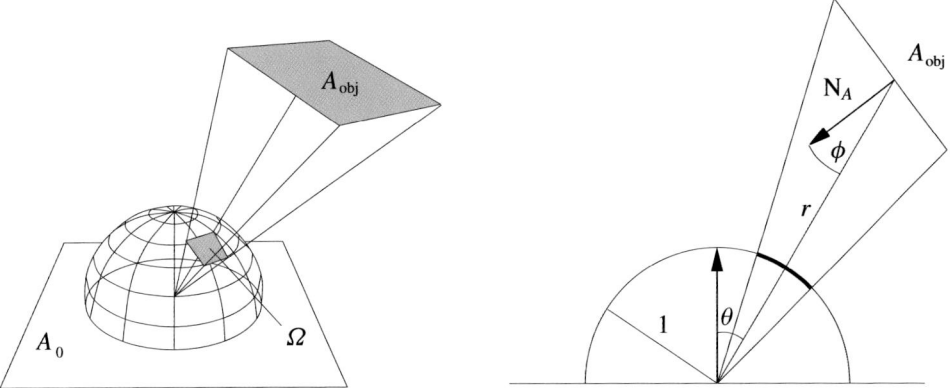

Figure 3.11 The solid angle is defined as the area of the projection of a planar area A on the unit hemisphere.

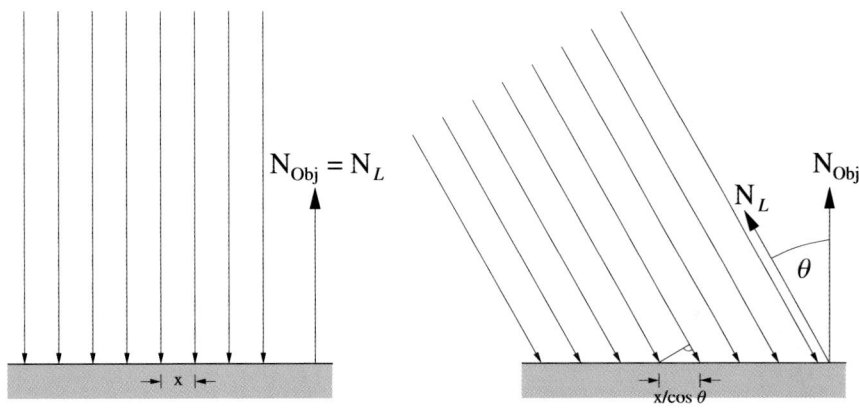

Figure 3.12 The irradiance falling onto a planar patch is reduced by a factor of $\cos\theta$, where θ is the angle between the direction of the light source and the normal vector of the planar patch.

of the patch relative to the viewer. The farther the patch is, the smaller the visual angle subtended by the patch. Also, the more tilted the patch is, the smaller it appears on the hemisphere. The projection of the patch on the unit hemisphere is maximally large if its surface normal is oriented toward the given point. As the patch is rotated away from this direction, the projection becomes smaller. The whole hemisphere has a solid angle of 2π.

Let L be the radiance given of by a light source. The irradiance falling on a planar patch depends on the angle θ between the normal vector of the patch and the direction of the light source as shown in Figure 3.12. Let $\mathbf{N}_{Obj}$ be the normal vector of the patch and let $\mathbf{N}_L$ be the normal vector that points to the direction of the light source. Then the

THEORY OF COLOR IMAGE FORMATION

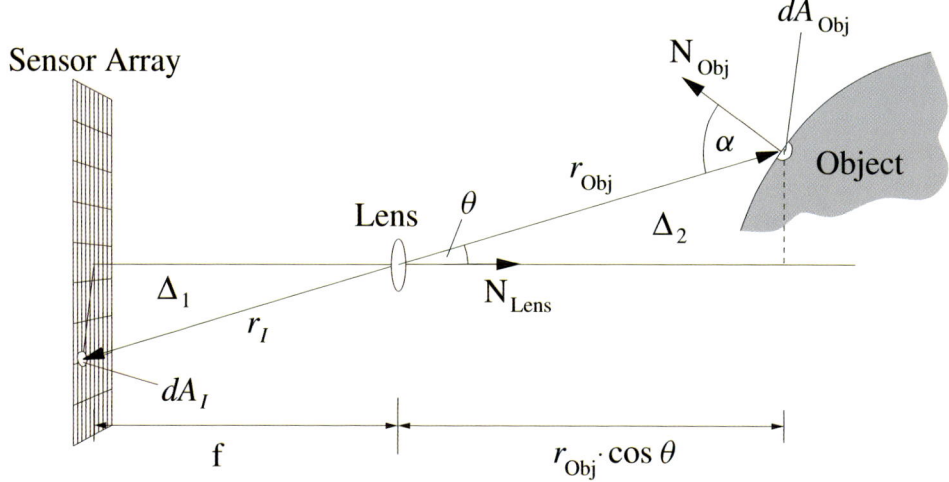

Figure 3.13 An array of sensors is located at a distance of f behind a lens. The center of the lens is assumed to be positioned at the origin. A small patch with normal vector $\mathbf{N}_{Obj}$ and area dA_{Obj} is projected through the lens onto the sensor array. The size of the projection on the sensor array is dA_I.

irradiance E is scaled by a factor of $\cos\theta$ where θ is the angle between $\mathbf{N}_{Obj}$ and $\mathbf{N}_L$.

$$E = L\cos\theta = L\mathbf{N}_{Obj}\mathbf{N}_L. \tag{3.5}$$

The irradiance at an image point depends on the radiance at a point on the object. In calculating the irradiance, we follow Horn (1986). We assume that the lens of the camera is located at the origin. The sensor array is located at a distance f behind the lens as shown in Figure 3.13. Each sensor measures the irradiance at a particular point of the image. A point in the upper left of the coordinate system of the camera is projected onto the lower right of the sensor array. Therefore, the image is inverted and needs to be transformed appropriately before display.

Let the area of the sensor be dA_I, then the irradiance measured by the sensor is given by

$$E = \frac{dP_{Lens}}{dA_I} \tag{3.6}$$

where dP_{Lens} is the power of the light that has passed through the lens and which originated at a small planar patch on the object. Let the radiance of the patch be L. Since radiance is defined as power per area per solid angle, we need to calculate the solid angle the lens subtends as seen from the planar patch in order to determine the power given off by the patch toward the lens. The power dP_{Lens} that passes through the lens, is given by

$$dP_{Lens} = L\Omega_{Lens}dA_{Obj}\cos\alpha \tag{3.7}$$

where Ω_{Lens} is the visual angle of the lens as seen from the planar patch and α is the angle between the normal vector $\mathbf{N}_{Obj}$ of the patch and the vector $\mathbf{r}_{Obj}$, which points from the

THEORY OF COLOR IMAGE FORMATION

lens to the planar patch. The factor $\cos\alpha$ takes care of the fact that the area of the patch appears foreshortened when viewed from the lens. Let the diameter of the lens be d, then we obtain for the solid angle

$$\Omega_{\text{Lens}} = \pi \frac{d^2}{4} \frac{\cos\theta}{r_{\text{Obj}}^2} \tag{3.8}$$

where θ is the angle between the normal vector $\mathbf{N}_{\text{Lens}}$ of the lens and the vector $\mathbf{r}_{\text{Obj}}$, which points from the center of the lens in the direction of the patch. Therefore, we obtain for the irradiance

$$E = \frac{dP_{\text{Lens}}}{dA_I} = L\Omega_{\text{Lens}} \frac{dA_{\text{Obj}}}{dA_I} \cos\alpha = L \frac{dA_{\text{Obj}}}{dA_I} \pi \frac{d^2}{4r_{\text{Obj}}^2} \cos\theta \cos\alpha. \tag{3.9}$$

The area of the projection of the patch on the image plane dA_I can be calculated using the fact that the image plane is located at a distance of f behind the lens. The light rays from the patch pass through the center of the lens and then onto the sensor array. The solid angle subtended by the patch as seen from the lens and the solid angle subtended by the projection on the sensor array have to be equal. Remember, we are trying to determine the amount of light falling onto a patch on the sensor. A particular patch is projected onto a corresponding area of the sensor, hence, the two solid angles of the patch and the solid angle of the sensor have to be equal. Let Ω_I be the solid angle of the projection of the patch on the sensor array and Ω_{Obj} be the solid angle of the patch as seen from the lens. Then we have,

$$\Omega_I = \Omega_{\text{Obj}} \tag{3.10}$$

$$\frac{dA_I \cos\theta}{r_I^2} = \frac{dA_{\text{Obj}} \cos\alpha}{r_{\text{Obj}}^2} \tag{3.11}$$

where $\mathbf{r}_I$ is the vector pointing from the center of the lens to the projection of the patch on the sensor array. We solve this equation to calculate the fraction dA_{Obj}/dA_I.

$$\frac{dA_{\text{Obj}}}{dA_I} = \frac{r_{\text{Obj}}^2 \cos\theta}{r_I^2 \cos\alpha}. \tag{3.12}$$

We now substitute this expression into the equation that describes the irradiance falling on the projection of the patch on the sensor array.

$$E = \pi L \frac{r_{\text{Obj}}^2 \cos\theta}{r_I^2 \cos\alpha} \frac{d^2}{4r_{\text{Obj}}^2} \cos\theta \cos\alpha = \pi L \frac{d^2 \cos^2\theta}{4r_I^2}. \tag{3.13}$$

The distance from the lens to the sensor $\mathbf{r}_I$ can be computed by considering the two similar triangles $\triangle_1$ and $\triangle_2$ which are shown in Figure 3.13.

$$\mathbf{r}_I = -f \frac{\mathbf{r}_{\text{Obj}}}{\mathbf{r}_{\text{Obj}} \mathbf{N}_{\text{Lens}}} = -f \frac{\mathbf{r}_{\text{Obj}}}{r_{\text{Obj}} \cos\theta}. \tag{3.14}$$

Therefore, we obtain for the irradiance E

$$E = \pi L \frac{d^2}{4f^2} \cos^4\theta. \tag{3.15}$$

The irradiance E on the projection of the patch on the sensor array is proportional to scene radiance L. The irradiance is also affected by the focal length f of the camera and the diameter d of the lens. Because of the factor $\cos^4 \theta$, the irradiance becomes smaller, the larger the angle θ between the patch and the optical axis.

3.4 Reflectance Models

So far, we have seen that the irradiance falling on the sensor array is directly proportional to the scene radiance. In other words, the sensors do measure scene radiance. We will now see how scene radiance depends on the surface properties of the object and on the light that illuminates the object. The light from a light source falls on a small patch as shown in Figure 3.14. Some of the light is absorbed by the patch, the remainder is reflected in different directions. The amount of reflected light may depend on the direction of the light source as well as the direction of the viewer. Matte surfaces reflect light equally in all directions. In this case, the amount of reflected light depends only on the direction of the light source relative to the normal of the surface. For specular surfaces, both, the direction of the light source and the direction of the viewer relative to the patch, have to be taken into account.

Let $\mathbf{N}_{Obj}$ be the normal vector of the patch, let $\mathbf{N}_L$ be the normal vector that points to the direction of the light source, and let $\mathbf{N}_V$ be the normal vector that points to the direction of the viewer. Let $E(\mathbf{N}_L)$ be the irradiance falling onto the patch, and $L(\mathbf{N}_V)$ be the radiance leaving the patch in the direction of the viewer. The fraction of the radiance relative to the irradiance falling onto the patch is called the *bidirectional reflectance distribution function*

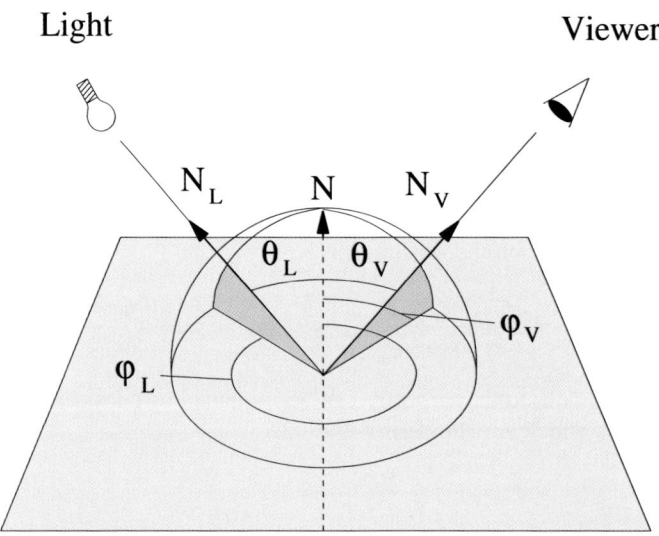

Figure 3.14 The bidirectional reflectance distribution function (BRDF) specifies how much of the incident light coming from direction (θ_L, ϕ_L) is reflected into the direction (θ_V, ϕ_V).

THEORY OF COLOR IMAGE FORMATION

or *BRDF* for short (Horn 1986; Lengyel 2002; Watt 2000).

$$f(\mathbf{N}_L, \mathbf{N}_V) = \frac{L(\mathbf{N}_V)}{E(\mathbf{N}_L)}. \tag{3.16}$$

The BRDF specifies how much of the incoming light is reflected for any given combination of light source direction $\mathbf{N}_L$ and viewer direction $\mathbf{N}_V$. If we fix the position of the light source relative to the patch, then the BRDF specifies how the reflected light is distributed around the patch. All BRDFs must fulfill the following requirement.

$$f(\mathbf{N}_L, \mathbf{N}_V) = f(\mathbf{N}_V, \mathbf{N}_L). \tag{3.17}$$

That is, if we exchange the positions of the light source and the viewer, then the BRDF must have the same form. It must be symmetric in this respect, i.e. bidirectional.

Ideal matte and specular surfaces have a very simple BRDF. In general, however, the BRDF does not have a simple structure. For example, cloth or brushed metal has a BRDF which reflects most of the light along a preferred direction. Sample BRDFs for aluminum, magnesium oxide ceramic, and sandpaper can be found in (He et al. 1991). If we know the BRDF for a particular material, then we can compute the radiance given off by an infinitesimal small patch of this material. The radiance leaving the patch is given by

$$L(\mathbf{N}_V) = f(\mathbf{N}_L, \mathbf{N}_V) E(\mathbf{N}_L). \tag{3.18}$$

Point Light Source

Let us first consider a point light source and let L be the radiance given off by this light source. If we assume that the light source is far away from the patch, then the irradiance $E(\mathbf{N}_L)$ is simply $L\mathbf{N}_{\text{Obj}}\mathbf{N}_L$. In this case, we obtain for the radiance $L(\mathbf{N}_V)$ given off by the patch in the direction of the viewer

$$L(\mathbf{N}_V) = f(\mathbf{N}_L, \mathbf{N}_V) L \mathbf{N}_{\text{Obj}} \mathbf{N}_L. \tag{3.19}$$

If we have multiple point light sources, we need to add the components from the different light sources. Let L_i be the radiance given off by light source i, then the radiance $L(\mathbf{N}_V)$ given off by the patch in the direction of the viewer is

$$L(\mathbf{N}_V) = \sum_{i=1}^{n} f(\mathbf{N}_{L_i}, \mathbf{N}_V) L_i \mathbf{N}_{\text{Obj}} \mathbf{N}_{L_i} \tag{3.20}$$

where n is the number of light sources and $\mathbf{N}_{L_i}$ is a normal vector pointing to the direction of the light source i.

Extended Light Source

If we have an extended light source, such as the sky, then we need to integrate the irradiance over the hemisphere centered at the patch. We use two angles θ and ϕ (polar angle and azimuth) to describe the direction of the light as seen from the patch. The solid angle ω of an infinitesimal patch on the hemisphere can be computed as shown in Figure 3.15. The solid angle ω of an infinitesimal patch on the hemisphere is given by

$$\omega = \sin\theta d\theta d\phi. \tag{3.21}$$

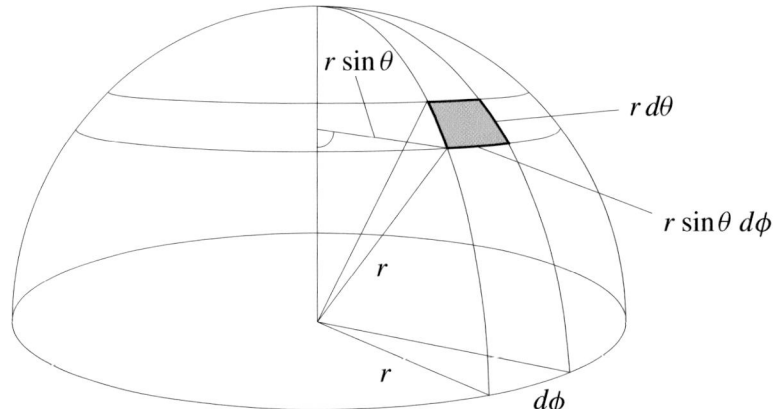

Figure 3.15 Computation of the size of an infinitesimal patch on a hemisphere.

Let $E(\theta, \phi)$ be the irradiance coming from direction (θ, ϕ). Then the irradiance falling on the small planar patch from (θ, ϕ) is $E(\theta, \phi) \cos \theta$ where the factor $\cos \theta$ takes care of the fact that the irradiance is reduced by $\cos \theta$ if it falls on the patch at an angle. We integrate over all infinitesimal patches of the hemisphere.

$$E = \int_0^{2\pi} \int_0^{\pi/2} E(\theta, \phi) \sin \theta \cos \theta \, d\theta \, d\phi. \tag{3.22}$$

This is the total irradiance falling onto the patch. The amount of light reflected in each direction depending on the direction of the irradiance is described by the BRDF. Therefore, the radiance given off by the patch into the direction of the viewer is given by

$$L(\mathbf{N}_V) = \int_0^{2\pi} \int_0^{\pi/2} f(\theta, \phi, \mathbf{N}_V) E(\theta, \phi) \sin \theta \cos \theta \, d\theta \, d\phi. \tag{3.23}$$

If we know the BRDF of the material of the viewed object, then we can compute the radiance using the preceding equation. We will now have a look at the BRDF for two idealized surfaces, a Lambertian surface and a perfect mirror. Let us address the BRDF of a perfect mirror first. If the incident light is coming from direction (θ_L, ϕ_L), then all of this light is reflected into the direction $(\theta_L, \phi_L + \pi)$. In this case, we have

$$L(\theta_V, \phi_V) = E(\theta_L, \phi_L + \pi). \tag{3.24}$$

Therefore, the BRDF is given by

$$f(\theta_L, \phi_L, \theta_V, \phi_V) = \frac{L(\theta_V, \phi_V)}{E(\theta_L, \phi_L)} = \frac{\delta(\theta_V - \theta_L)\delta(\phi_V - \phi_L - \pi)}{\sin \theta_L \cos \theta_L}. \tag{3.25}$$

where δ is the Dirac delta function (see Appendix A for a definition).

Let us now turn to a Lambertian surface. A Lambertian surface reflects the incident light equally in all directions. The radiance given off by matte objects can be approximated using the BRDF of a Lambertian surface. If the irradiance is reflected equally in all directions,

THEORY OF COLOR IMAGE FORMATION

the BRDF must be a constant. If none of the irradiance is absorbed, then the radiance given off by the surface must be equal to the irradiance. Let L be the radiance of a point light source. Let $\mathbf{N}_L$ be the normal vector in the direction of the light source, then the irradiance falling on the surface is $E = L\mathbf{N}_{\text{Obj}}\mathbf{N}_L$. Let f be the constant BRDF, then the radiance L_{Obj} given off in each direction (θ, ϕ) is

$$L_{\text{Obj}}(\theta, \phi) = fE = fL\mathbf{N}_{\text{Obj}}\mathbf{N}_L = fL\cos\alpha \qquad (3.26)$$

where α is the angle between the normal vector of the patch and the direction of the light source. If we integrate over the hemisphere above the patch, we obtain the total radiance given off by the patch. Note, that we must take into account a foreshortening factor of $\cos\theta$.

$$\int_0^{2\pi}\int_0^{\pi/2} L_{\text{Obj}}(\theta, \phi)\cos\theta\sin\theta\, d\theta\, d\phi = \int_0^{2\pi}\int_0^{\pi/2} fL\cos\alpha\cos\theta\sin\theta\, d\theta\, d\phi \qquad (3.27)$$

$$= \pi fL\cos\alpha \int_0^{\pi/2} 2\sin\theta\cos\theta\, d\theta \qquad (3.28)$$

$$= \pi fL\cos\alpha [\sin^2\theta]_0^{\pi/2} \qquad (3.29)$$

$$= \pi fL\cos\alpha. \qquad (3.30)$$

If none of the light is absorbed, then this must be equal to the radiance falling on the patch, i.e. we must have

$$\pi fL\cos\alpha = L\cos\alpha. \qquad (3.31)$$

Therefore, the BRDF of a Lambertian surface is

$$f(\mathbf{N}_L, \mathbf{N}_V) = \frac{1}{\pi}. \qquad (3.32)$$

This BRDF allows us to compute the radiance given off by a Lambertian surface illuminated by a point light source.

$$L_{\text{Obj}} = \frac{1}{\pi}L\cos\alpha. \qquad (3.33)$$

This is the result for a single point light source. If we have n point light sources and the radiance given off by each light source is L_i, then the radiance L_{Obj} given off by the patch is

$$L_{\text{Obj}} = \sum_{i=1}^{n} \frac{1}{\pi} L_i \mathbf{N}_{\text{Obj}} \mathbf{N}_{L_i} \qquad (3.34)$$

where n is the number of light sources and $\mathbf{N}_{L_i}$ is a normal vector pointing to the direction of the light source i. If the radiance of all light sources is approximately equal, we obtain

$$L_{\text{Obj}} = \sum_{i=1}^{n} \frac{1}{\pi} L \mathbf{N}_{\text{Obj}} \mathbf{N}_{L_i} \qquad (3.35)$$

$$= \frac{1}{\pi} L \mathbf{N}_{\text{Obj}} \sum_{i=1}^{n} \mathbf{N}_{L_i} \qquad (3.36)$$

$$= \frac{n}{\pi} L \mathbf{N}_{\text{Obj}} \mathbf{N}_{L\text{avg}} \qquad (3.37)$$

$$= \frac{n}{\pi} L \cos\alpha \qquad (3.38)$$

where α is the angle between the normal vector of the surface and a vector pointing to the direction of the averaged light vectors with

$$\mathbf{N}_{Lavg} = \frac{1}{n} \sum_{i=1}^{n} \mathbf{N}_{L_i}. \qquad (3.39)$$

Illumination on a Cloudy Day

Another frequently occurring case is the illumination given on a cloudy day or a brightly lit room with white walls and indirect illumination. In both cases, we assume that the radiance falling onto a small patch is constant over the hemisphere above ground. Let α be the angle between the normal vector of the patch and the vector pointing to the zenith of the hemisphere. The irradiance E is given by

$$E(\theta, \phi) = \begin{cases} E & \text{if } 0 \leq \theta \leq \frac{\pi}{2} \quad \text{and} \quad 0 \leq \phi \leq \pi \\ E & \text{if } 0 \leq \theta \leq \frac{\pi}{2} - \alpha \quad \text{and} \quad \pi \leq \phi \leq 2\pi \\ 0 & \text{otherwise} \end{cases} \qquad (3.40)$$

That is, except for a small segment of the hemisphere, the patch is uniformly illuminated as shown in Figure 3.16. Given the irradiance, we can now compute the radiance L given off by the patch. For the following calculation, we assume that $\alpha \leq \frac{\pi}{2}$.

$$L = \int_0^{2\pi} \int_0^{\pi/2} \frac{1}{\pi} E(\theta, \phi) \sin\theta \cos\theta \, d\theta \, d\phi \qquad (3.41)$$

$$= \int_0^{\pi} \int_0^{\pi/2} \frac{1}{\pi} E \sin\theta \cos\theta \, d\theta \, d\phi + \int_0^{\pi} \int_0^{\pi/2-\alpha} \frac{1}{\pi} E \sin\theta \cos\theta \, d\theta \, d\phi \qquad (3.42)$$

$$= \frac{E}{2} \int_0^{\pi/2} 2 \sin\theta \cos\theta \, d\theta + \frac{E}{2} \int_0^{\pi/2-\alpha} 2 \sin\theta \cos\theta \, d\theta \qquad (3.43)$$

$$= \frac{E}{2} [\sin^2\theta]_0^{\pi/2} + \frac{E}{2} [\sin^2\theta]_0^{\pi/2-\alpha} \qquad (3.44)$$

$$= \frac{E}{2} + \frac{E}{2} \sin^2\left(\frac{\pi}{2} - \alpha\right) \qquad (3.45)$$

$$= \frac{E}{2} \left(1 + \cos^2\alpha\right). \qquad (3.46)$$

The difference between a single point light source above the ground and a uniformly illuminated hemisphere is shown in Figure 3.17.

3.5 Illuminants

So far, we have only considered light of a single wavelength. Illuminants are described by their power spectrum. The power spectrum of several different illuminants are shown in Figure 3.18. The data was measured by Funt et al. (1998) with a Photoresearch PR-650 spectrometer. Figure 3.18 shows the power spectrum of a Sylvania 75 W halogen bulb, a

THEORY OF COLOR IMAGE FORMATION

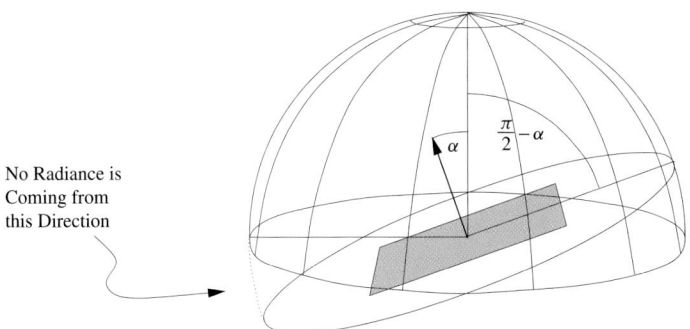

Figure 3.16 A tilted patch with uniform illumination for a hemisphere above ground. If the angle between the normal vector of the patch and the vector pointing to the zenith is α, then the patch is uniformly illuminated except for a small segment of the hemisphere above the patch.

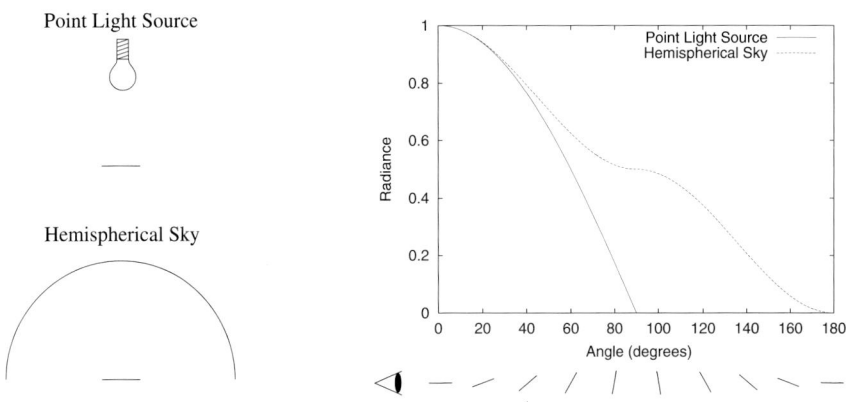

Figure 3.17 The radiance given off by a patch illuminated by a point light source is different from the radiance given off by a patch, which is illuminated by a uniformly illuminated hemisphere.

Macbeth 5000 K fluorescent, a Philips Ultralume fluorescent, and a Sylvania Cool White fluorescent tube. Some color constancy algorithms try to find out what type of illuminant produced the particular color sensation, which was measured by the sensor. If we measure the entire power spectrum for a particular patch of our object and also know the type of illuminant used, then it is easy to compute the BRDF. Let $L(\lambda)$ be the measured power spectrum and let $E(\lambda)$ be the power spectrum of the illuminant. Assuming a Lambertian surface, where the BRDF is independent of the normal vector of the patch $\mathbf{N}_{\text{Obj}}$, and also independent of the normal vector that points to the direction of the light source $\mathbf{N}_L$, we

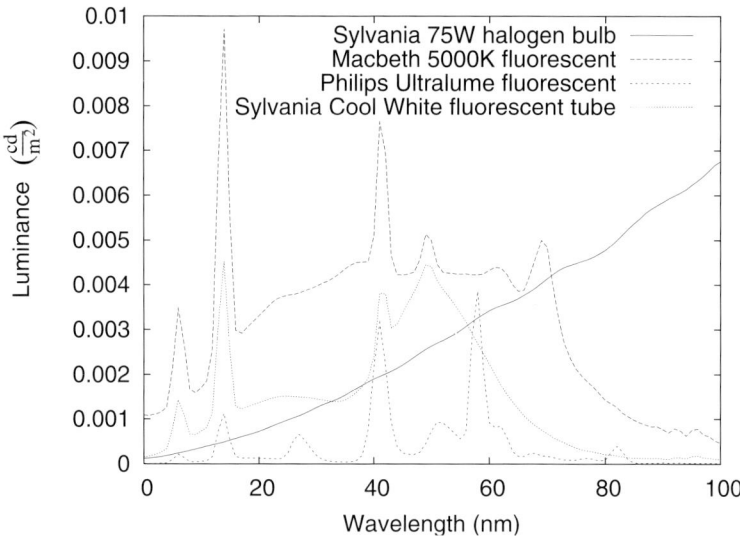

Figure 3.18 Power spectrum of four different illuminants: a Sylvania 75 W halogen bulb, a Macbeth 5000 K fluorescent, a Philips Ultralume fluorescent, and a Sylvania Cool White fluorescent tube. The power spectra were measured by Funt et al. (1998) with a Photoresearch PR-650 spectrometer.

obtain

$$f(\lambda) = \frac{L(\lambda)}{E(\lambda)}. \tag{3.47}$$

An important concept in this respect is the black-body radiator. Many natural light sources can be approximated by a black-body radiator. Therefore, this concept can be used to constrain the infinitely large set of possible illuminants. A black-body radiator is a body that absorbs all radiation (Haken and Wolf 1990; Jähne 2002). According to the laws of thermodynamics, the absorbed radiation also has to be emitted again. A black-body radiator is experimentally realized by creating a concavity whose walls are kept at a constant temperature. Radiation can only escape through a little hole. The hole has to be so small that radiation entering or leaving the whole does not alter the temperature of the black-body radiator. Radiation is measured in front of the hole using a spectrometer. The power spectrum $L(\lambda, T)$ of a black-body depends on its temperature T. It is given off equally in all directions. Such a radiator is called a *Lambertian radiator*. The power spectrum can be described by the following equation (Haken and Wolf 1990; Jähne 2002):

$$L(\lambda, T) = \frac{2hc^2}{\lambda^5} \frac{1}{(e^{\frac{hc}{k_B T \lambda}} - 1)} \tag{3.48}$$

where T is the temperature of the black-body measured in Kelvin, $h = 6.626176 \cdot 10^{-34} Js$ is Planck's constant, $k_B = 1.3806 \cdot 10^{-23} \frac{J}{K}$ is Boltzmann's constant, and $c = 2.9979 \cdot 10^8 \frac{m}{s}$ is the speed of light. The power spectra of a black-body radiator at different temperatures

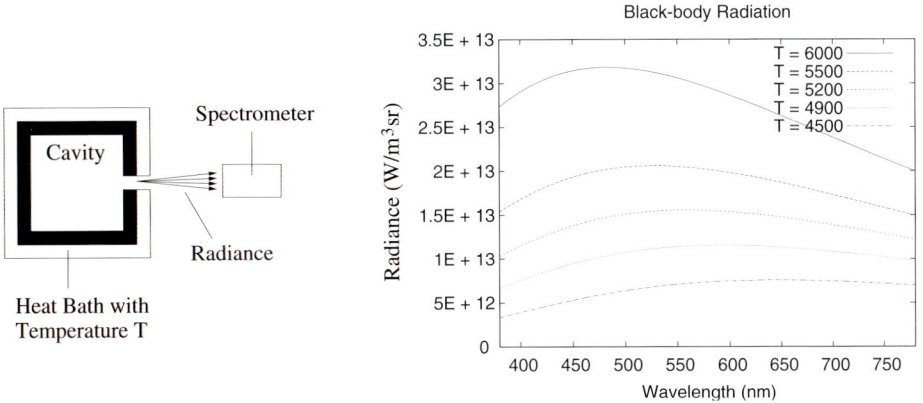

Figure 3.19 Power spectra of a black-body radiator for different temperatures.

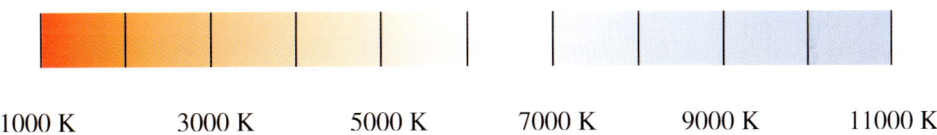

Figure 3.20 Color of a black-body radiator at different temperatures.

are shown in Figure 3.19. As the temperature of the black-body radiator increases, the maximum shifts to lower wavelengths. The wavelength of maximum emission is described by Wien's law:

$$\lambda_{max} \approx \frac{2.898 \cdot 10^{-3} K m}{T}. \tag{3.49}$$

Many natural light sources such as the flame of a candle, a light bulb, or sunlight can be approximated using the concept of a black-body radiator. A flame burning at low temperature emits more light in the red part of the spectrum. As the fire becomes hotter, the flame will turn white or even blue if the fire is very hot, i.e. the fire of a cutting torch. The observed color of the radiance given off by a black-body radiator is shown in Figure 3.20 for the temperature range from 1000 to 11000 K. Note that the power spectra of real objects is not equivalent to the power spectrum of a black-body radiator. Real objects emit less radiance than a black-body radiator.

The International Commission on Illumination (CIE) International Commission on Illumination (CIE) has defined a set of standard illuminants to be used for colorimetry (International Commission on Illumination 1996). Figure 3.21 shows the CIE illuminants A, C, D_{50}, D_{55}, D_{65}, and D_{75}. Illuminant A represents the power spectrum of light from a black-body radiator at approximately 2856 K. If this type of light is required for experiments, a gas-filled tungsten filament lamp that operates at a temperature of 2856 K is to be used. Illuminant D_{65} represents a phase of daylight with a correlated color temperature of approximately 6500 K. The CIE recommends to use this illuminant wherever possible.

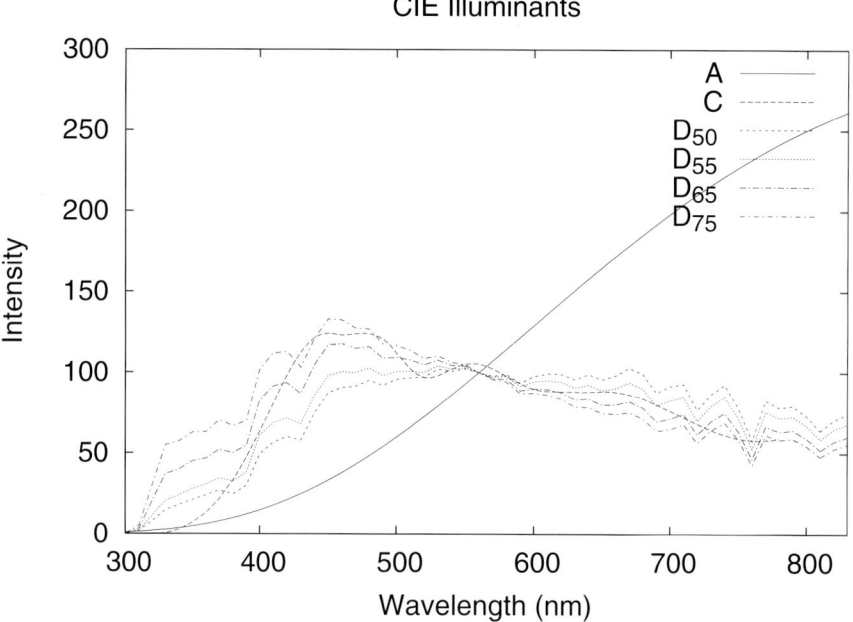

Figure 3.21 CIE illuminants A, C, D_{50}, D_{55}, D_{65}, and D_{75} (data from International Commission on Illumination 1996).

Similarly, illuminants D_{50}, D_{55}, and D_{75} with correlated color temperatures of approximately 5000 K, 5500 K, and 7500 K respectively were also defined. These illuminants should be used whenever D_{65} cannot be used.

3.6 Sensor Response

Measuring the power spectrum reflected from an object patch would allow us to accurately compute the reflectances of the corresponding object patch if we also knew the type of illuminant used. Not having an entire power spectrum of the reflected light makes it more difficult to compute the reflectances. In Chapter 2, we have already seen how the receptors of the human visual system respond to light. The cones are responsible for color vision. There are three types of cones, one primarily responds to light in the red part of the spectrum, one primarily responds to light in the green part of the spectrum, and one primarily responds to light in the blue part of the spectrum. They respond to a whole range of wavelengths and not only to one particular wavelength. Similarly, the sensors in a digital camera also respond to a whole range of wavelengths. The response curves of the Sony DXC-930 digital camera as measured by Barnard et al. (2002c) are shown in Figure 3.22.

In order to obtain the output of a sensor, we need to integrate the response over all possible wavelengths. Let $E(\lambda, \mathbf{x}_I)$ be the irradiance of wavelength λ falling onto an infinitesimal small patch on the sensor array located at position $\mathbf{x}_I$. Let $\mathbf{S}(\lambda)$ be a vector of the sensor's response function. That is, if we have three different types of sensors, one

THEORY OF COLOR IMAGE FORMATION

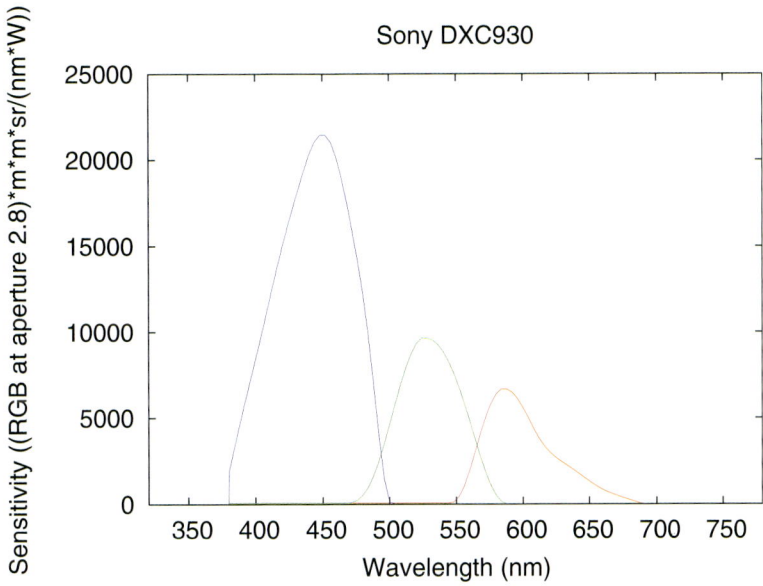

Figure 3.22 Sensor response curves of the Sony DXC-930 digital camera. The data was derived by Barnard et al. (2002c).

each for red, green and blue, then $S_i(\lambda)$ describes the response curve of the i-th sensor with $i \in \{r, g, b\}$. The intensities **I** measured by the sensor are given by

$$\mathbf{I}(\mathbf{x}_I) = \int E(\lambda, \mathbf{x}_I) S(\lambda) d\lambda \qquad (3.50)$$

where the integration is done over all wavelengths.

In Section 3.3 we have seen that the irradiance at the sensor array is proportional to the radiance given off by the corresponding object patch. In the following equation, we assume a scaling factor of one for simplicity, i.e. $E(\lambda, \mathbf{x}_I) = L(\lambda, \mathbf{x}_{Obj})$ where $\mathbf{x}_{Obj}$ is the position of the object patch that is projected onto $\mathbf{x}_I$ on the sensor array. Then we obtain the following expression for the intensity measured by the sensor.

$$\mathbf{I}(\mathbf{x}_I) = \int L(\lambda, \mathbf{x}_{Obj}) S(\lambda) d\lambda. \qquad (3.51)$$

Now we can use the reflectance models from Section 3.4. We assume that the object patch is a Lambertian surface. For a Lambertian surface, the BRDF is constant. However, the constant depends on the wavelength of the light. It defines how much of the incident light is reflected for any given wavelength λ. We denote this constant by $R(\lambda, \mathbf{x})$, the reflectance for each point $\mathbf{x}$ on the surface. The radiance given off by a Lambertian surface is equivalent to the product between the radiance given off by the light source and the reflectance, except for a scaling factor that depends on the angle α between the normal vector of the surface $\mathbf{N}_{Obj}$ and the direction of the light source $\mathbf{N}_L$. Let $G(\mathbf{x}_{Obj})$ be the scaling factor at position

$\mathbf{x}_{\text{Obj}}$. For a Lambertian surface illuminated with a point light source, we have

$$G(\mathbf{x}_{\text{Obj}}) = \cos \alpha = \mathbf{N}_{\text{Obj}} \mathbf{N}_L \qquad (3.52)$$

where α is the angle between the normal vector of the surface and a vector that points to the direction of the light source at position $\mathbf{x}_{\text{Obj}}$. We then obtain for the radiance $L(\lambda, \mathbf{x}_{\text{Obj}})$ given off by the patch

$$L(\lambda, \mathbf{x}_{\text{Obj}}) = R(\lambda, \mathbf{x}_{\text{Obj}}) L(\lambda) G(\mathbf{x}_{\text{Obj}}) \qquad (3.53)$$

where $L(\lambda)$ is the radiance given off by the light source. This gives us

$$\mathbf{I}(\mathbf{x}_I) = \int R(\lambda, \mathbf{x}_{\text{Obj}}) L(\lambda) G(\mathbf{x}_{\text{Obj}}) \mathbf{S}(\lambda) d\lambda \qquad (3.54)$$

$$= G(\mathbf{x}_{\text{Obj}}) \int R(\lambda, \mathbf{x}_{\text{Obj}}) L(\lambda) \mathbf{S}(\lambda) d\lambda \qquad (3.55)$$

for the intensity $\mathbf{I}$ measured by the sensor. This model of color image formation is used as a basis for many color constancy algorithms (Buchsbaum 1980; Finlayson 1996; Finlayson and Hordley 2001a,b; Finlayson et al. 1995, 1994b, 2004, 2002; Finlayson and Schaefer 2001; Finlayson et al. 1998; Forsyth 1988, 1992; Novak and Shafer 1992).

The sensor's response characteristics can often be approximated by delta functions (Finlayson and Hordley 2001b). Even though each sensor responds to a range of wavelengths, in practice they are either close enough to delta functions or they can be sharpened (Barnard et al. 2001; Finlayson and Funt 1996; Finlayson et al. 1994a,b) such that this assumption is approximately correct. Assuming that the sensor's response characteristics can be described by delta functions, we have

$$S_i(\lambda) = \delta(\lambda - \lambda_i) \qquad (3.56)$$

with $i \in \{r, g, b\}$. Sensor S_r responds to a single wavelength λ_r in the red part of the spectrum, sensor S_g responds to a single wavelength λ_g in the green part of the spectrum and S_b responds to a single wavelength λ_b in the blue part of the spectrum. Given this assumption for the sensor, we obtain

$$I_i(\mathbf{x}_I) = G(\mathbf{x}_{\text{Obj}}) \int R(\lambda, \mathbf{x}_{\text{Obj}}) L(\lambda) S_i(\lambda) d\lambda \qquad (3.57)$$

$$= G(\mathbf{x}_{\text{Obj}}) \int R(\lambda, \mathbf{x}_{\text{Obj}}) L(\lambda) \delta(\lambda - \lambda_i) d\lambda \qquad (3.58)$$

$$= G(\mathbf{x}_{\text{Obj}}) R(\lambda_i, \mathbf{x}_{\text{Obj}}) L(\lambda_i) \qquad (3.59)$$

for the intensity $I_i(\mathbf{x}_I)$ measured by sensor i at position $\mathbf{x}_I$ on the sensor array.

If we use these simplifying assumptions, the intensity of each color channel of the image pixel is proportional to the product of the reflectance at the corresponding object position times the intensity of the light for the particular wavelength. The result is also scaled by a factor that depends on the scene geometry. The preceding calculation only applies to matte, i.e. Lambertian, surfaces. In this case, the amount of reflected light does not depend on the orientation of the surface relative to the viewer.

At each image point, we only measure three values. We obtain the intensities from the sensor in the red, green, and blue part of the spectrum. This leads to the RGB color

THEORY OF COLOR IMAGE FORMATION   63

space where each point is identified by just three numbers (Gonzalez and Woods 1992; Hanbury and Serra 2003). Each number denotes the intensity of the corresponding channel. The measurements are, of course, dependent on the type of sensor used. Standardized color spaces have been defined to allow for exchange of images that were obtained from different devices. Accurate color reproduction is addressed in the next chapter.

Given the three measurements from the cones of the retina, we are somehow able to discount the illuminant and to compute color constant descriptors. In solving the problem of color constancy, we need to estimate the reflectances of the object. That is, we need to determine at least six unknowns (reflectance at the corresponding object position and the irradiance falling onto the object) from only three input values. Note that discriminating between two different reflectances is actually quite easy. This important point was made by Young (1987). Even if a particular set of reflectances viewed through one sensor cannot be distinguished, it may well be true that the same set of reflectances can be easily distinguished using a sensor with a different response characteristic.

3.7 Finite Set of Basis Functions

Reflectances can be approximated using a finite set of basis functions (Brill 1978; Brill and West 1981; D'Zmura and Lennie 1986; Finlayson et al. 1994a; Funt and Drew 1988; Funt et al. 1991, 1992; Healey and Slater 1994; Ho et al. 1992; Maloney 1992; Maloney and Wandell 1992, 1986; Novak and Shafer 1992; Tominaga 1991; Wandell 1987). If we decompose the space of reflectances into a finite set of n_R basis functions $\hat{\mathbf{R}}_i$ with $i \in \{1, \ldots, n_R\}$, then the reflectance $R(\lambda)$ for a given wavelength λ can be written as

$$R(\lambda) = \sum_i R_i \hat{\mathbf{R}}_i(\lambda) \qquad (3.60)$$

where R_i are the coefficients of the basis functions. The basis vectors can be computed by principal component analysis. Similarly, the illuminant can also be decomposed onto a set of basis functions. Most surface reflectances can be modeled using only three basis functions (Ho et al. 1992; Maloney 1992). Three to five basis functions are sufficient to model most daylight spectra. Thus, in practice, it suffices to work with just three sensor responses. It does not matter if an illuminant contains some small sharp peaks, since the visual system integrates over the available light.

Funt et al. (1991, 1992) use a finite dimensional linear model to recover ambient illumination and the surface reflectance by examining mutual reflection between surfaces. Ho et al. (1992) show how a color signal spectrum can be separated into reflectance and illumination components. They compute the coefficients of the basis functions by finding a least squares solution, which best fits the given color signal. However, in order to do this, they require that the entire color spectrum and not only the measurements from the sensors is available. Ho et al. suggest to obtain the measured color spectrum from chromatic aberration. Novak and Shafer (1992) suggest to introduce a color chart with known spectral characteristics to estimate the spectral power distribution of an unknown illuminant.

The standard model of color image formation was already described in the previous section. Let us now assume that the reflectances are approximated by a set of basis functions.

If we also assume that the illuminant is uniform over the entire image and $G(\mathbf{x}_{Obj}) = 1$, we obtain

$$\mathbf{I}(\mathbf{x}_I) = \int L(\lambda) \mathbf{S}(\lambda) \sum_i R_i(\mathbf{x}_{Obj}) \hat{\mathbf{R}}_i(\lambda) d\lambda \tag{3.61}$$

$$= \sum_i \int L(\lambda) \mathbf{S}(\lambda) R_i(\mathbf{x}_{Obj}) \hat{\mathbf{R}}_i(\lambda) d\lambda \tag{3.62}$$

$$= \mathbf{AR} \tag{3.63}$$

where $\mathbf{A}$ is a $n_R \times n_S$ matrix for a sensor with n_S different spectral sensitivities and $\mathbf{R} = [R_1, ..., R_{n_R}]^T$ is a vector with the reflectance coefficients. Wandell (1987) calls the matrix $\mathbf{A}$ the *lighting* matrix. The elements of $\mathbf{A}(L)$ for illuminant L are given by

$$A_{ji}(L) = \int L(\lambda) S_j(\lambda) \hat{\mathbf{R}}_i(\lambda) d\lambda. \tag{3.64}$$

Thus, the measurements made by the sensor can be viewed as a linear transformation of the reflectances. Let us consider only the three wavelengths λ_i with $i \in \{r, g, b\}$, then the lighting matrix would be given by

$$\mathbf{A} = \begin{pmatrix} S_1(\lambda_r)L(\lambda_r) & S_1(\lambda_g)L(\lambda_g) & S_1(\lambda_b)L(\lambda_b) \\ S_2(\lambda_r)L(\lambda_r) & S_2(\lambda_g)L(\lambda_g) & S_2(\lambda_b)L(\lambda_b) \\ S_3(\lambda_r)L(\lambda_r) & S_3(\lambda_g)L(\lambda_g) & S_3(\lambda_b)L(\lambda_b) \end{pmatrix} \tag{3.65}$$

assuming $\hat{\mathbf{R}}_i(\lambda_i) = 1$ and $\hat{\mathbf{R}}_i(\lambda) = 0$ for $\lambda \neq \lambda_i$. If we have an estimate of the illuminant $L(\lambda_i)$ and we knew the response functions of the sensor $\mathbf{S}(\lambda)$, we could estimate the reflectances $\mathbf{R}$ by computing

$$\mathbf{R} = \mathbf{A}^{-1} \mathbf{I}. \tag{3.66}$$

Note, that this can only be done if the illuminant is uniform.

Suppose that we have two different illuminants. Each illuminant defines a local coordinate system inside the three-dimensional space of receptors as shown in Figure 3.23. A diagonal transform, i.e. a simple scaling of each color channel, is not sufficient to align the coordinate systems defined by the two illuminants. A simple scaling of the color channels can only be used if the response functions of the sensor are sufficiently narrow band, i.e. they can be approximated by a delta function.

If we measure the light reflected from a scene illuminated by two different illuminants, we obtain two measurements $\mathbf{I}$ and $\mathbf{I}'$.

$$\mathbf{I} = \mathbf{A}(L)\mathbf{R} \tag{3.67}$$

$$\mathbf{I}' = \mathbf{A}(L')\mathbf{R}. \tag{3.68}$$

The two measurements are related by the transform $\mathbf{M}$.

$$\mathbf{I} = \mathbf{A}(L)\mathbf{R} \tag{3.69}$$

$$= \mathbf{A}(L)\mathbf{A}^{-1}(L)\mathbf{I}' \tag{3.70}$$

$$= \mathbf{MI}' \tag{3.71}$$

with $\mathbf{M} = \mathbf{A}(L)\mathbf{A}^{-1}(L)$.

THEORY OF COLOR IMAGE FORMATION 65

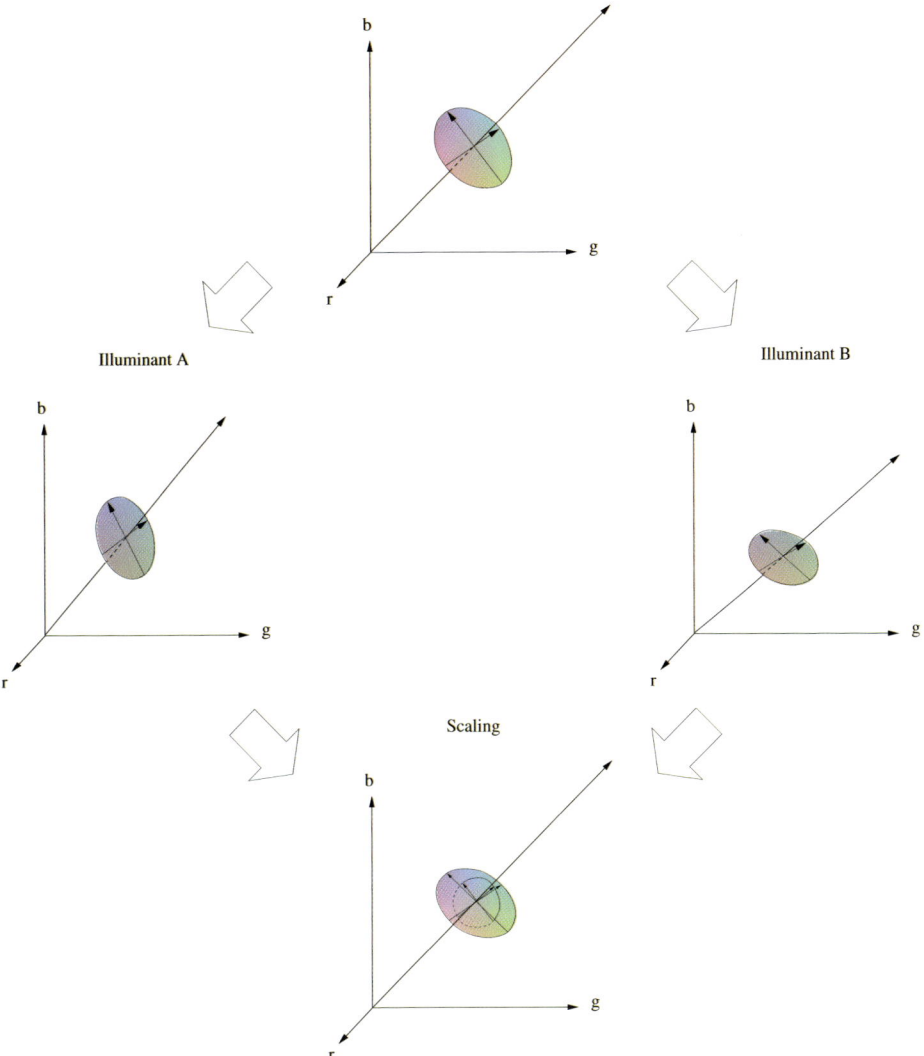

Figure 3.23 Two different illuminants define two coordinate systems within the space of receptors. A simple diagonal transform, i.e. scaling of the color channels, is not sufficient to align the two coordinate systems.

In developing color constancy algorithms, there are basically two roads to follow (Finlayson et al. 1994a). One is the accurate estimation of reflectances. This is of particular importance for object recognition. Object recognition becomes much easier if the reflectances are known. Another possibility would be to compute colors as they would appear under a canonical, e.g. white, illuminant. These are two different objectives and both are justified in their own right. In the first case (computation of reflectances), color constant descriptors **D** are given by

$$\mathbf{D} = \mathbf{A}^{-1}(L)\mathbf{I} = \mathbf{A}^{-1}(L)\mathbf{A}(L)\mathbf{R} = \mathbf{R}. \quad (3.72)$$

For the latter case (computation of colors under a canonical illuminant L_c), color constant descriptors are given by

$$\mathbf{D} = \mathbf{A}(L_c)\mathbf{A}^{-1}(L)\mathbf{I} = \mathbf{A}(L_c)\mathbf{A}^{-1}(L)\mathbf{A}(L)\mathbf{R} = \mathbf{A}(L_c)\mathbf{R}. \quad (3.73)$$

Another question of interest is to compute colors similar to the colors perceived by the photographer who took the image. Even though perceived color correlates with object reflectances, as we have seen in Section 2.5, there are also situations when this does not hold. This is described in Section 14.1. Creating an accurate model of human color perception would be a great step forward for consumer photography.

In order to solve the problem of color constancy, some assumptions have to be made. One popular assumption is that the illuminant is either constant or is varying smoothly over the objects and that abrupt changes are caused by a change in the reflectance, i.e. color of the object. Many methods also assume that, on average, the world is gray. The gray world assumption is probably the most well-known algorithm for color constancy. In Chapter 6 and Chapter 7, we will look at several different algorithms for color constancy. Once the correct colors are determined, they have to be reproduced somewhere either by displaying the image on a computer screen or by printing the image. Accurate color reproduction is discussed in the next chapter.

4

Color Reproduction

The human retina contains only three types of receptors, which respond mainly to light in the red, green, and blue parts of the spectrum. Suppose that we equip a digital camera with filters that model the response characteristics of the receptors found in the retina. If the processing done by the retina and by the brain were completely understood, we could compute the same color constant descriptors as are computed by the human brain. This method could then be used for object recognition based on color.

For areas such as consumer photography, however, we also have to display the processed image somewhere. We could print the image or show it on a color monitor or thin film transistor (TFT) display. Care must be taken that the measured colors are reproduced accurately using the given output device. If the illuminant is assumed to be constant over the entire scene, then the illuminant can be measured using a spectrometer. Once the illuminant is known, we can correct the color of the illuminant. However, accurate color reproduction is difficult even if the type of illuminant is known. Our requirement is that a given color sensation be exactly recreated by the display device. A number of standards have been developed to ensure accurate color reproduction. In order to display colors accurately, we have to use a properly defined coordinate space for the measured colors, i.e. a standardized color space such as CIE XYZ, CIE $L^*a^*b^*$ or sRGB. The CIE XYZ color space is described in the subsequent text. Additional color spaces are described in the next chapter. Standards such as the sRGB also define the viewing conditions (Stokes et al. 1996). If the actual viewing condition deviates from the standard viewing condition, then the result may not be as intended. Accurate color reproduction is also obviously important for the manufacturing industry (Harold 2001).

In order to reproduce the color sensation of an observer, we could reproduce exactly the same spectral characteristic for every point of the image. This sensation would be indistinguishable from the original sensation. However, we do not usually have a spectrometer to measure the entire spectrum for every image point, let alone a method to create a color spectrum given such measurements. However, quite a large number of colors can be reproduced using only three color primaries.

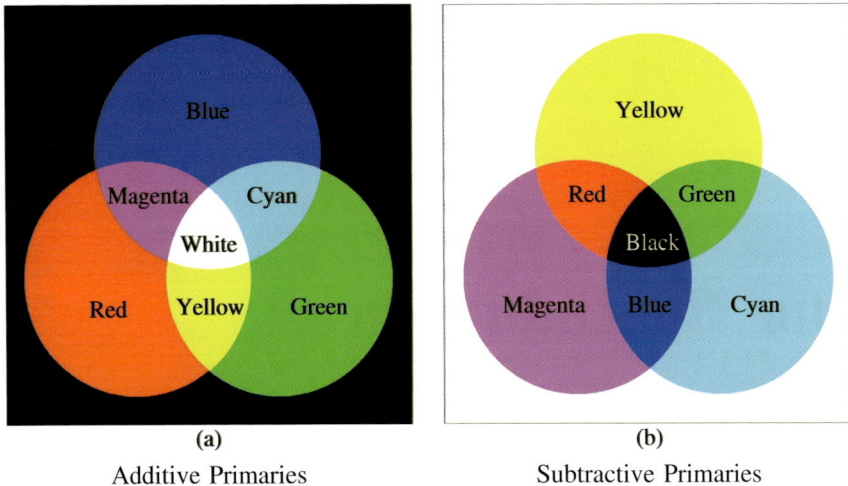

Figure 4.1 Additive and subtractive color generation.

4.1 Additive and Subtractive Color Generation

Two different modes of color generation can be distinguished. One is additive and the other is subtractive (Gonzalez and Woods 1992). For instance, a television set, flat panel display or beamer uses an additive color model. Colors are generated by adding three primary colors red, green, and blue in different quantities. The space of possible colors that can be produced by this process is called the *gamut* of the display device. Secondary colors are produced by adding two primary colors. Adding red and green produces yellow, red and blue produces magenta, and green and blue produces cyan. White is created by combining red, green, and blue. Alternatively, white can be created by mixing a secondary color with its opposite primary color. Black is the default color. This is illustrated in Figure 4.1(a).

A subtractive color model is used for printing. A printed document or image only reflects the light that illuminates it. Some light is absorbed and the remainder is reflected. The light reflected by a point on the surface determines the color of the point. Colors are created by depositing pigments that absorb one primary color. The three primaries in this case are yellow, magenta, and cyan. A yellow point absorbs blue and reflects red and green. A magenta point absorbs green and reflects red and blue. A cyan point absorbs red and reflects green and blue. If yellow and magenta are combined, green and blue are absorbed and only red remains. Therefore, red can be created by combining yellow and magenta. Green can be created by combining yellow and cyan. Yellow absorbs blue while cyan absorbs red. Finally, blue can be created by combining magenta and cyan. Magenta absorbs green, while cyan absorbs red. Therefore, the secondary colors are red, green, and blue. Black is created by combining yellow, magenta, and blue. Black can also be created by mixing a secondary color with its opposite primary color. White is the default color. The subtractive color model is shown in Figure 4.1(b).

4.2 Color Gamut

If images are displayed on a monitor, then the measured intensities are used to modulate the output of an electron gun (Foley et al. 1996; Möller and Haines 1999). This electron gun, a cathode, is located behind the screen of the monitor. This is the reason why a monitor is often referred to as a cathode-ray tube. The ray of electrons coming from a cathode sweeps over the screen from left to right. After a line is completed, the ray advances to the next line. At the screen, the electrons hit a layer of phosphor. The phosphor emits light when hit by the electrons. Since the output of the electron gun is modulated using the measured intensities, a gray-level image appears on the computer monitor. The image is redrawn at least 30 times/s. Most modern monitors achieve frame rates of over 100 Hz.

Some color monitors have three electron guns: one for each of the red, green, and blue components while others use a single electron gun for all three colors (Kalawsky 1993). The electron gun produces little flashes of red, green, and blue light on the computer screen. The individual pixels are so small that the eye sees a composite pixel whose color depends on the power applied to the electron gun. Similarly, a flat panel display or liquid crystal display (LCD) uses liquid crystals in combination with a backlight and color filters to create the colors. The question now is, how much of the three primary colors do we have to add to create a particular color sensation. For accurate color reproduction, we would need three primary colors with which we can reproduce the color sensation for any given wavelength. However, it is not possible to generate all color sensations using just three primary colors. The CIE has developed standards that address this problem.

4.3 Computing Primary Intensities

In Chapter 2 we have seen that the human eye contains three different types of cones. Color sensations can be produced by varying the amount of light in the sections of the visible spectrum to which the cones respond. Let S_i with $i \in \{r, g, b\}$ be the response curves of the red, green, and blue cones. Then the response Q_i of each cone i is given by

$$Q_i = \int S_i(\lambda) L(\lambda) \, d\lambda \qquad (4.1)$$

where $L(\lambda)$ defines the illuminant for all wavelengths λ. If we choose three color primaries $L_1(\lambda)$, $L_2(\lambda)$, and $L_3(\lambda)$, then the illuminant L is a linear combination of these primaries:

$$L(\lambda) = XL_1 + YL_2 + ZL_3 \qquad (4.2)$$

with $X, Y, Z \in \mathbb{R}_+$. If monochromatic colors are used as primaries, then the response simplifies to a sum over the three primaries. In this case, we have

$$L_1 = \delta(\lambda - \lambda_1) \qquad (4.3)$$

$$L_2 = \delta(\lambda - \lambda_2) \qquad (4.4)$$

$$L_3 = \delta(\lambda - \lambda_3) \qquad (4.5)$$

where λ_j with $j \in \{1, 2, 3\}$ are the monochromatic wavelengths of the three primaries. The response of the receptor i with $i \in \{r, g, b\}$ is then given by

$$Q_i = \int S_i(\lambda) L(\lambda) \, d\lambda \tag{4.6}$$

$$= \int S_i(\lambda) \left(X L_1 + Y L_2 + Z L_3 \right) d\lambda \tag{4.7}$$

$$= \int S_i(\lambda) \left(X \delta(\lambda - \lambda_1) + Y \delta(\lambda - \lambda_2) + Z \delta(\lambda - \lambda_3) \right) d\lambda \tag{4.8}$$

$$= X S_i(\lambda_1) + Y S_i(\lambda_2) + Z S_i(\lambda_3) \tag{4.9}$$

The three primaries span a three-dimensional space of colors. By varying the parameters X, Y, and Z we can create different color sensations.

Suppose that we want to create the color sensation created by monochromatic light at wavelength $\lambda = 500$ nm. The response of the three cones are given by $S_i(\lambda)$ with $i \in \{r, g, b\}$. In order to create the same color sensation using the monochromatic primaries L_j with $j \in \{1, 2, 3\}$, we have to solve the following set of equations:

$$\begin{bmatrix} S_r(\lambda) \\ S_g(\lambda) \\ S_b(\lambda) \end{bmatrix} = \begin{pmatrix} S_r(\lambda_1) & S_r(\lambda_2) & S_r(\lambda_3) \\ S_g(\lambda_1) & S_g(\lambda_2) & S_g(\lambda_3) \\ S_b(\lambda_1) & S_b(\lambda_2) & S_b(\lambda_3) \end{pmatrix} \begin{bmatrix} X \\ Y \\ Z \end{bmatrix}. \tag{4.10}$$

In other words, the response of the three receptors at the given wavelength λ have to be equivalent to the response to a linear combination of the three monochromatic primary colors for some X, Y, and Z. If any one of the parameters X, Y, or Z should become negative, then we cannot create the particular color sensation using only these three primaries.

Suppose that we have a set of monochromatic primaries located at 620 nm, 530 nm, and 400 nm. What would be the corresponding parameters X, Y, and Z for these primaries if we were to create the same color sensation as light of wavelength $\lambda = 500$ nm. If we solve the set of equations for $\lambda = 500$ nm using the absorbance characteristics (shown in Figure 2.4) of Dartnall et al. (1983) for the sensors, we obtain $X = -0.355954$, $Y = 0.831075$, and $Z = 0.120729$. Obviously it is not possible to create this particular color sensation because one of the three primaries has to be subtracted from the other two primaries. Since this is not physically possible, this particular color sensation cannot be created. The inability to create certain color sensations is caused by the overlap of the response curves of the visual system. Figure 4.2 shows the required intensities for the wavelengths 400 nm through 640 nm.

4.4 CIE XYZ Color Space

The International Commission on Illumination (CIE) has defined a standard observer to be used for accurate color reproduction (International Commission on Illumination 1983, 1990, 1996). In Chapter 2 we have seen that the rods mediate vision when very little light is available. This type of vision is called *scotopic* vision. The cones mediate high acuity vision in bright light conditions. This type of vision is called *photopic* vision. The sensitivities for a standard observer as defined by the CIE for scotopic and photopic vision are shown in Figure 4.3. The scotopic function is denoted by $V'(\lambda)$. The photopic

COLOR REPRODUCTION

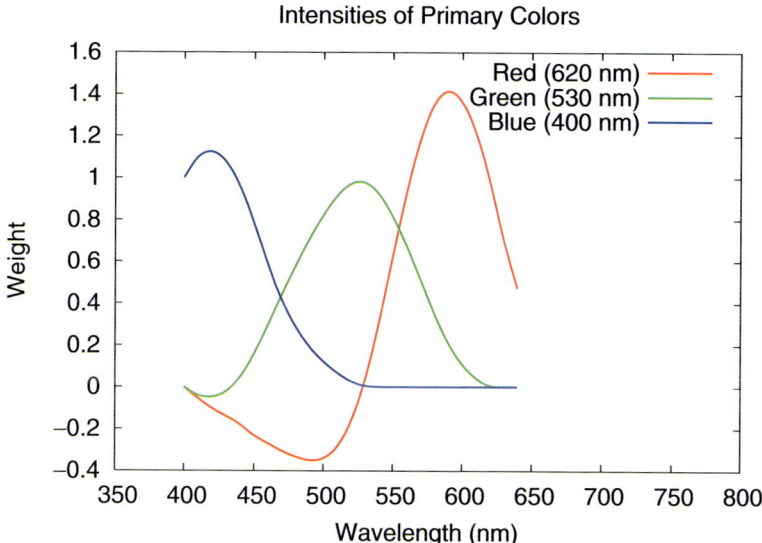

Figure 4.2 Weights for monochromatic primaries located at 620 nm, 530 nm, and 400 nm for an observer whose response characteristics are described by the absorbance curves of the retinal curves as measured by Dartnall et al. (1983).

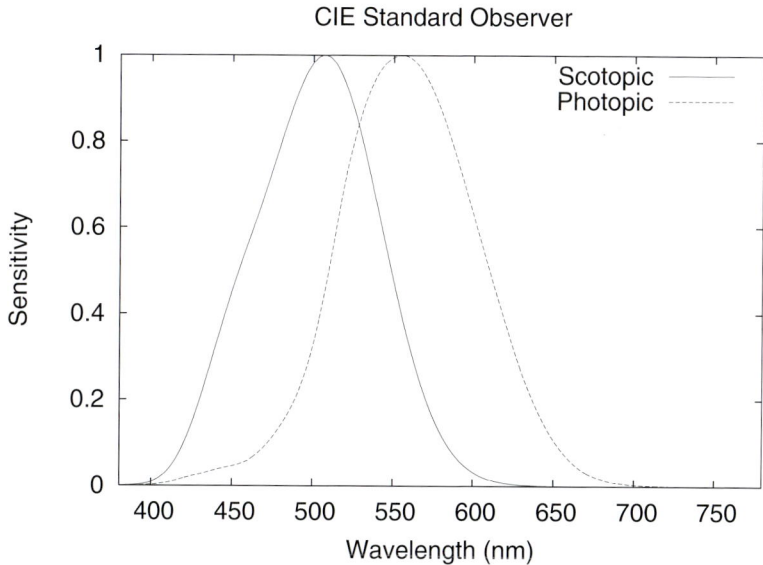

Figure 4.3 Sensitivities as defined by CIE, which are used to model scotopic $V'(\lambda)$ and photopic vision $V_M(\lambda)$ of a standard observer (International Commission on Illumination 1983, 1990, 1996) (data from International Commission on Illumination 1988).

function was originally denoted by $V(\lambda)$. In 1988, a slight modification to the efficiency function for photopic vision was made (International Commission on Illumination 1990). This modified version is denoted by $V_M(\lambda)$. The two functions $V(\lambda)$ and $V_M(\lambda)$ differ only for wavelengths below 460 nm.

Using these sensitivities, we can calculate the luminous flux of any light source (International Commission on Illumination 1983; Jähne 2002). Let $\Phi(\lambda)$ be the radiant flux at wavelength λ of the light source. Let $V'(\lambda)$ be the sensitivity of scotopic vision and $V(\lambda)$ be the sensitivity of photopic vision, then the intensity of the light source is

$$\Phi'_v = K'_m \int_{380 \text{ nm}}^{780 \text{ nm}} \Phi(\lambda) V'(\lambda)\, d\lambda \qquad \text{for scotopic vision, and} \qquad (4.11)$$

$$\Phi_v = K_m \int_{380 \text{ nm}}^{780 \text{ nm}} \Phi(\lambda) V(\lambda)\, d\lambda \qquad \text{for photopic vision} \qquad (4.12)$$

with $K'_m = 1700.06 \frac{\text{lm}}{\text{W}}$ and $K_m = 683.002 \frac{\text{lm}}{\text{W}}$.

The CIE also defined a set of imaginary primaries that can be added using only positive weights X, Y, and Z to create all possible colors (Colourware Ltd 2001; International Commission on Illumination 1996; Jähne 2002; Wyszecki and Stiles 2000). The weights that would be used by a standard or average observer to create a given color using the imaginary primaries are shown in Figure 4.4. This graph tells us how much of the three primaries we have to add to produce the color sensation for a particular wavelength λ. Two sets of weights exist: one for a 2° observer and one for a 10° observer. The weights

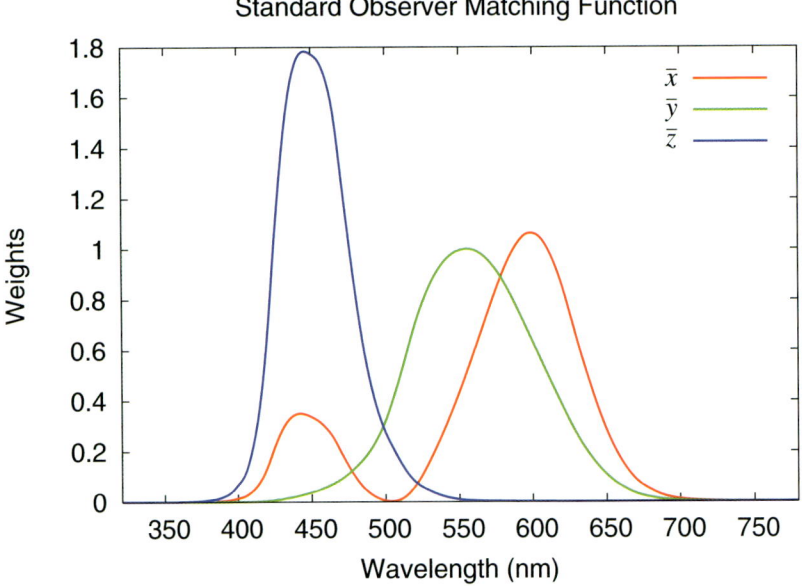

Figure 4.4 Weights used by a standard observer to create any given color sensation. Note that all of the weights are positive (data from International Commission on Illumination 1988).

COLOR REPRODUCTION

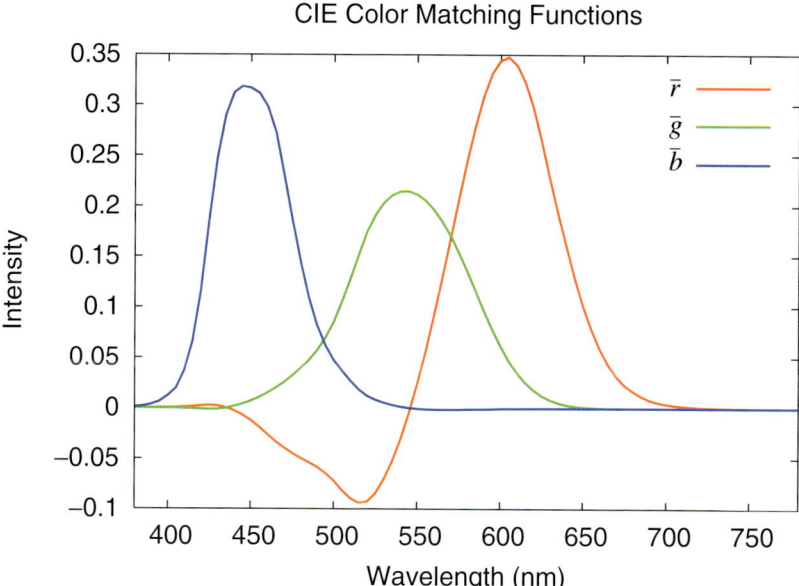

Figure 4.5 Set of weights for three monochromatic stimuli at wavelengths 700 nm, 546.1 nm, and 435.8 nm to match a color stimulus of a given wavelength (data from International Commission on Illumination 1996).

for the 2° observer were computed from color-matching experiments, where an observer had to match stimuli of wavelengths 700 nm, 546.1 nm, and 435.8 nm to a given stimulus of monochromatic wavelength. The experimental setup was done in such a way that the stimuli activated only an area of the retina of 2° of visual angle. This resulted in a set of weights $\bar{r}(\lambda)$, $\bar{g}(\lambda)$, and $\bar{b}(\lambda)$ for the three monochromatic stimuli. These weights are shown in Figure 4.5. The agreement between Figure 4.5 and Figure 4.2 is striking. Note that the values in Figure 4.2 were arrived at using the absorbance characteristics of the retinal cones while that in Figure 4.5 were determined experimentally. In order to derive the weights for the standard observer, a linear transformation is applied such that all values are positive and the second coordinate corresponds to the photopic luminous efficiency function

$$\begin{bmatrix} x(\lambda) \\ y(\lambda) \\ z(\lambda) \end{bmatrix} = \begin{pmatrix} 0.49000 & 0.31000 & 0.20000 \\ 0.17697 & 0.81240 & 0.01063 \\ 0.00000 & 0.01000 & 0.99000 \end{pmatrix} \begin{bmatrix} \bar{r}(\lambda) \\ \bar{g}(\lambda) \\ \bar{b}(\lambda) \end{bmatrix}. \quad (4.13)$$

After the linear transformation has been applied, the weights $\bar{x}$, $\bar{y}$, and $\bar{z}$ are computed as follows:

$$\bar{x}(\lambda) = \frac{x(\lambda)}{y(\lambda)} V(\lambda) \quad (4.14)$$

$$\bar{y}(\lambda) = V(\lambda) \quad (4.15)$$

$$\bar{z}(\lambda) = \frac{z(\lambda)}{y(\lambda)} V(\lambda) \quad (4.16)$$

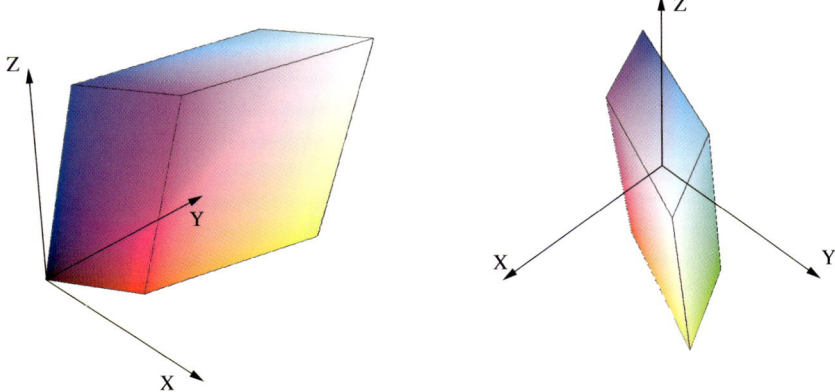

Figure 4.6 sRGB color cube drawn in XYZ color space.

where $V(\lambda)$ is the photopic luminous efficiency function shown in Figure 4.3. The linear transformation followed by the multiplication with $\frac{V(\lambda)}{y(\lambda)}$ results in all positive weights $\bar{x}$, $\bar{y}$, and $\bar{z}$ and also makes the second coordinate equivalent to the photopic luminous efficiency function. Figure 4.6 shows the shape of the sRGB color cube transformed to XYZ space.

Tristimulus values X, Y, and Z are computed for a primary light source with power spectrum $L(\lambda)$ from the color-matching functions $\bar{x}$, $\bar{y}$, and $\bar{z}$ as follows:

$$X = k \int_{380 \text{ nm}}^{830 \text{ nm}} L(\lambda)\bar{x}(\lambda)\,d\lambda \tag{4.17}$$

$$Y = k \int_{380 \text{ nm}}^{830 \text{ nm}} L(\lambda)\bar{y}(\lambda)\,d\lambda \tag{4.18}$$

$$Z = k \int_{380 \text{ nm}}^{830 \text{ nm}} L(\lambda)\bar{z}(\lambda)\,d\lambda \tag{4.19}$$

where $k = 683.002 \frac{\text{lm}}{W}$. Since $\bar{y}(\lambda) = V(\lambda)$, the Y coordinate measures the luminance of the object. If tristimulus values X, Y, and Z are to be calculated for a reflecting or transmitting object, then the following equations are to be used:

$$X = k \int_{380 \text{ nm}}^{830 \text{ nm}} R(\lambda)L(\lambda)\bar{x}(\lambda)\,d\lambda \tag{4.20}$$

$$Y = k \int_{380 \text{ nm}}^{830 \text{ nm}} R(\lambda)L(\lambda)\bar{y}(\lambda)\,d\lambda \tag{4.21}$$

$$Z = k \int_{380 \text{ nm}}^{830 \text{ nm}} R(\lambda)L(\lambda)\bar{z}(\lambda)\,d\lambda \tag{4.22}$$

COLOR REPRODUCTION

Here $R(\lambda)$ denotes the reflectance of the object. In this case, the constant k is set to

$$k = \frac{100}{\int_{380 \text{ nm}}^{830 \text{ nm}} L(\lambda)\bar{y}(\lambda)\,d\lambda} \tag{4.23}$$

which results in $Y = 100$ for an object that reflects all incident light over the visible spectrum.

Another set of weights exists for a 10° observer. These are denoted by $\bar{x}_{10}$, $\bar{y}_{10}$, and $\bar{z}_{10}$. The second set of weights was defined because it was found that the 2° observer is not appropriate for large-field visual color judgments. Thus, if large-field color judgments are to be made, i.e. the visual stimulus subtends an area of the retina of 10° of visual angle, this set of weights is to be used. The set of weights for the 10° observer was defined in a similar way as that for the 2° observer except that monochromatic stimuli of 645.2 nm, 526.3 nm, and 444.4 nm were used for the matching experiments.

If the energy of each primary color is increased by the same percentage, then the relative responses of the receptors will not be any different. Therefore, colors are usually specified using normalized colors, or chromaticities. The chromaticities of the CIE color space are given by

$$\hat{x} = \frac{X}{X+Y+Z}, \quad \hat{y} = \frac{Y}{X+Y+Z}, \quad \hat{z} = \frac{Z}{X+Y+Z}. \tag{4.24}$$

The normalization projects all colors onto the plane at $X+Y+Z = 1$. Therefore, the third coordinate is redundant and we have to only specify the coordinates $\hat{x}$ and $\hat{y}$. The third coordinate $\hat{z}$ is then given by $\hat{z} = 1 - \hat{x} - \hat{y}$. Since the weights as defined by the standard observer are all positive, the coordinates $\hat{x}$, $\hat{y}$, and $\hat{z}$ will also be positive. Owing to the constraint $\hat{x} + \hat{y} + \hat{z} = 1$, they will all be smaller than one. Instead of specifying a color by the triplet X, Y, and Z, we can also specify the color by the chromaticity coordinates $\hat{x}$ and $\hat{y}$ as well as the absolute luminance Y. The absolute coordinates X and Z are then given as (Poynton 2003)

$$X = \hat{x}\frac{Y}{\hat{y}} \tag{4.25}$$

$$Z = (1 - \hat{x} - \hat{y})\frac{Y}{\hat{y}}. \tag{4.26}$$

If one looks at the chromaticity coordinates of the colors produced by light in the visible range from 400 nm to 700 nm, one obtains the graph shown in Figure 4.7. As we pass through the wavelengths from 400 nm to 700 nm, the curve forms a horseshoe-shaped region. This region contains all possible colors. Blue is located at the lower left of the graph, green is at the upper left, and red is at the lower right of the horseshoe-shaped region. The straight line at the bottom, the so-called purple line, connects blue at the lower left with red at the lower right. White is located at the point of equal energy [1/3, 1/3] at the center of the graph. Colors can be described by their hue, saturation, and intensity (Jähne 2002). These perceptual terms were originally introduced by Munsell. As we move along the border of the graph, we vary the hue of the color. Monochromatic colors are fully saturated. As we move towards the point of equal energy, i.e. add more white, colors become less saturated. If we draw a line from the center of the graph to any point inside the horseshoe-shaped region, all points along the line will have the same hue. They differ only in their saturation.

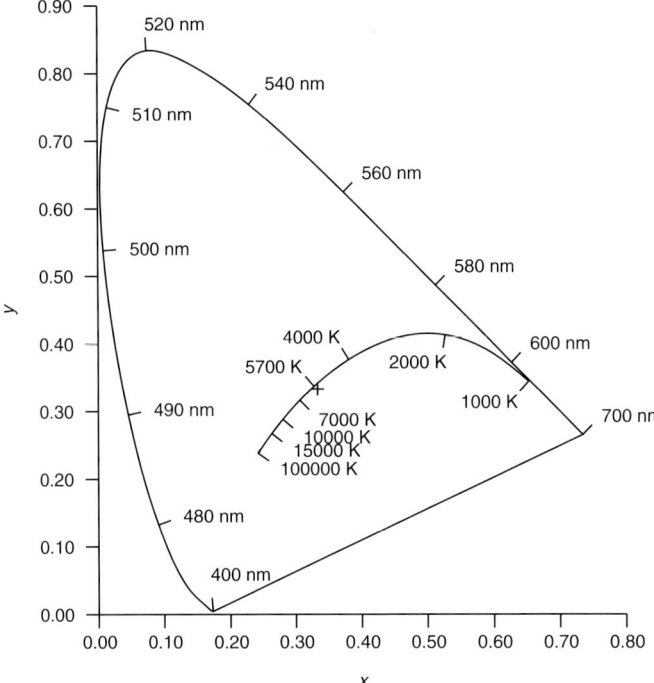

Figure 4.7 Gamut of colors in CIE XYZ space. The chromaticities of light in the visible range from 400 nm to 700 nm form a horseshoe-shaped region. The chromaticities for a black-body radiator at different temperatures is shown in the center. The color of a black-body radiator passes from red at low temperatures through white and on to blue at higher temperatures. The point of equal energy is located at the center of the graph.

In Section 3.5 we have already seen that many natural light sources can be described by a black-body radiator. The radiation given off by a black-body depends on its temperature. If we observe the color of the black-body for different temperatures, one obtains the curve in the center of the graph. The curve starts off at low temperatures, which are mostly red, passes close to the point of equal energy, which is white, and on to blue colors at higher temperatures. The shape of the black-body radiator closely resembles a parabola in CIE chromaticity space and can be approximated by a quadratic equation. The chromaticities of typical daylight follows the curve of the black-body radiator closely but stays slightly on the green side (Judd et al. 1964). Like the curve of the black-body radiator, the coordinates of daylight (D) can be approximated by a simple quadratic equation. The approximation given by CIE (International Commission on Illumination 1996) for the chromaticity coordinates of daylight $(\hat{x}_D, \hat{y}_D)$ is as follows:

$$\hat{y}_D = -3.000\hat{x}_D^2 + 2.870\hat{x}_D - 0.275 \qquad (4.27)$$

where $\hat{x}_D$ is a function of the temperature T. Two different formulas are used to compute $\hat{x}_D$ depending on the temperature T. For temperatures from approximately 4000 to 7000K,

COLOR REPRODUCTION

$\hat{x}_D$ is given by

$$\hat{x}_D = -4.6070\frac{10^9}{T^3} + 2.9678\frac{10^6}{T^2} + 0.09911\frac{10^3}{T} + 0.244063. \quad (4.28)$$

For temperatures from approximately 7000 to 25000K, $\hat{x}_D$ is given by

$$\hat{x}_D = -2.0064\frac{10^9}{T^3} + 1.9018\frac{10^6}{T^2} + 0.24748\frac{10^3}{T} + 0.237040. \quad (4.29)$$

From the graph shown in Figure 4.7, it is obvious that if any three monochromatic colors are chosen along the horseshoe-shaped region, we create a triangular subregion. This is shown in Figure 4.8. This subregion is the color gamut of the display device. Colors that are located outside this region cannot be created using the chosen primary colors. Note that different devices may have different gamuts. Figure 4.9 shows the gamut of colors for three different methods of color reproduction: film, a color monitor, and printing ink. It may well be true that a given color that can be displayed by one device cannot be displayed using another device. It is also clear that different illuminants may fall onto the same point inside this color system and that different reflectances illuminated by the same illuminant may be mapped onto the same point. Colors that appear to be the same under one illuminant

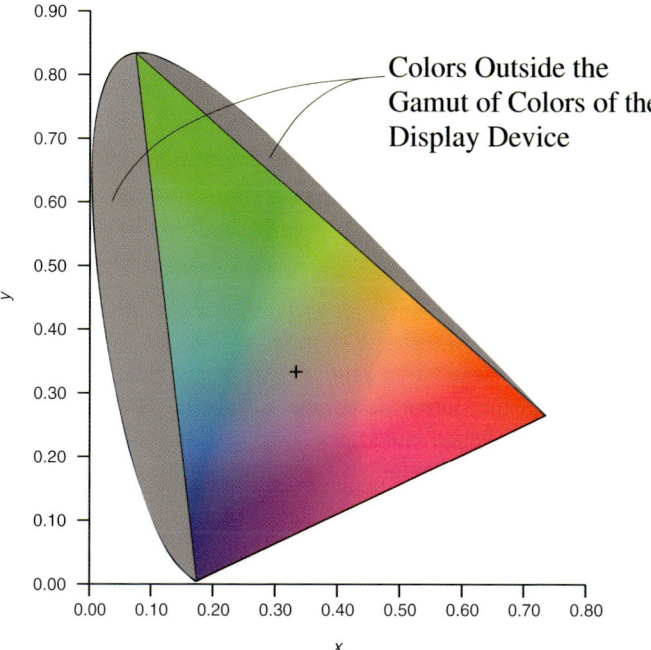

Figure 4.8 If three monochromatic primary colors are chosen for the display device then some colors cannot be created by the display device. The triangular-shaped subregion assumes three monochromatic primary colors at 400 nm, 520 nm, and 700 nm. Colors that are located inside the two gray regions lie outside the gamut of colors of the display device.

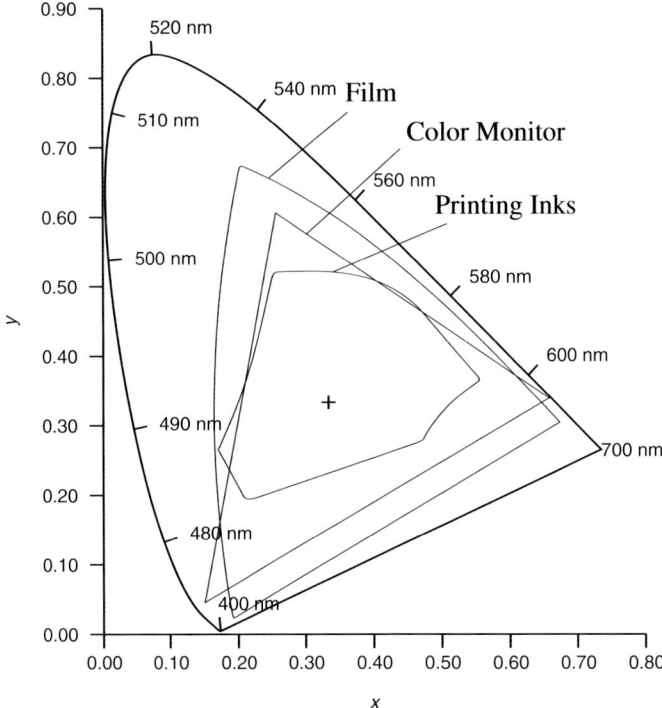

Figure 4.9 Color gamuts of film, a color monitor, and printing inks (Reproduced by permission of Pearson Education from 3D Computer Graphics Third Edition, Alan Watt, Pearson Education Limited, © Pearson Education Limited 2000).

and different under a second illuminant are called *metamers* (Colourware Ltd 2001; Jähne 2002; Mallot 1998; Wyszecki and Stiles 2000).

On a display system that uses three primary colors, we have to transform the output color to the color space of the output device before display. Digital cameras are usually using three types of narrow-shaped band pass filters that pass light in the red, green, and blue parts of the spectrum. Digital monitors are using light from three different phosphors as additive primaries. The light given off by the phosphors may not be exactly equivalent to the shape of the filters of the digital camera. So, if we obtain a digital image using a digital camera and want to display it on a computer monitor, then we have to transform the colors from the coordinate system of the digital camera to the coordinate system of the computer monitor. Today, flat panel displays are ubiquitous. Such displays use filtered white light to create color sensations. A backlight is located at the back of the display and three types of filters (red, green, and blue) are used to create the color primaries. If one would use the same type of narrow-band filters on the camera as well as on the display device, then we would not need an additional transformation between the two color spaces.

If the gamut of two display or measurement devices is not equivalent, we can nevertheless exchange data between the two by using the CIE XYZ coordinate system as a

COLOR REPRODUCTION

reference. Provided that the coordinate system of the display device is linear, we can write the transformation between the display device's coordinate system and XYZ color space as (Watt 2000)

$$\begin{bmatrix} X \\ Y \\ Z \end{bmatrix} = \begin{pmatrix} X_r & X_g & X_b \\ Y_r & Y_g & Y_b \\ Z_r & Z_g & Z_b \end{pmatrix} \begin{bmatrix} R \\ G \\ B \end{bmatrix} = \mathbf{T} \begin{bmatrix} R \\ G \\ B \end{bmatrix} \quad (4.30)$$

where $[R, G, B]^T$ are the coordinates of the display or measurement device and $[X, Y, Z]^T$ are the corresponding coordinates in XYZ color space. Let $\mathbf{T}_1$ be the transform for the first display device and $\mathbf{T}_2$ be the transform for the second display device. In order to view the data originally intended for display device 1 on display device 2, we have to apply the transform $\mathbf{T}_{2 \leftarrow 1}$

$$\mathbf{T}_{2 \leftarrow 1} = \mathbf{T}_2^{-1} \mathbf{T}_1 \quad (4.31)$$

to the original data. Note that some coordinates may be outside of the gamut of the destination display device. In this case, the coordinates have to be either clipped or transformed in some other way before viewing, e.g. by reducing the intensity or the saturation of the colors to avoid clipping.

The matrix $\mathbf{T}$ which transforms the linear RGB coordinates of the display device to XYZ space can be computed as follows (Watt 2000): In case of a color monitor, let $[\hat{x}_i, \hat{y}_i]$ with $i \in \{r, g, b\}$ be the chromaticity coordinates of the phosphors red, green, and blue. Note that $\hat{z}_i = 1 - \hat{x}_i - \hat{y}_i$. We can rewrite the matrix $\mathbf{T}$ as

$$\mathbf{T} = \begin{pmatrix} X_r & X_g & X_b \\ Y_r & Y_g & Y_b \\ Z_r & Z_g & Z_b \end{pmatrix} = \begin{pmatrix} k_r \hat{x}_r & k_g \hat{x}_g & k_b \hat{x}_b \\ k_r \hat{y}_r & k_g \hat{y}_g & k_b \hat{y}_b \\ k_r \hat{z}_r & k_g \hat{z}_g & k_b \hat{z}_b \end{pmatrix} \quad (4.32)$$

with $k_i = X_i + Y_i + Z_i$ and $\hat{x}_i = \frac{X_i}{k_i}$. Now we factor $\mathbf{T}$ into two matrices, one with the chromaticities of the display device and the other with the coefficients k_i.

$$\mathbf{T} = \begin{pmatrix} \hat{x}_r & \hat{x}_g & \hat{x}_b \\ \hat{y}_r & \hat{y}_g & \hat{y}_b \\ \hat{z}_r & \hat{z}_g & \hat{z}_b \end{pmatrix} \begin{pmatrix} k_r & 0 & 0 \\ 0 & k_g & 0 \\ 0 & 0 & k_b \end{pmatrix}. \quad (4.33)$$

We assume that $[R, G, B]^T = [1, 1, 1]^T$ should be mapped to standard white D_{65}. The chromaticity coordinates in XYZ space of standard white for a standard colorimetric observer are $[X_w, Y_w, Z_w]^T = [95.05, 100.00, 108.88]^T$ (International Commission on Illumination 1996). This gives us

$$\begin{bmatrix} X_w \\ Y_w \\ Z_w \end{bmatrix} = \begin{pmatrix} \hat{x}_r & \hat{x}_g & \hat{x}_b \\ \hat{y}_r & \hat{y}_g & \hat{y}_b \\ \hat{z}_r & \hat{z}_g & \hat{z}_b \end{pmatrix} \begin{bmatrix} k_r \\ k_g \\ k_b \end{bmatrix} \quad (4.34)$$

from which we can compute k_i by inverting the matrix.

4.5 Gamma Correction

A color television set contains an array of electron-sensitive phosphors. Three different phosphors are used, one emits red light, another emits green light, and the remaining one emits blue light. The electron gun inside the television set is modulated to produce the desired intensity. However, the correspondence between the power applied to the electron gun and the amount of light given off by the phosphor I, is not linear. The nonlinearity is caused by the function of the cathode and the electron gun. It is not caused by the phosphor (Poynton 1998, 2003). The relationship is best described by a power function (Möller and Haines 1999; Stokes et al. 1996)

$$I \approx A(k_1 x + k_2)^\gamma \tag{4.35}$$

where I is the luminance of the pixel on the computer screen, x is the intensity of the color channel, A is the maximum intensity given off by the monitor, k_1 and k_2 are system gain and offset respectively, and γ is a constant which is specific for a given monitor. The constant k_2 describes the black level. Both k_1 and k_2 can be changed by adjusting the contrast and brightness knobs. The performance is best when the black level is set to zero. In this case, black pixels are located at the point where such pixels are just about to give off light. If k_2 is assumed to be zero, then the equation is simplified to

$$I \approx A x^\gamma. \tag{4.36}$$

The gamma values of computer monitors are in the range of 2.3 to 2.6. The National Television Standards Committee (NTSC) standard specifies a gamma value of 2.2. This relationship between luminance and the voltage, which is applied to the electron guns, is shown in Figure 4.10. The graph assumes a gamma of 2.2. Modern cathode-ray tubes have a gamma of 2.5 (Poynton 2003).

If one applies a power proportional to the measured luminance to the electron guns, then the image will appear too dark. Suppose that we want to display three pixels, one with intensity 0, one with intensity 0.5, and one with intensity 1.0. Let us assume a black level of zero, i.e. $k_2 = 0$, $A = 1$, and a gamma value of 2.2. If we now apply voltages of 0, 0.5, and 1.0 for the three pixels, then the result would be measured intensities of 0,

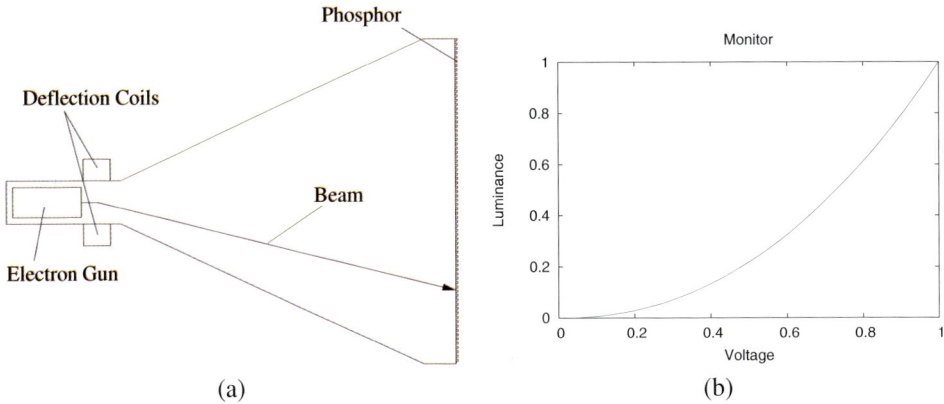

Figure 4.10 A cathode-ray tube consists of an electron gun and a phosphor covered screen. Deflection coils are used to sweep the electron beam over the entire screen (a). The correspondence between the power applied to the electron gun and the light given off by the phosphor is nonlinear (b).

COLOR REPRODUCTION

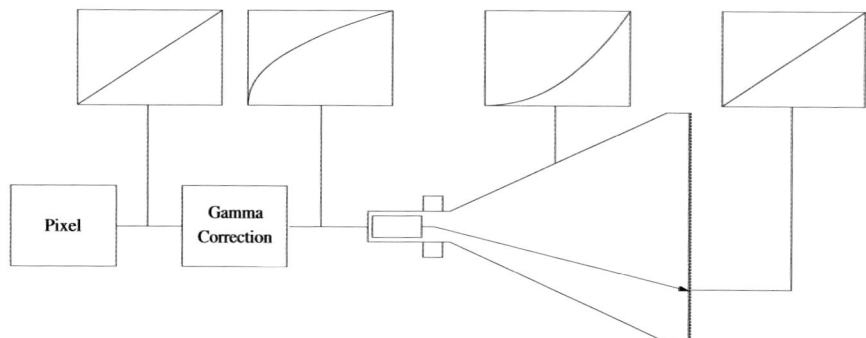

Figure 4.11 In order to achieve a linear relationship between pixel values and amount of light given off by the phosphor, one has to apply a gamma correction.

0.22, and 1.0. In order to obtain output images that appear to be equivalent to the measured ones, one has to apply a so-called gamma correction. The relationship between the output voltage and the pixel intensity has to be linearized. Let c be the pixel intensity that is used to drive the electron gun. Then, we obtain

$$c = I^{\frac{1}{\gamma}} \tag{4.37}$$

as gamma-corrected pixel values for our image. This function is also shown in Figure 4.11. Poynton (2003) denotes the nonlinear, i.e. gamma-corrected values using prime symbols. In other words, R', G', and B' denote the gamma-corrected values of linear R, G, and B. If we use gamma-corrected values, we apply 0, 0.73, and 1.0 instead of 0, 0.22, 1.0, which produces the desired intensities of 0, 0.5, and 1.0. Note that the necessary gamma correction depends on the display device. Quite often, the gamma factor is assumed to be 2.2 and a gamma correction of 1/2.2 is applied such that the image appears correct on a display device with a gamma factor of 2.2. If the actual gamma factor of the display device differs from 2.2, then the image will appear too bright or too dark.

Sometimes a linear transfer function is used for small intensities. This has the advantage that the inverse of the function can still be computed using integer arithmetic. Figure 4.12 shows two such functions. The International Telecommunication Union ITU Rec. 709 gamma transfer function (defined in International Telecommunication Union-Radiocommunication ITU-R Recommendation BT.709) is used for motion video images, e.g. high definition television (HDTV), whereas the International Electrotechnical Commission (IEC) sRGB gamma transfer function is used for images viewed in an office environment (Poynton 2003). The two functions differ because perception of brightness variations changes with the type of surrounding in which the images are viewed. It is assumed that television is viewed in a dim environment compared to an office environment, which is usually much brighter. The IEC sRGB gamma transfer function has a linear section for $I < 0.0031308$ (Poynton 2003).

$$\text{gamma}_{\text{sRGB}}(x) = \begin{cases} 12.92x & \text{if } x \leq 0.0031308 \\ 1.055 x^{\frac{1}{2.4}} - 0.055 & \text{if } x > 0.0031308 \end{cases}. \tag{4.38}$$

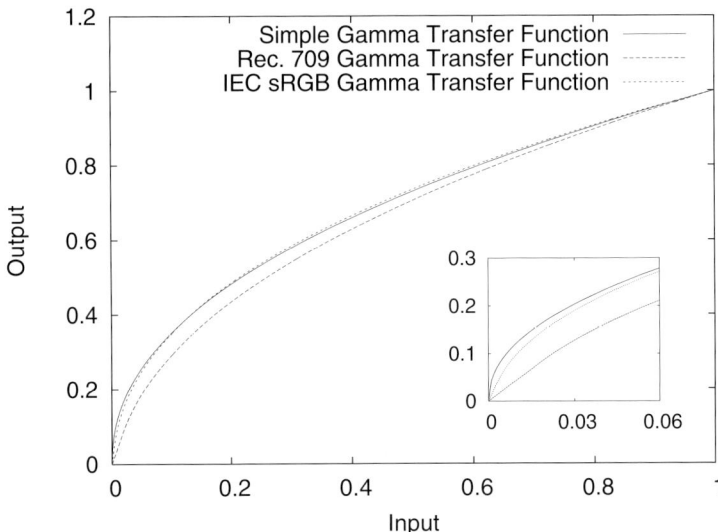

Figure 4.12 The ITU Rec. 709 gamma transfer function and the IEC sRGB gamma transfer function have a linear section. The inset shows the transfer function for small intensities.

Even though the sRGB standard contains a power function with an exponent of 2.4, the overall curve is best described as a power function with an exponent of $\frac{1}{2.22}$. If a display device with a gamma of 2.5 is used, the end-to-end gamma will be 1.125, which is suitable for an office environment. Because of the exponent of 1.125, the image will be slightly brighter. The ITU Rec. 709 gamma transfer function has a linear section for $I < 0.018$ (Henderson 2003; Poynton 1998, 2003).

$$\text{gamma}_{709}(x) = \begin{cases} 4.5x & \text{if } x \leq 0.018 \\ 1.099x^{0.45} - 0.099 & \text{if } x > 0.018 \end{cases}. \quad (4.39)$$

This gamma function contains a power function with an exponent of 0.45; however, the overall curve is best described as a power function with an exponent of 0.5. The standard assumes that the display device has a gamma of 2.5. This gives an end-to-end gamma of 1.25. Since the end-to-end gamma is not unity, the image will be slightly brighter, which is suitable for a dim environment.

Use of a gamma correction is also very important for computer graphics (Foley et al. 1996; Möller and Haines 1999; Watt 2000). In computer graphics, one computes the color of each image pixel based on physical reflectance models. The computations are carried out independently for the three color bands red, green, and blue. In order to accurately display such images on a computer display, a gamma correction has to be applied. Computer images are often stored with a gamma correction already applied. For instance, many digital cameras produce JPEG images using the sRGB color space. This color space assumes a gamma factor of 2.2.

Gamma correction is also important for image-processing applications. Suppose that we want to apply a blur filter to a stored image. We have two pixel intensities of 0.6 and

COLOR REPRODUCTION

1.0. When we compute the average of the two values, we obtain 0.8. If the intensities were stored using the sRGB color space, then the values 0.6 and 1.0 actually correspond to linear intensities $(0.6)^{2.2} = 0.325$ and 1.0. Therefore, the correct result of the averaging would be $(0.325 + 1)/2 = 0.663$. If we want to store the result using the same color space, then we have to apply the gamma correction again. In this case, we would have to store the value $(0.663)^{\frac{1}{2.2}}$ in the output image. Thus, when working with such images, it is very important to linearize the images first before applying any image-processing algorithms. Many imaging operations, such as smoothing or averaging operations, assume that we are operating in a linear space. Therefore, we have to first undo the gamma correction implicit in the image format and then perform the desired image-processing operation. In order to display the images, we have to apply the gamma correction again, which of course depends on the display device.

4.6 Von Kries Coefficients and Sensor Sharpening

In Chapter 3 we have seen that sensors that have very narrow band-response functions simplify the functions which relate the energy measured by the sensor to the geometry of the object, the reflectance of the object patch, and the irradiance falling onto the patch. If the response function is very narrow, it can be approximated by a delta function. In this case, the energy measured by a sensor I at sensor position $\mathbf{x}_I$ is given by

$$I(\mathbf{x}_I) = G(\mathbf{x}_{\text{Obj}}) R(\lambda, \mathbf{x}_{\text{Obj}}) L(\lambda, \mathbf{x}_{\text{Obj}}) \qquad (4.40)$$

where λ is the single wavelength to which the sensor responds, $G(\mathbf{x}_{\text{Obj}})$ is a factor due to the scene geometry at position $\mathbf{x}_{\text{Obj}}$, $R(\lambda, \mathbf{x}_{\text{Obj}})$ is the reflectance for wavelength λ at the corresponding object position $\mathbf{x}_{\text{Obj}}$, and $L(\lambda, \mathbf{x}_{\text{Obj}})$ is the irradiance for wavelength λ at position $\mathbf{x}_{\text{Obj}}$.

From this equation, it is clear that the illuminant only scales the response of the sensor. In order to adjust the color of the illuminant, we have to divide by $L(\lambda, \mathbf{x}_{\text{Obj}})$ to obtain an image taken under white light. A simple diagonal transform suffices. Suppose that we have a sensor that responds to the three different wavelengths red (λ_r), green (λ_g), and blue (λ_b). This sensor is used to create a computer image. Let $\mathbf{c} = [c_r, c_g, c_b]^T$ be the intensity values stored in the image. If we assume a gamma factor of one, then we have

$$c_i(x, y) = I_i(x, y) = G(x, y) R_i(x, y) L_i(x, y) \qquad (4.41)$$

with $i \in \{r, g, b\}$. It is assumed here, that object position $\mathbf{x}_{\text{Obj}}$ is imaged at image pixel (x, y). The difficult part is to obtain an accurate estimate of the illuminant for every object position. Algorithms for estimating the illuminant will be covered in the following chapters.

If the illuminant is assumed to be uniform over the entire image, we obtain an even simpler equation.

$$I_i(x, y) = G(x, y) R_i(x, y) L_i. \qquad (4.42)$$

In order to obtain an output image that is corrected for the color of the illuminant we only need to divide each channel by L_i. The output image will then appear to have been taken under a white light. This can be done by applying a diagonal transform. Let $\mathbf{S}$ be a diagonal

84   COLOR REPRODUCTION

transform
$$\mathbf{S}(s_r, s_g, s_b) = \begin{pmatrix} s_r & 0 & 0 \\ 0 & s_g & 0 \\ 0 & 0 & s_b \end{pmatrix}. \tag{4.43}$$

An image corrected for the color of the illuminant can be computed by applying the linear transform $\mathbf{S}(\frac{1}{L_r}, \frac{1}{L_g}, \frac{1}{L_b})$ to each image pixel. The output color $\mathbf{o} = [o_r, o_g, o_b]^T$ is given by

$$\begin{bmatrix} o_r \\ o_g \\ o_b \end{bmatrix} = \begin{pmatrix} \frac{1}{L_r} & 0 & 0 \\ 0 & \frac{1}{L_g} & 0 \\ 0 & 0 & \frac{1}{L_b} \end{pmatrix} \begin{bmatrix} I_r \\ I_g \\ I_b \end{bmatrix} = \begin{bmatrix} G(x,y)R_r(x,y) \\ G(x,y)R_g(x,y) \\ G(x,y)R_b(x,y) \end{bmatrix}, \tag{4.44}$$

which is independent of the illuminant as desired.

Now, suppose that we have a given image and we want to transform this image such that it appears that the image was taken under a different illuminant. Let $L(\lambda)$ and $L'(\lambda)$ be the two illuminants. Since we are working with narrow-shaped sensors, only the intensities L_i and L_i' at wavelength λ_i with $i \in \{r, g, b\}$ are relevant. Let us call illuminant L the canonical illuminant. If the image was taken using L', then we only have to apply the transformation $\mathbf{S}(\frac{L_r}{L_r'}, \frac{L_g}{L_g'}, \frac{L_b}{L_b'})$ to each image pixel. The result will be an image that seems to have been taken under illuminant L. Let $I_i' = G(x,y)R_r(x,y)L_r'$, then we obtain

$$\begin{bmatrix} o_r \\ o_g \\ o_b \end{bmatrix} = \begin{pmatrix} \frac{L_r}{L_r'} & 0 & 0 \\ 0 & \frac{L_g}{L_g'} & 0 \\ 0 & 0 & \frac{L_b}{L_b'} \end{pmatrix} \begin{bmatrix} I_r' \\ I_g' \\ I_b' \end{bmatrix} = \begin{bmatrix} G(x,y)R_r(x,y)L_r \\ G(x,y)R_g(x,y)L_g \\ G(x,y)R_b(x,y)L_b \end{bmatrix}. \tag{4.45}$$

The coefficients $K_i = \frac{L_i}{L_i'}$ with $i \in \{r, g, b\}$ are known as von Kries coefficients (Richards and Parks 1971). If the sensors are not narrow band and we are given two measurements that were taken under different illuminants L_i and L_i' at the same position of the image sensor, then the von Kries coefficients are given by

$$K_i = \frac{I_i}{I_i'} = \frac{\int G(x,y)R_r(x,y,\lambda)L(\lambda)S_i(\lambda)\,d\lambda}{\int G(x,y)R_r(x,y,\lambda)L'(\lambda)S_i(\lambda)\,d\lambda}. \tag{4.46}$$

The second illuminant L' can be considered to be an illuminant to which the eye has adapted itself. Necessary and sufficient conditions for von Kries chromatic adaptation to provide color constancy were derived by West and Brill (1982).

Finlayson et al. (1994a) have shown that diagonal transforms suffice even if the sensors are not narrow-band sensors. A similar argument was also made by Forsyth (1992). Finlayson et al. showed this under the assumption that illumination can be approximated using three basis functions and reflectances can be approximated using two basis functions. The same also holds if reflectances are assumed to be three-dimensional and illuminants are assumed to be two-dimensional. In general, if we assume that color constancy is achieved using a linear symmetric transformation, then color constancy can always be achieved using a diagonal transform. Let

$$\mathbf{o} = \mathbf{Ac} \tag{4.47}$$

be a linear transform with **A** being a 3 × 3 matrix. If the matrix **A** is symmetric, it can also be written as

$$\mathbf{A} = \mathbf{TDT}^T \tag{4.48}$$

where **T** is an orthogonal matrix and **D** is a diagonal matrix (Bronstein et al. 2001). In other words, a change of the coordinate system allows us to apply a diagonal transform to perform color constancy even if the sensors are not narrow band.

5

Color Spaces

A large number of different color spaces exist. Most of the color spaces are just a linear transformation of the coordinate system. Each color space was created for a certain task. For instance, some color spaces were devised for paint programs, while other color spaces were devised for color perception. Poynton (1997) gives a good overview of many of the different color spaces. See also Wyszecki and Stiles (2000) for a description of the CIE color spaces.

5.1 RGB Color Space

In the RGB color space, color is defined with respect to a unit cube (Foley et al. 1996). The cube is defined using three axes, red, green, and blue, as shown in Figure 5.1. Each point inside the cube defines a unique color. Let $\mathbf{c} = [R, G, B]$ with $R, G, B \in [0, 1]$ be a point inside the color cube. Then the individual components r, g, b specify the intensity of the red, green, and blue components that are used to produce a single pixel on the screen. The eight corners of the cube can be labeled with the colors black, red, green, blue, yellow, magenta, cyan, and white. The gray scale is located inside the cube. It starts at the origin with the black color and extends all the way through the cube to the opposite corner of the cube to the white color.

5.2 sRGB

The sRGB standard was proposed by Hewlett-Packard and Microsoft (Stokes et al. 1996). It is based on a calibrated calorimetric RGB color space (Figure 5.2). As we have seen in the previous chapter, we need to apply different gamma factors depending on the type of display device used. Some computers such as the Apple Macintosh and SGI workstations have internal lookup tables that can be used to adjust the gamma value of a given image. If an artist creates a digital image on one computer and then transfers this image to another computer, then the image will look different depending on the type of computer used to view the image. Today, images are transferred frequently over the Internet using the World

---

Color Constancy M. Ebner
© 2007 John Wiley & Sons, Ltd

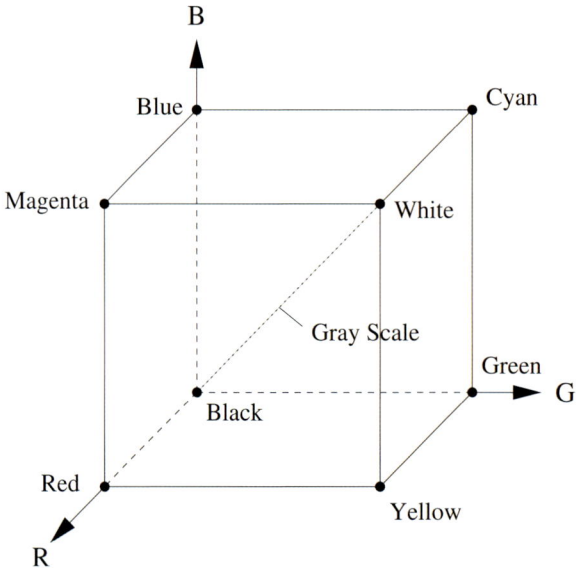

Figure 5.1 RGB color space. Color is defined with respect to a unit cube.

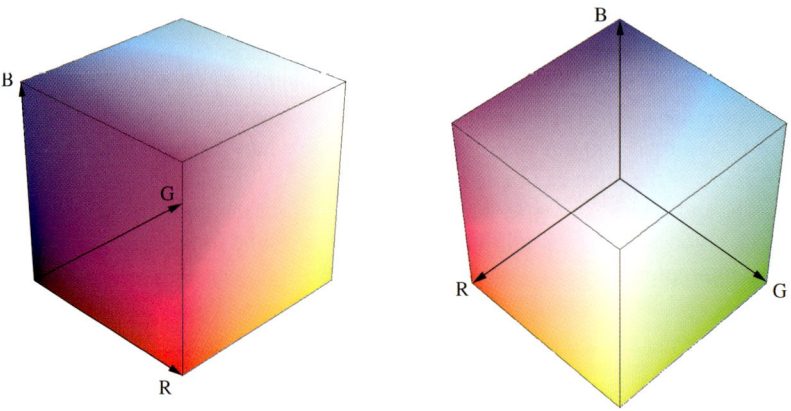

Figure 5.2 sRGB color space.

Wide Web. In order to achieve accurate color reproduction on different systems, a color profile (International Color Consortium 2003) of the generating system has to be attached to every image. However, the associated color profile of the system that was used, when the image was created, is not always stored with the image. Also, sometimes it may be more convenient not to save this information if the this profile can be inferred in some way.

COLOR SPACES

For older images, the data may simply be not available. The sRGB standard was created to fill this void.

The sRGB standard assumes a display gamma value of 2.2. It allows for the unambiguous communication of color without the overhead of storing a color profile with every image. Parameters for a reference viewing environment are also defined. The transformation from CIE XYZ values to sRGB space is given by

$$\begin{bmatrix} R_{sRGB} \\ G_{sRGB} \\ B_{sRGB} \end{bmatrix} = \begin{pmatrix} 3.2410 & -1.5374 & -0.4986 \\ -0.9692 & 1.8760 & 0.0416 \\ 0.0556 & -0.2040 & 1.0570 \end{pmatrix} \begin{bmatrix} X \\ Y \\ Z \end{bmatrix}. \quad (5.1)$$

Tristimulus sRGB values that lie outside the range [0, 1] are clipped to the range [0, 1]. Gamma correction is performed linearly for values smaller than or equal to 0.0031308 and nonlinearly for values larger than 0.0031308 (Poynton 2003).

$$\text{gamma}(x) = \begin{cases} 12.92x & \text{if } x \leq 0.0031308 \\ 1.055 x^{\frac{1}{2.4}} - 0.055 & \text{if } x > 0.0031308 \end{cases} \quad (5.2)$$

This function closely fits the standard gamma function of $\text{gamma}(x) = x^{\frac{1}{2.2}}$. The two sections (one linear and the other nonlinear) were introduced to allow for invertability using integer arithmetic. Finally the gamma-corrected RGB values are mapped to the range [0, 255].

The inverse transformation is performed as follows. RGB values in the range [0, 255] are mapped to the range [0, 1]. The inverse to the gamma function just defined is given by

$$\text{gamma}^{-1}(x) = \begin{cases} \frac{1}{12.92}x & \text{if } x \leq 0.0031308 \\ \left(\frac{1}{1.055}(x + 0.055)\right)^{2.4} & \text{if } x > 0.0031308. \end{cases} \quad (5.3)$$

Finally, the transform to CIE XYZ coordinates is given by

$$\begin{bmatrix} X \\ Y \\ Z \end{bmatrix} = \begin{pmatrix} 0.4124 & 0.3576 & 0.1805 \\ 0.2126 & 0.7152 & 0.0722 \\ 0.0193 & 0.1192 & 0.9505 \end{pmatrix} \begin{bmatrix} R_{sRGB} \\ G_{sRGB} \\ B_{sRGB} \end{bmatrix}. \quad (5.4)$$

This is just the inverse of the above matrix.

The sRGB standard has the advantage that it is compatible with older equipment and older images. Older images will be reproduced very well by assuming that they are sRGB. Today, many manufacturers of displays, monitors, scanners, and digital cameras support the sRGB standard. Advantages of the sRGB standard are discussed in the whitepaper by Starkweather (1998).

5.3 CIE $L^*u^*v^*$ Color Space

In 1976, the CIE defined a three-dimensional color space that is perceptually more uniform than the CIE XYZ color space (International Commission on Illumination 1996). The

$L^*u^*v^*$ color space is intended for self-luminous colors (Colourware Ltd 2001). The three coordinates L^*, u^*, and v^* are computed from the tristimulus values X, Y, and Z as follows:

$$L^* = \begin{cases} 116 \left(\frac{Y}{Y_n}\right)^{\frac{1}{3}} - 16 & \text{if } \frac{Y}{Y_n} > 0.008856 \\ 903.3 \left(\frac{Y}{Y_n}\right) & \text{if } \frac{Y}{Y_n} \leq 0.008856 \end{cases} \quad (5.5)$$

$$u^* = 13L^*(u' - u'_n) \quad (5.6)$$

$$v^* = 13L^*(v' - v'_n) \quad (5.7)$$

where Y_n, u'_n, and v'_n describe a specified white object color stimulus and u' and v' are computed as follows:

$$u' = \frac{4x}{x + 15y + 3z} \quad (5.8)$$

$$v' = \frac{9y}{x + 15y + 3z}. \quad (5.9)$$

The coordinates u' and u' have a range of approximately $[-100, 100]$ (Poynton 2003). The three-dimensional color space basically transforms the XYZ coordinates such that L^* denotes luminance, i.e. radiance weighted by the spectral sensitivity function. The lightness curve has a linear section for values smaller than 0.008856. The overall lightness curve is best approximated by a 0.4-power function. Thus, CIE lightness approximates the response to reflectance. At this point, the reader should have another look at Figure 2.22, where we considered different approximations such as a logarithmic, cube root, and square root approximation to the lightness–reflectance relationship. The CIE has basically standardized the 0.4-power function. A transformation of the sRGB color cube to the $L^*u^*v^*$ color space is shown in Figure 5.3.

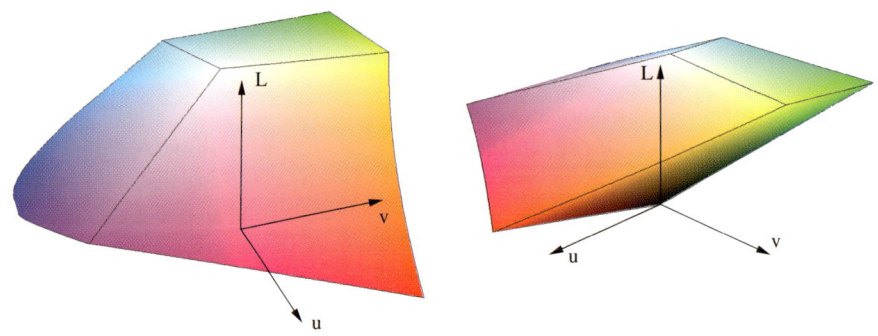

Figure 5.3 sRGB cube visualized in $L^*u^*v^*$ color space.

COLOR SPACES

Let $[L_1, u_1, v_1]$ and $[L_2, u_2, v_2]$ be two different colors inside the $L^*u^*v^*$ color space. Then the color difference between them is simply the Euclidean distance

$$\Delta E^*_{uv} = \sqrt{(L_1 - L_2)^2 + (u_1 - u_2)^2 + (v_1 - v_2)^2}. \tag{5.10}$$

A difference ΔE^*_{uv} of less than 1 is said to be imperceptible (Poynton 2003). Differences between 1 and 4 may or may not be perceptible. This depends on the region of the color space where the two colors are from. If the difference is larger than 4, it is likely that the difference can be perceived. Correlates of lightness, saturation, chroma, and hue are given by

$$L^* = 116 \left(\frac{Y}{Y_n} \right)^{\frac{1}{3}} - 16 \quad \text{if} \quad \frac{Y}{Y_n} > 0.008856 \quad \text{(lightness)} \tag{5.11}$$

$$s_{uv} = 13\sqrt{(u' - u'_n)^2 + (v' - v'_n)^2} \quad \text{(saturation)} \tag{5.12}$$

$$C^*_{uv} = \sqrt{u^{*2} + v^{*2}} = L^* s_{uv} \quad \text{(chroma)} \tag{5.13}$$

$$h_{uv} = \tan^{-1} \left(\frac{v^*}{u^*} \right) \quad \text{(hue angle).} \tag{5.14}$$

Differences in the hue of two colors $[L_1, u_1, v_1]$ and $[L_2, u_2, v_2]$ are computed as

$$\Delta H^*_{uv} = \sqrt{(\Delta E^*_{uv})^2 - (\Delta L^*)^2 - (\Delta C^*_{uv})^2}. \tag{5.15}$$

The main advantage of such a color system is that color is decoupled from intensity. For instance, if we view a red sphere illuminated by a single light source, then illumination varies over the surface of the sphere (Figure 5.4). Because of this, one can use this information to extract the three-dimensional shape of objects given a single image. This research area is known as *shape from shading* (Horn 1986; Jain et al. 1995; Mallot 1998; Marr 1982). But suppose we wanted to segment the image. We want to locate all pixels that belong to the sphere. This can be easily done using a color space where color is specified by lightness, saturation, and hue. Knowing that the sphere is red, one transforms the RGB values into such a color space. All pixels that have a red hue are assumed to belong to the sphere.

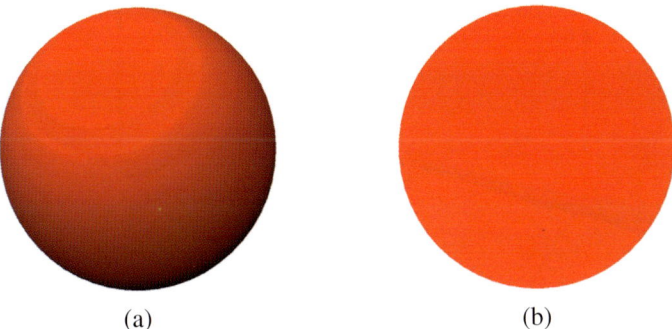

(a)   (b)

Figure 5.4 Sphere illuminated by a single light source (a). Segmentation is simpler if only hue is considered (b).

5.4 CIE $L^*a^*b^*$ Color Space

Another perceptually approximately uniform color space is the CIE $L^*a^*b^*$ color space (Colourware Ltd 2001; International Commission on Illumination 1996). This color space was intended for use with surface colors. The three coordinates L^*, a^*, and b^* are computed from the tristimulus values X, Y, and Z as follows:

$$L^* = \begin{cases} 116 \left(\frac{Y}{Y_n}\right)^{\frac{1}{3}} - 16 & \text{if } \frac{Y}{Y_n} > 0.008856 \\ 903.3 \left(\frac{Y}{Y_n}\right) & \text{if } \frac{Y}{Y_n} \leq 0.008856 \end{cases} \quad (5.16)$$

$$a^* = 500 \left(f\left(\frac{X}{X_n}\right) - f\left(\frac{Y}{Y_n}\right) \right) \quad (5.17)$$

$$b^* = 200 \left(f\left(\frac{Y}{Y_n}\right) - f\left(\frac{Z}{Z_n}\right) \right) \quad (5.18)$$

where X_n, Y_n, and Z_n describe a specified white object color stimulus and the function f is defined as

$$f(X) = \begin{cases} X^{\frac{1}{3}} & \text{if } X > 0.008856 \\ 7.787 X + \frac{16}{116} & \text{if } X \leq 0.008856 \end{cases}. \quad (5.19)$$

L^* describes lightness and extends from 0 (black) to 100 (white). The a coordinate represents the redness–greenness of the sample. The b coordinate represents the yellowness–blueness. The coordinates a and b have a range of approximately $[-100, 100]$. Notice the cube root in the above equation. The cube root was introduced in order to obtain a homogeneous, isotropic color solid (Glasser et al. 1958). Visual observations correspond closely to color differences calculated in this uniform coordinate system. A transformation of the sRGB color cube to the $L^*a^*b^*$ color space is shown in Figure 5.5.

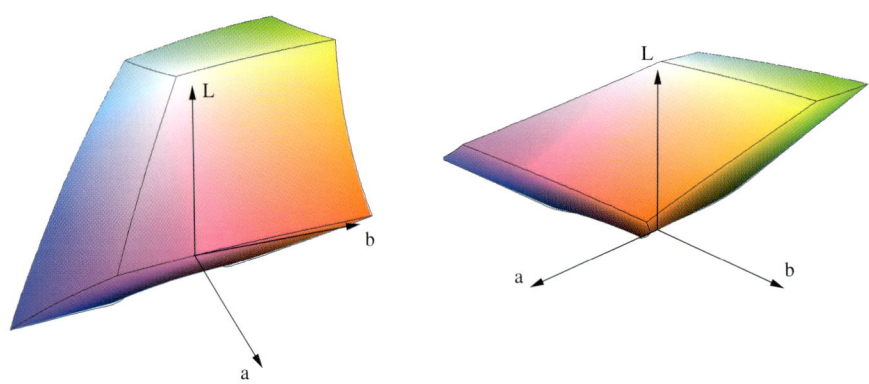

Figure 5.5 sRGB cube visualized in $L^*a^*b^*$ color space.

Again, color differences are computed using the Euclidean distance between the two stimuli. Let $[L_1, a_1, b_1]$ and $[L_2, a_2, b_2]$ be two different colors. Then the color difference between them is given by

$$\Delta E_{ab}^* = \sqrt{(L_1 - L_2)^2 + (a_1 - a_2)^2 + (b_1 - b_2)^2}. \tag{5.20}$$

Correlates of lightness, chroma, and hue are given by

$$L^* = 116 \left(\frac{Y}{Y_n}\right)^{\frac{1}{3}} - 16 \quad \text{if} \quad \frac{Y}{Y_n} > 0.008856 \quad \text{(lightness)} \tag{5.21}$$

$$C_{ab}^* = \sqrt{a^{*2} + b^{*2}} \quad \text{(chroma)} \tag{5.22}$$

$$h_{ab} = \tan^{-1}\left(\frac{b^*}{a^*}\right) \quad \text{(hue angle)}. \tag{5.23}$$

Differences in hue of two colors $[L_1, a_1, b_1]$ and $[L_2, a_2, b_2]$ are computed as

$$\Delta H_{ab}^* = \sqrt{(\Delta E_{ab}^*)^2 - (\Delta L^*)^2 - (\Delta C_{ab}^*)^2}. \tag{5.24}$$

5.5 CMY Color Space

The CMY color space is used for color printing (Foley et al. 1996). A color printer deposits colored pigments on paper. The three primary colors are cyan, magenta, and yellow. Each of these colored pigments absorbs a single primary color. Suppose that we want to print a color with RGB values in the range [0, 1]. The corresponding CMY values can be calculated by subtracting the RGB values from [1, 1, 1]:

$$\begin{bmatrix} C \\ M \\ Y \end{bmatrix} = \begin{bmatrix} 1 \\ 1 \\ 1 \end{bmatrix} - \begin{bmatrix} R \\ G \\ B \end{bmatrix}. \tag{5.25}$$

The more the red light that should be reflected, the lesser the amount of cyan that should be deposited. Similarly, the more the green light that should be reflected, the lesser the amount of magenta that should be deposited. Finally, the more the blue light should be reflected, the lesser the amount of yellow that is required. Sometimes black pigments are used in addition to cyan, magenta, and yellow. This is called the *CMYK* color model. The exact transformation which is required to transform sRGB to CMY or CMYK of course depends on the spectral absorbance curves of the color pigments which are used by the printer.

5.6 HSI Color Space

Color in the HSI color space is represented as three components: hue, saturation, and intensity (Gonzalez and Woods 1992; Jain et al. 1995; Smith 1978). Any point in the RGB color space can also be specified using the HSI color model. The attribute hue defines the color of the point. Saturation specifies the amount of white added to the pure color. Finally, intensity specifies the brightness of the point. It is not clear if the HSI components are defined for linear RGB or nonlinear RGB space (Poynton 1997, 1998). Therefore, it is not possible to determine whether the components of the HSI color space refer to a physical

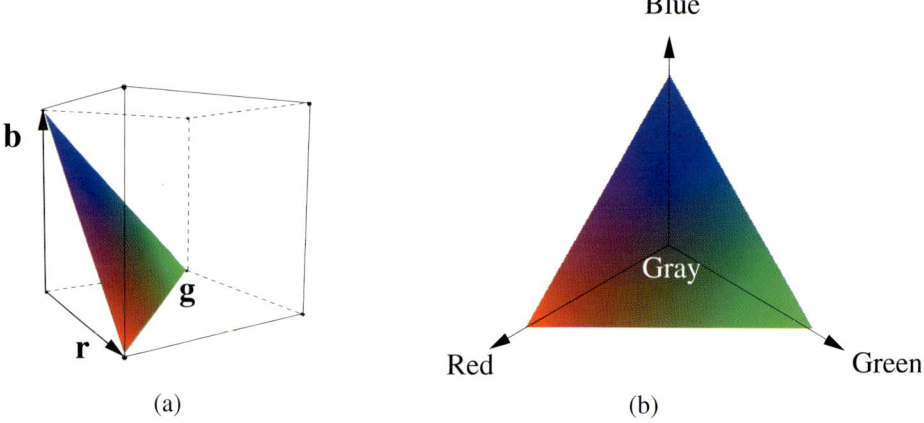

Figure 5.6 The HSI triangle is located at a distance of $\frac{1}{\sqrt{3}}$ from the origin. The triangle is defined by the points $[1, 0, 0]$, $[0, 1, 0]$, and $[0, 0, 1]$.

quantity or a perceptual quantity. Since the HSI transformation is simply a transformation of the coordinate system, it may be used with either linear RGB or nonlinear RGB values.

We will now look at the conversion from RGB color space to that of HSI and back. In order to define the HSI color space, we place a color triangle inside the unit cube (Figure 5.6). The triangle is defined by the three points $\mathbf{r} = [1, 0, 0]$, $\mathbf{g} = [0, 1, 0]$, and $\mathbf{b} = [0, 0, 1]$. It is located at a distance of $\frac{1}{\sqrt{3}}$ from the origin with the normal vector pointing in the direction of the gray vector. Fully saturated colors are located at the border of the triangle. Less saturated colors are located toward the center. Hue is defined as the angle between the vector from the center of the triangle to the red corner and the vector pointing from the center of the triangle to the projection of the given color on the triangle. Saturation is defined as the amount of white added to a fully saturated color. Intensity is defined as the brightness of a pixel.

Let $\mathbf{c} = [R, G, B]$ be a color in RGB space. The transformation from RGB space to HSI space is given by (Gonzalez and Woods 1992)

$$I = \frac{1}{3}(R + G + B) \tag{5.26}$$

$$H = \begin{cases} \frac{360° - \alpha}{360°} & \text{if } b > g \\ \frac{\alpha}{360°} & \text{otherwise.} \end{cases} \tag{5.27}$$

$$\text{with} \quad \alpha = \cos^{-1} \frac{2R - G - B}{\sqrt{R^2 + G^2 + B^2 - RG - GB - RB}}$$

$$S = 1 - \frac{\min\{R, G, B\}}{I} \tag{5.28}$$

COLOR SPACES

The transformation from HSI space to RGB space is given by

$$b = \frac{1}{3}(1 - S) \qquad (5.29)$$

$$r = \frac{1}{3}\left(1 + \frac{S \cos \alpha}{\cos(\alpha - 60°)}\right) \qquad (5.30)$$

$$g = 1 - b - r \qquad (5.31)$$

if $0 < \alpha \leq 120°$. It is given by

$$r = \frac{1}{3}(1 - S) \qquad (5.32)$$

$$g = \frac{1}{3}\left(1 + \frac{S \cos \alpha'}{\cos(\alpha' - 60°)}\right) \qquad (5.33)$$

$$b = 1 - r - g \qquad (5.34)$$

if $120° < \alpha \leq 240°$ with $\alpha' = \alpha - 120°$. It is given by

$$g = \frac{1}{3}(1 - S) \qquad (5.35)$$

$$b = \frac{1}{3}\left(1 + \frac{S \cos \alpha'}{\cos(\alpha' - 60°)}\right) \qquad (5.36)$$

$$r = 1 - g - b \qquad (5.37)$$

if $240° < \alpha \leq 360°$ with $\alpha' = \alpha - 240°$. We have

$$r = \frac{R}{R + G + B} = \frac{R}{3I} \qquad (5.38)$$

where I is the intensity of the color point. Similarly, $g = \frac{G}{3I}$ and $b = \frac{B}{3I}$. This lets us calculate R, G, and B as

$$R = 3Ir \qquad (5.39)$$

$$G = 3Ig \qquad (5.40)$$

$$B = 3Ib. \qquad (5.41)$$

Note that care must be taken not to perform the transformation for HSI values that would specify RGB values outside of the unit cube. Suppose we start with a red color $[R, G, B] = [1, 0, 0]$. If we transform this color to HSI space, we obtain $[H, S, I] = [0, 1, \frac{1}{3}]$. This is a maximally saturated color of intensity $I = \frac{1}{3}$. It is not possible to further increase the intensity to 1 while maintaining the same saturation.

A generalization of the HSI color space, where the white point may be located at an arbitrary position on the triangle, is given by Smith (1978). This color space is known as the *HSL* color space.

5.7 HSV Color Space

The HSV color space was defined by Smith (1978) (see also Foley et al. (1996) and Watt (2000)). This color space was defined to model an artist's method of mixing colors. Therefore, the HSV color space may be most useful in painting or in drawing programs. As in the HSI color space, a color is represented using three components: hue (H), saturation (S), and value (V). Colors are defined given an RGB color cube. The color cube is oriented such that the value component is oriented along the gray axis, which runs from black to white. Value is defined as the maximum intensity of the three RGB components red, green, and blue. Saturation is defined in terms of the maximum intensity and the minimum intensity of the RGB stimulus. The minimum intensity specifies how much white has been added to achieve the given color. When the RGB color cube is projected along the gray axis onto a plane that is oriented perpendicular to the gray axis, one obtains a hexagonal disk as shown in Figure 5.7. This can be done for different sizes of the RGB color space. The scaling is determined by the value component. If this is done for all values V, one obtains a hexagonal cone. Therefore, this color model is also known as the *hexcone* model. Figure 5.8 shows the resulting hexagonal disks for different values of V.

The saturation S and hue H components specify a point inside the hexagonal disk. Saturation for a given level of V is defined as the relative length of the vector that points to the given color to the length of the vector that points to the corresponding color on the border or the hexagonal disk. This results in a set of loci of constant S as shown in Figure 5.9. Hue H is defined as the linear length along the loci of constant S beginning

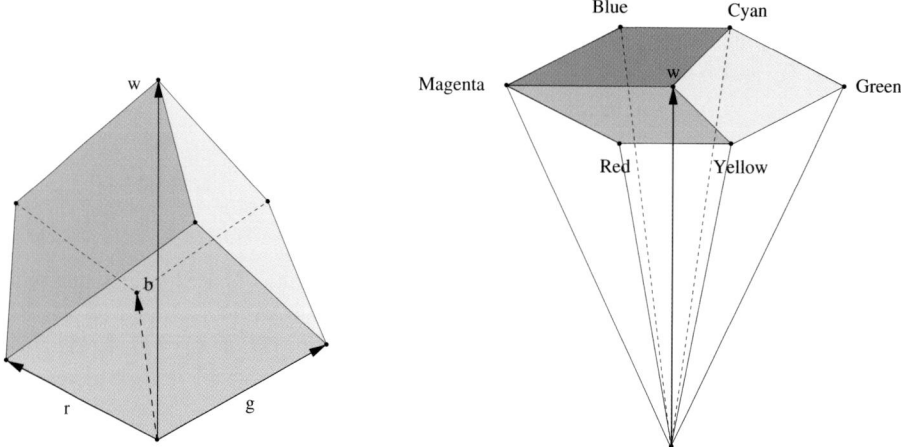

Figure 5.7 HSV color space. The RGB cube is projected along onto a plane that is perpendicular to the gray vector. The result is a hexagonal disk.

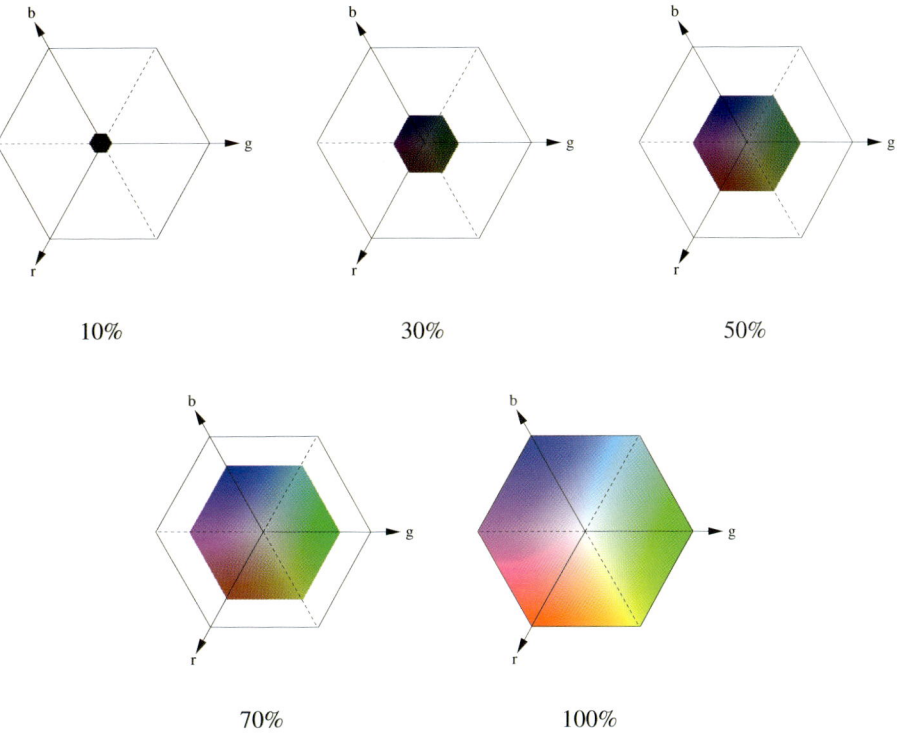

Figure 5.8 HSV color space at $V = 0.1$, $V = 0.3$, $V = 0.5$, $V = 0.7$, and $V = 1.0$.

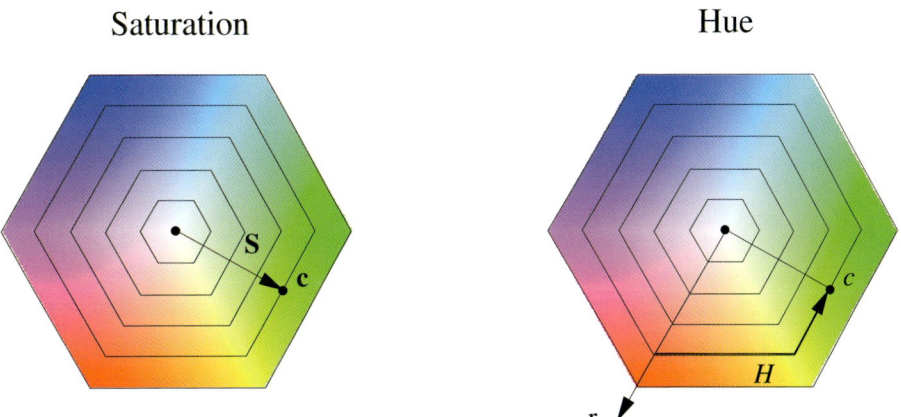

Figure 5.9 Saturation and hue of a color **c** as defined in the HSV color space.

at the red vector. Note that the definitions of hue and saturation of the HSV color space are different from the definitions of hue and saturation of the HSI color space. In the HSI color space, hue is defined as an angle around the gray vector. In the HSV color space, hue is defined as the linear distance around the loci of constant saturation.

The transformation from RGB color space to HSV color space is given by

$$V = \max\{R, G, B\} \tag{5.42}$$

$$S = \frac{\max - \min}{\max} \tag{5.43}$$

$$H = \begin{cases} \frac{1}{6} \frac{G-B}{\max - \min} & \text{if } R = \max \\ \frac{1}{6} \left(2 + \frac{B-R}{\max - \min}\right) & \text{if } G = \max \\ \frac{1}{6} \left(4 + \frac{R-G}{\max - \min}\right) & \text{if } B = \max \end{cases} \tag{5.44}$$

where $\max = \max\{R, G, B\}$ and $\min = \min\{R, G, B\}$. All three components V, S, and H are in the range $[0, 1]$. The transformation from HSV back to RGB is given by

$$[R, G, B] = [V, V, V] \tag{5.45}$$

if the saturation S is equal to zero. If the saturation is not zero, then the RGB components are given by

$$[R, G, B] = [V, K, M] \quad \text{if } 0 \leq H < \frac{1}{6} \tag{5.46}$$

$$[R, G, B] = [N, V, M] \quad \text{if } \frac{1}{6} \leq H < \frac{2}{6} \tag{5.47}$$

$$[R, G, B] = [M, V, K] \quad \text{if } \frac{2}{6} \leq H < \frac{3}{6} \tag{5.48}$$

$$[R, G, B] = [M, N, V] \quad \text{if } \frac{3}{6} \leq H < \frac{4}{6} \tag{5.49}$$

$$[R, G, B] = [K, M, V] \quad \text{if } \frac{4}{6} \leq H < \frac{5}{6} \tag{5.50}$$

$$[R, G, B] = [V, M, N] \quad \text{if } \frac{5}{6} \leq H < 1 \tag{5.51}$$

where M, N, and K are defined as

$$M = V(1 - S) \tag{5.52}$$

$$N = V(1 - SF) \tag{5.53}$$

$$K = V(1 - S(1 - F)) \tag{5.54}$$

and $F = 6H - \lfloor I \rfloor$.

The HSV as well as the HSI coordinate system has several drawbacks (Hanbury and Serra 2003). Colors with a low intensity may be described as being fully saturated. For instance, the color $[0.01, 0, 0]$ has a saturation of 1 even though for all practical purposes it is indistinguishable from black, which has a saturation of zero. Thus, dark regions of an image usually have a noisy saturation.

5.8 Analog and Digital Video Color Spaces

Video color spaces transform the measured RGB signals into a perceptually uniform color space. One signal is used to transmit the luminance information and two signals are used to transmit the color information. The reasons for this are twofold. In the early days of television, this had the advantage that only two additional signals had to be transmitted. Old black-and-white equipment would just decode the luminance information. Another reason is that the two color signals may be subsampled without any perceptible loss of detail when viewed at normal viewing distance (Poynton 2003). Color acuity of vision is considered to be poor compared to the discrimination between different luminance levels. As we have seen in Section 2.4, the perception of lightness is nonlinear. Roughly, the human visual system cannot distinguish two luminance levels if the ratio between them is less than about 1.01 (Poynton 2003). If luminance was coded using a linear color space, 11 bits or more would be required in order to describe a perceptually uniform color space. However, many of the coded values would be redundant. Recall that the perception of lightness can be approximated using a power function. Indeed, the CIE $L^*a^*b^*$ color space uses a power function of approximately 0.4. Therefore, it makes sense to apply a power function with an exponent of 0.4. In essence, lesser bits are spent for the very dark areas of the image. Figure 5.10 illustrates this process.

Since the color RGB signals are eventually displayed on a monitor, a gamma correction has to be applied. We have already discussed gamma correction in Section 4.5. Note that the exponent 2.5 of the decoding stage and the exponent 0.4 of the gamma correction would cancel were it not for the color transform that is applied in between. In order to avoid the application of two power functions, the power functions are swapped with the application of the color transform as shown in Figure 5.11. The end result is a much simpler decoder. However, this comes at a cost. The color transform is applied to nonlinear signals. Hence, the encoded signal is not equivalent to luminance. This nonlinear quantity is called *luma* and is denoted by Y' (Poynton 2003). In general, nonlinear quantities are denoted with a prime.

The encoding process used for standard definition television (SDTV) is shown in Figure 5.12. First, the nonlinear quantity luma is computed from the nonlinear RGB values. Then the color differences between the red channel and luma as well as the blue channel and luma are computed. These two differences are called *chroma*. The $R' - Y'$ channel is scaled by 0.564 and is denoted by P_R. The $B' - Y'$ channel is scaled by 0.713 and is denoted by P_B. This is called the $Y'P_BP_R$ color space. The transform shown in Figure 5.12 can be

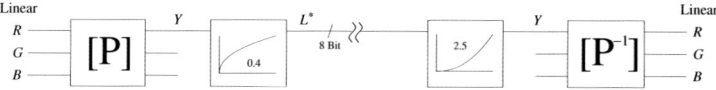

Figure 5.10 Encoding and decoding of the luminance signal using only eight bits. Linear RGB signals are encoded by transforming them to a color space that is spanned by lightness and two color signals. Next, a 0.4-power function is applied. This process is inverted for decoding. (Reprinted from Digital Video and HDTV. Algorithms and Interfaces, Charles Poynton, Morgan Kaufmann Publishers, San Francisco, CA, Copyright 2003, with permission from Elsevier.)

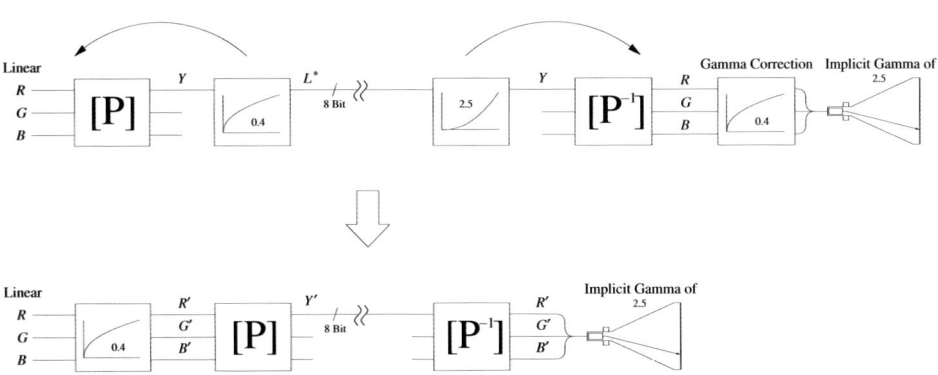

Figure 5.11 Encoding and decoding of the luminance signal. A gamma correction has to be applied to undo the nonlinearity of the display device. A simplified decoder results if the power functions are swapped with the application of the color transform.

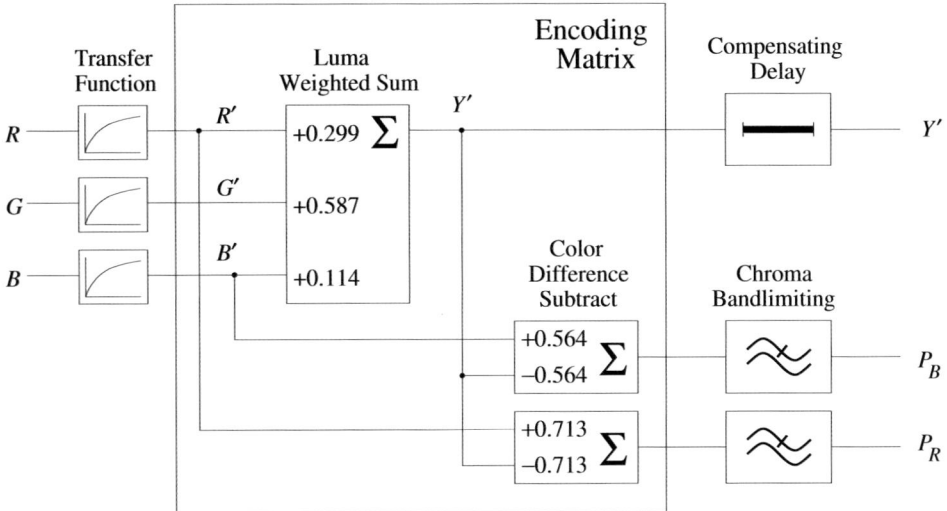

Figure 5.12 SDTV luma/color difference encoder. Luma Y' is computed as a weighted sum from nonlinear RGB values. $R' - Y'$ and $B' - Y'$ color difference values denoted by P_R and P_B are also computed. The color difference signals are low-pass filtered and a compensating delay is introduced for the luma signal. (Reprinted from Digital Video and HDTV. Algorithms and Interfaces, Charles Poynton, Morgan Kaufmann Publishers, San Francisco, CA, Copyright 2003, with permission from Elsevier.)

summarized as a single matrix transform $\mathbf{P}$, which is applied to nonlinear RGB tristimulus values $[R', G', B']$.

$$\begin{bmatrix} Y' \\ P_B \\ P_R \end{bmatrix} = \mathbf{P} \begin{bmatrix} R' \\ G' \\ B' \end{bmatrix} = \begin{pmatrix} 0.299 & 0.587 & 0.114 \\ -0.169 & -0.331 & 0.5 \\ 0.5 & -0.419 & -0.081 \end{pmatrix} \begin{bmatrix} R' \\ G' \\ B' \end{bmatrix}. \quad (5.55)$$

COLOR SPACES

The decoding matrix of SDTV is given by

$$\begin{bmatrix} R' \\ G' \\ B' \end{bmatrix} = \mathbf{P}^{-1} \begin{bmatrix} Y' \\ P_B \\ P_R \end{bmatrix} = \begin{pmatrix} 1 & 0 & 1.402 \\ 1 & -0.344 & -0.714 \\ 1 & 1.772 & 0 \end{pmatrix} \begin{bmatrix} Y' \\ P_B \\ P_R \end{bmatrix}. \tag{5.56}$$

Slightly different luma coefficients are used for high definition television (HDTV). That aside, different primary chromaticities and transfer functions have been standardized for the different systems. Poynton (2003) introduced a standard notation to clearly define the color space used. Historically, the notation $Y'C_BC_R$ defines essentially the same color space for digital systems such as HDTV. The two color spaces $Y'P_BP_R$ and $Y'C_BC_R$ differ only in scaling. Eight-bit systems use the range [0, 255] together with an offset of 128 for the two color channels. Some space is allowed for head and footroom. Thus, the luma channel may actually only use the range [16, 235]. JPEG uses the entire range. The transformation from nonlinear RGB in the range [0, 255] to full range $Y'C_BC_R$ is given by

$$\begin{bmatrix} Y' \\ C_B \\ C_R \end{bmatrix} = \frac{1}{256} \begin{pmatrix} 76.544 & 150.272 & 29.184 \\ -43.027 & -84.471 & 127.498 \\ 127.498 & -106.764 & -20.734 \end{pmatrix} \begin{bmatrix} R' \\ G' \\ B' \end{bmatrix}. \tag{5.57}$$

The luma channel has the range [0, 255], and the two chroma components C_B and C_R have the range [−128, 128]. The decoding matrix is given by

$$\begin{bmatrix} R' \\ G' \\ B' \end{bmatrix} = \frac{1}{256} \begin{pmatrix} 256 & 0 & 357.510 \\ 256 & -87.755 & -182.105 \\ 256 & 451.860 & 0 \end{pmatrix} \begin{bmatrix} Y' \\ C_B \\ C_R \end{bmatrix}. \tag{5.58}$$

6

Algorithms for Color Constancy under Uniform Illumination

Numerous algorithms have been proposed for color constancy. We now describe in detail how these algorithms work. We also see how the algorithms perform in practice. The performance of the algorithms will be shown on two sample images. One image shows a scene with a uniform illumination and the other shows a scene with a nonuniform illumination.

The two sample images are shown in Figure 6.1. The images were taken with a Canon 10D using the sRGB color space. The first image (a) shows a table with plates, coffee cups, spoons, and so on. The image looks very yellow because the room was illuminated by a yellow illuminant. The second image (b) shows an office scene with a desk and several utensils on top of the desk. The image was taken on a bright sunny day. The blue window curtains were closed and the desk lamp was switched on. The blue window curtains created the blue background illumination. The spotlight effect on the table was created by the desk lamp.

In this chapter, we will assume that the color is uniform across the scene. The intensity $\mathbf{I}$ measured by a sensor at position $\mathbf{x}_I$ can be modeled as

$$\mathbf{I}(\mathbf{x}_I) = G(\mathbf{x}_{\text{Obj}}) \int R(\lambda, \mathbf{x}_{\text{Obj}}) L(\lambda) \mathbf{S}(\lambda) \, d\lambda \tag{6.1}$$

where $G(\mathbf{x}_{\text{Obj}})$ is a scaling factor due to the geometry of the patch at position $\mathbf{x}_{\text{Obj}}$, $R(\lambda, \mathbf{x}_{\text{Obj}})$ denotes the reflectance at position $\mathbf{x}_{\text{Obj}}$, $L(\lambda)$ is the radiance given off by the light source, and $\mathbf{S}(\lambda)$ describes the sensitivity of the sensors.

For all of the following algorithms, we will assume that the response functions of the sensors are very narrow-band, i.e. they can be approximated by delta functions. Let λ_i with $i \in \{r, g, b\}$ be the wavelengths to which the sensors respond. We will now denote the sensor coordinates by (x, y). The intensity measured by the sensor at position (x, y) is then given by

$$I_i(x, y) = G(x, y) R_i(x, y) L_i \tag{6.2}$$

where $G(x, y)$ is a factor that depends on the scene geometry at the corresponding object position, $R_i(x, y)$ is the reflectance for wavelength λ_i, and L_i is the irradiance at wavelength

Color Constancy M. Ebner
© 2007 John Wiley & Sons, Ltd

104 ALGORITHMS FOR COLOR CONSTANCY UNDER UNIFORM ILLUMINATION

(a)   (b)

Figure 6.1 Two sample images. The scene in (a) shows a table illuminated by a yellow illuminant. The scene in (b) shows a desk illuminated by sunlight falling through a blue curtain. This creates a blue background illumination. The lamp on the desk was also switched on.

λ_i. We have already derived this equation in Section 3.6. The algorithms discussed in this chapter assume that the illuminant is uniform across the entire image, i.e. the irradiance does not depend on the coordinates (x, y). Algorithms that do not make these assumptions are discussed in the next chapter.

6.1 White Patch Retinex

The white patch retinex algorithm is basically just a simplified version of the retinex algorithm (Cardei and Funt 1999; Funt et al. 1998, 1996; Land and McCann 1971), which is described in the next chapter. The retinex algorithm relies on having a bright patch somewhere in the image. The idea is that, if there is a white patch in the scene, then this patch reflects the maximum light possible for each band. This will be the color of the illuminant, i.e. if $R_i(x, y) = 1$ for all $i \in \{r, g, b\}$ and $G(x, y) = 1$, then

$$I_i(x, y) = L_i. \tag{6.3}$$

If one assumes a linear relationship between the response of the sensor and pixel colors, i.e. $c_i(x, y) = I_i(x, y)$, and one also assumes that the sensor's response characteristic is similar to delta functions, then the light illuminating the scene simply scales the product of the geometry term G and the reflectance R_i of the object.

$$c_i(x, y) = G(x, y) R_i(x, y) L_i \tag{6.4}$$

Therefore, we can rescale all color bands once we have located such a bright patch. In practice, one does not look for a white patch but looks for the maximum intensity of each color channel. Let $L_{i,\max}$ be the maximum of each band over all pixels.

$$L_{i,\max} = \max_{x,y} \{c_i(x, y)\} \tag{6.5}$$

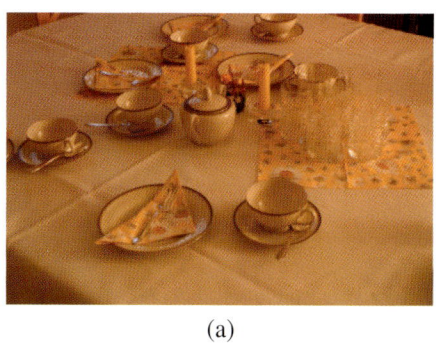 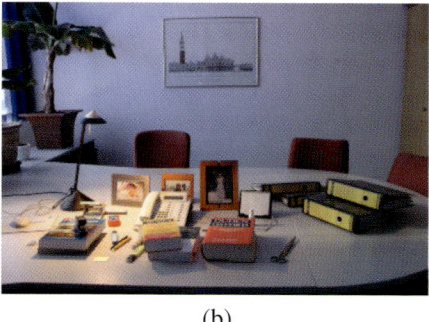

(a) (b)

Figure 6.2 The white patch retinex algorithm is able to perform some color adjustments for the image shown in (a) but the image still looks very yellow. The white patch retinex algorithm does not work very well for the image shown in (b) because of the nonuniform illuminant.

This maximum is then used to scale each color band of the pixels back to the range [0, max]

$$o_i(x, y) = \frac{c_i(x, y)}{L_{i,\max}} = G(x, y) R_i(x, y) \qquad (6.6)$$

where $\mathbf{o}(x, y) = [o_r(x, y), o_g(x, y), o_b(x, y)]^T$ is the color of the output pixel.

Figure 6.2 shows the results achieved with the white patch retinex algorithm. The problem with the white patch retinex algorithm is that it assumes that a single uniform illuminant illuminates the scene. Therefore, it does not work if we have a nonuniform illumination as is the case for the image shown in (b). It also performs poorly on the image shown in (a). This image has some bright specularity in the glassware on the table. The resulting image is still too yellow.

A drawback of this simple version of the white patch retinex algorithm is that a single bright pixel can lead to a bad estimate of the illuminant. If we have a highlight in the image caused by an object that does not reflect the color of the illuminant uniformly, then the estimate will not be equivalent to the actual color of the illuminant. Noise in the image will also be a problem. The white patch retinex algorithm is also highly susceptible to clipped pixels (Funt et al. 1998). If one or more color channels are clipped, then the color of the illuminant cannot be reliably estimated from the brightest pixel.

The white patch retinex algorithm can be made more robust by computing a histogram H_i for each color channel i. The histogram tells us how many image pixels have a particular intensity for a given color channel. Let n_b be the number of buckets of the histogram and let $H_i(j)$ be the number of pixels of color channel i that have intensity j. Instead of choosing the pixel with the maximum intensity for each color channel, one can choose the intensity such that all pixels with intensity higher than the one chosen account for some percentage of the total number of pixels. This method is also used as an automatic white balance by some scanning software. Finlayson et al. (2002) also use it in their method for shadow removal. Let p be the percentage, i.e. 1% or a similar small value, and n be the total number of pixels in the image. Let $c_i(j)$ be the intensity of color channel i represented by bucket j of the histogram H_i. Then the estimate of the illuminant is given by

$$L_i = c_i(j_i) \tag{6.7}$$

with j_i chosen such that

$$pn \leq \sum_{k=j_i}^{n_b} H_i(k) \quad \text{and} \quad pn \geq \sum_{k=j_i+1}^{n_b} H_i(k). \tag{6.8}$$

Results for this algorithm are shown in Figure 6.3. The image of the coffee table looks much better now.

6.2 The Gray World Assumption

The gray world assumption was proposed by Buchsbaum (1980). It estimates the illuminant using the average color of the pixels. That the illuminant could be estimated by computing some kind of average of the light received by the observer was known for a long time and was also suggested by Land (see Judd 1960). However, Buchsbaum was the first to formalize the method. The gray world assumption is probably one of the best-known algorithms for color constancy. Many algorithms have been proposed, which use the gray world assumption in one way or another (Ebner 2002, 2003a,c, 2004c,d; Finlayson et al. 1998; Gershon et al. 1987; Moore et al. 1991; Paulus et al. 1998; Pomierski and Groß 1995; Rahman et al. 1999; Tominaga 1991). These algorithms are all based on the assumption that, on average, the world is gray. Buchsbaum's algorithm estimates the illuminant by assuming that a certain standard spatial spectral average exists for the total visual field. This average is used to estimate the illuminant, which is then used to estimate the reflectances. Results were only shown for simulated data. The derivation of the gray world assumption given here differs from the one given by Buchsbaum. Buchsbaum considers overlapping response characteristics of the sensor array. We assume nonoverlapping response characteristics because little may be gained by using a more general transform (Barnard et al. 2001; Finlayson et al. 1994b). In addition, we also include geometry information in the reflection model whereas Buchsbaum used a simple reflection model without any geometry information.

From the theory of color image formation, we have seen that the intensity $I_i(x, y)$ measured by a sensor i with $i \in \{r, g, b\}$ at position (x, y) on the sensor array can be approximated by

$$I_i(x, y) = G(x, y) R_i(x, y) L_i(x, y) \tag{6.9}$$

if we assume that the sensor sees a surface that reflects light equally in all directions, i.e. it is a Lambertian surface. Each sensor determines the intensity of the light of a particular wavelength λ_i. Here, $G(x, y)$ is a factor that depends on scene geometry at the corresponding object point that is shown at position (x, y) and the type of lighting model used, $R_i(x, y)$ is the amount of light reflected at the corresponding object position (x, y) for wavelength λ_i, and $L_i(x, y)$ is the intensity of the illuminant at the corresponding object position.

For the derivation of this equation, we have also assumed ideal sensors that only respond to light of a single wavelength, i.e. the sensor can be described by a delta function. This is an

ALGORITHMS FOR COLOR CONSTANCY UNDER UNIFORM ILLUMINATION

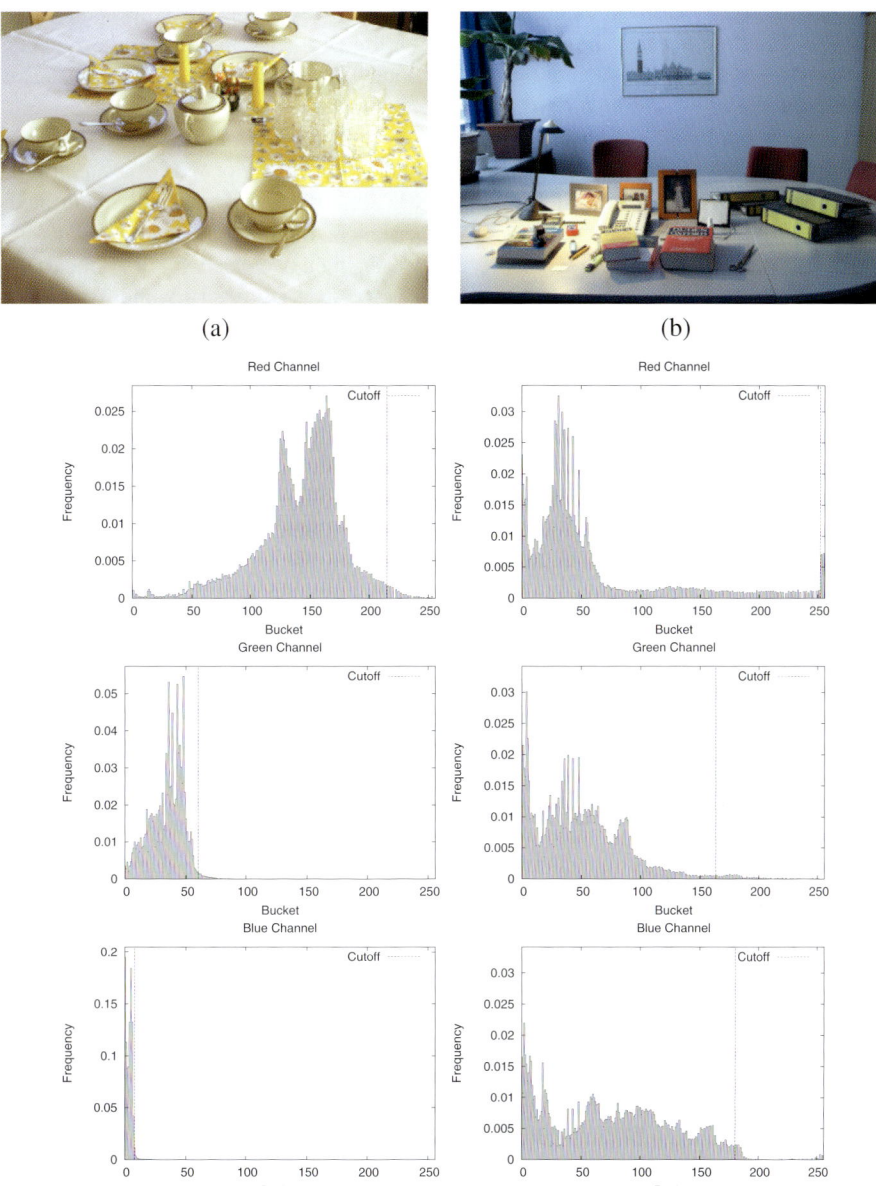

Figure 6.3 Results for the white patch retinex algorithm where histograms were used to find a white patch. The histograms for all the three color bands of the input image are also shown. The cutoff value is marked with a vertical line. All the three cutoff values represent an estimate of the illuminant.

assumption frequently made to achieve color constancy. From the preceding equation, we see that the illuminant scales the product between the geometry factor and the reflectance of the object. Thus, one can achieve color constancy by independently scaling the three color bands if the light illuminating the scene was known.

For display purposes, the measured intensities are often gamma corrected and transformed to the range [0, 1] or [0, 255]. In order to apply the gray world assumption, we need to linearize pixel colors. This point is very important and was not mentioned by Buchsbaum. If we process stored images, for instance, images that are stored as TIFF, JPEG, or PPM files (Murray and van Ryper 1994), then we need to first linearize the intensity values. If the pixel colors are gamma corrected, we need to undo this gamma correction. If the colors are gamma corrected with a factor of 1/2.2, then we linearize the pixel colors by applying a gamma correction with a factor of 2.2. The problem is that quite often the gamma factor of the original gamma correction is not known. In this case, it is assumed that images use the sRGB color space, which assumes a gamma of 2.2. See Section 4.5 for details on how to transform a given image to linear intensity space. Let

$$\mathbf{c}(x, y) = [c_r(x, y), c_g(x, y), c_b(x, y)]^T \quad (6.10)$$

be the linearized color of the input image at position (x, y). Let $[0, 1]$ be the range of the intensities for each color channel.

If we assume that the colors of the objects in view are uniformly distributed over the entire color range and we have a sufficient number of objects with different colors in the scene, then the average color computed for each channel will be close to $\frac{1}{2}$. To show this, we assume a linear mapping between sensor measurements and image pixel colors, i.e. $c_i(x, y) = I_i(x, y)$, and an illuminant that is uniform across the entire image, i.e. $L_i(x, y) = L_i$. In this case, space average color $\mathbf{a}$ of an image of size $n = n_x \times n_y$, where n_x is the width and n_y is the height of the image, is given by

$$a_i = \frac{1}{n} \sum_{x,y} c_i(x, y) \quad (6.11)$$

$$= \frac{1}{n} \sum_{x,y} G(x, y) R_i(x, y) L_i \quad (6.12)$$

$$= L_i \frac{1}{n} \sum_{x,y} G(x, y) R_i(x, y) \quad (6.13)$$

where $G(x, y)$ is a factor that depends on scene geometry, $R_i(x, y)$ is the reflectance of the object point displayed at position (x, y) in the image, L_i is the intensity of the light that illuminates the scene, and the index i denotes the corresponding color channel.

Let $E[GR_i]$ be the expected value of the geometry factor G multiplied by the reflectance R_i. Both can be considered as independent random variables, as there is no correlation between the shape and the color of an object. Let us assume that the reflectances are uniformly distributed, i.e. many different colors are present in the scene and each color is equally likely. Therefore, the reflectance can be considered to be a random variable drawn from the range [0, 1]. We obtain (Johnson and Bhattacharyya 2001)

$$E[GR_i] = E[G]E[R_i] = E[G] \left(\int_0^1 x \, dx \right) = E[G] \frac{1}{2}. \quad (6.14)$$

For large n, we have

$$a_i = L_i \frac{1}{n} \sum_{x,y} G(x,y) R_i(x,y) \qquad (6.15)$$

$$\approx L_i E[G R_i] \qquad (6.16)$$

$$= L_i E[G] \frac{1}{2}. \qquad (6.17)$$

Instead of assuming that the reflectances are uniformly distributed over the range [0, 1], we can also use the actual distribution of reflectances to compute $E[R_i]$ (Barnard et al. 2002a). In this case, we also need to know the shape of the response curves of the camera to compute the expected value. This value would then depend on the set of reflectances chosen for the surrounding and would also depend on the type of camera used.

We now see that we can use space average color to estimate the color of the illuminant as

$$L_i \approx \frac{2}{E[G]} a_i = f a_i \qquad (6.18)$$

where $f = \frac{2}{E[G]}$ is a factor that depends on the scene viewed. Note that this derivation is only valid if the relationship between measured sensor values and pixel colors is linear. Given the color of the illuminant, we can estimate the combined geometry factor and the reflectance of the object. With $c_i(x,y) = G(x,y) R_i(x,y) L_i$, we have,

$$o_i(x,y) = \frac{c_i(x,y)}{L_i} \approx \frac{c_i(x,y)}{f a_i} = G(x,y) R_i(x,y) \qquad (6.19)$$

where $\mathbf{o} = [o_r(x,y), o_g(x,y), o_b(x,y)]^T$ is the color of the output pixel. Thus, the combined geometry and reflectance factor can be estimated by dividing the color of the current pixel by the product of f and space average color. The factor f only scales all color channels equally and affects only the intensity of the colors. It can be set as $f = 2$. This assumes that $E[G] = 1$, i.e. there is a perpendicular orientation between the object and the camera. The factor f can also be estimated directly from the image. In this case, one first rescales each channel by dividing the intensity by the average value of the channel. Next all channels are rescaled equally such that, say, only 1% of all pixels are clipped.

Figure 6.4 shows the results achieved with the gray world assumption. In practice, we have found the gray world assumption to produce better results than the white patch retinex algorithm. The gray world assumption is based on the average of a large number of pixels. In this respect, it is much more robust than the white patch retinex, which is only based on the maximum pixel value. If one only has a single pixel that is very bright, then the white patch retinex algorithm will probably produce incorrect results. This holds especially if shiny objects are in the scene, which reflect all of the incident light, which may result in clipped pixels.

The gray world assumption works nicely if we only have a single illuminant. However, if we have multiple illuminants, then the gray world assumption does not work. We can see this in image (b) of Figure 6.4. This is not surprising, as one of the assumptions was that we

110 ALGORITHMS FOR COLOR CONSTANCY UNDER UNIFORM ILLUMINATION

Figure 6.4 Gray world assumption. The gray world assumption produces nice results if we only have a single illuminant (a). If we have multiple illuminants, then the gray world assumption does not work very well (b).

Figure 6.5 The gray world assumption will fail to produce correct colors if sufficiently large numbers of colors are not present in the scene. A leaf from a banana plant is shown in (a). The image in (b) shows the output image.

have a uniform illuminant. The gray world assumption requires that there be a sufficiently large number of different colors in the image. If this is not the case, then the gray world assumption will not work. Figure 6.5(a) shows a close-up of a leaf from a banana plant. The average of the channels for this image is [0.227827, 0.339494, 0.049392]. Thus, the gray world assumption will increase the influence of the blue channel to a great extent. The output image is shown in (b). Clearly, this is not what is desired.

The gray world assumption as well as the white patch retinex algorithm are used frequently for automatic white balance. The popular draw utility written by Coffin (2004) scales each color channel using the average as an automatic white-balance option. After rescaling, the white point is set at the 99th percentile. In other words, all channels are scaled equally such that only the top 1% of all pixels are clipped. This assumes that there are only a few highlights.

Another assumption is that the reflectances of the image are uniformly distributed over the range [0,1]. This assumption may not be correct in practice. Suppose that we have one uniformly colored object in the image and that the object covers most of the image pixels. In this case, space average color will be close to the color of the object irrespective of the background. Therefore, it may make sense to segment the image before performing the averaging operation. Gershon et al. (1987) suggest that the input image may be segmented into different regions. Let n_r be the number of different regions and let $\mathbf{a}(R_j) = [a_r(R_j), a_g(R_j), a_b(R_j)]^T$ be the average color of region $j \in \{1, ..., n_r\}$. We now calculate the average color by looping over the unique regions:

$$a_i = \frac{1}{n_r} \sum_{j=1}^{n_r} a_i(R_j) \qquad (6.20)$$

This space average color is then used to calculate the reflectances.

$$L_i \approx f a_i \qquad (6.21)$$

Obviously, if the input image is segmented before calculating the average, then each region contributes only once to the average. The result is independent of the number of pixels the different areas cover. Figure 6.6 shows a segmented image. For the image shown in (a), each region was colored using the average color of the pixels that belong to the particular region. To show the different regions of the segmented image better, the regions are shown with random colors in (b). Figure 6.7 shows the result when segmentation is used to compute the average.

Suppose that we have one large object that is covered by a second object in front of it. If both objects have a different color, a segmentation algorithm may segment the image into three regions even though there are only two unique colors. In order to solve this problem, instead of segmenting the image we could compute a color histogram. Such a color histogram is shown in Figure 6.8. Figure 6.8 (a) shows the input image. The plot in (b) shows the color histogram. To visualize the histogram better, each color channel was quantized into 10 different intensities. Since we have three color channels, the histogram has 1000 buckets. For each bucket a cube is drawn. The size of the cube is proportional to

(a)   (b)

Figure 6.6 Segmented input image. For the image shown in (a), each region was assigned the average color of the pixels that belong to that region. For the image in (b), random colors were used for each region.

112 ALGORITHMS FOR COLOR CONSTANCY UNDER UNIFORM ILLUMINATION

Figure 6.7 Output images produced by the algorithm of Gershon et al. (1987).

(a)

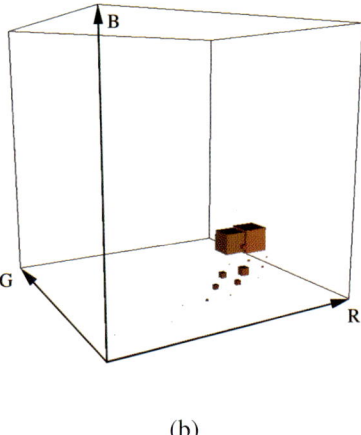
(b)

Figure 6.8 The input image is shown in (a). The graph in (b) shows the color histogram. A quantization of 10 was used for each color channel. Therefore, there are 1000 buckets in the histogram. Each bucket is represented by a cube where the size of the cube is proportional to the number of pixels of the same color in the original image.

the number of pixels having the corresponding color. The peak of the histogram is located at color [0.85, 0.45, 0.15], which is caused by the yellowish table cloth.

Let n_b be the total number of buckets. Let n_{nz} be the number of nonzero buckets and let $\mathbf{c}(j)$ be the color represented by bucket j of the histogram. The average color can then be calculated by summing up the colors of the buckets that are nonzero, i.e. averaging over all unique colors in the image.

$$a_i = \frac{1}{n_{nz}} \sum_{j=1}^{n_b} c_i(j) \qquad (6.22)$$

This space average color can then be used to estimate the illuminant, which in turn can be used to calculate the reflectances. Figure 6.9 shows the result of this algorithm.

Irrespective of the exact method used to estimate the illuminant, i.e. whether we estimate the illuminant from all pixels, segment the image, or compute the histogram, the gray world

Figure 6.9 Output images produced by estimating the illuminant from the histogram of the input image.

assumption relies on the assumption that there is only a single illuminant present. However, in practice this assumption does not hold.

6.3 Variant of Horn's Algorithm

Horn (1974, 1986) has developed an algorithm for color constancy under varying illumination. We discuss this algorithm in detail in Section 7.2. Horn's algorithm assumes that the illuminant is nonuniform across the image. He suggests that the logarithm of the input signal be first taken. Next, the Laplacian is applied. Then, a threshold operation is used to separate a change in reflectance from a change of the illuminant. Changes due to a change of the illuminant are removed. Finally, the output of the thresholded data is reintegrated to obtain the log reflectances. Let us now consider what happens if the illumination is constant over the entire image. If the illuminant is constant, then we can omit taking the Laplacian, thresholding, and reintegrating.

It is assumed that the image color $c_i(x, y)$ is basically the product of the reflectance $R_i(x, y)$ and the illuminant $L_i(x, y)$ at the corresponding object location for color channel i.

$$c_i(x, y) = R_i(x, y) L_i(x, y) \tag{6.23}$$

For a constant illumination $L_i(x, y) = L_i$, we have

$$c_i(x, y) = R_i(x, y) L_i. \tag{6.24}$$

If we now apply the logarithm, the product of reflectance and illuminant turns into a sum of logarithms.

$$\log(c_i(x, y)) = \log(R_i(x, y)) + \log L_i \tag{6.25}$$

The unknown constant $\log L_i$ can be removed by transforming the logarithm of the pixel colors independently for each color channel to the range [0, 1]. This constant does not depend on the coordinates (x, y). By transforming each channel to the range [0, 1], this constant will be subtracted from each channel and the result will be an image independent of the illuminant. After transforming the log-pixel values to the range [0, 1], each channel is a color constant descriptor, given by the log reflectances. In order to undo the logarithm, we can now exponentiate and then transform the result a second time to the range

114 ALGORITHMS FOR COLOR CONSTANCY UNDER UNIFORM ILLUMINATION

Figure 6.10 Results obtained using a simplified version of Horn's algorithm. It is assumed that the illuminant is constant across the entire image. The algorithm cannot be used if multiple illuminants are present.

[0, 1] independently for each color channel. In essence, this maps the brightest point of the image to white and the darkest to black. The result of this simple algorithm is shown in Figure 6.10. This is the method we have used for the experiments described in Chapter 13, as it produced the best results.

Note that this algorithm is also able to cope with any gamma correction that may have been applied to the input pixels. If the pixels are gamma corrected, i.e. $c = I^\gamma$ for some gamma value γ, we have

$$\log(c_i(x, y)) = \gamma \log(R_i(x, y)) + \gamma \log L_i. \tag{6.26}$$

Again, the constant due to the illuminant can be removed by transforming the result to the range [0, 1]. This will also remove the constant due to the gamma correction.

An alternative way to remove the constant $\log L_i$ is to normalize with respect to a white reference patch that is assumed to be in the image. This normalization was suggested by Horn (1974, 1986). A white patch reflects all incident light. It has a reflectance $R_i = 1$. Thus, all we need to do is determine the maximum value for each channel and then subtract this value from the pixel data for each channel. This will normalize the data from the brightest patch to have a reflectance of $R_i = 1$. We compute

$$o_i = \exp(\log(c_i(x, y)) - \text{Max}_i) = R_i(x, y) \tag{6.27}$$

where $\text{Max}_i = \max_{x,y}\{\log(c_i(x, y))\}$. This variant is identical to the white patch retinex algorithm. Figure 6.11 shows the results if this type of normalization is used.

Instead of normalizing to white, we can also use the gray world assumption. If we compute the average of the n pixel values after the logarithm has been applied, we obtain

$$\frac{1}{n} \sum_{x,y} \log(c_i) = \frac{1}{n} \sum_{x,y} (\log L_i + \log R_i) \tag{6.28}$$

$$= \log L_i + \frac{1}{n} \sum_{x,y} \log R_i \tag{6.29}$$

$$= \log L_i + \log \left(\prod_{x,y} R_i^{\frac{1}{n}} \right). \tag{6.30}$$

Figure 6.11 Results obtained using a simplified version of Horn's algorithm. A constant has been added to each color band such that the maximum of each color channel is 1 after exponentiation.

The second term is the logarithm of the geometric mean of the reflectances. If we assume that the reflectances are uniformly distributed over the range $[0, 1]$, then we can rewrite the second term as

$$\log\left(\prod_{x,y} R_i^{\frac{1}{n}}\right) \approx \log\left(\prod_{i=1}^{n} \frac{i}{n}^{\frac{1}{n}}\right) = \log\left(\frac{n!^{\frac{1}{n}}}{n}\right). \tag{6.31}$$

Using Stirling's formula (Weisstein 1999a)

$$\lim_{n \to \infty} \frac{(n!)^{\frac{1}{n}}}{n} = \frac{1}{e}, \tag{6.32}$$

we obtain

$$\log\left(\prod_{x,y} R_i^{\frac{1}{n}}\right) \approx -1. \tag{6.33}$$

This second term would produce an offset of 1 if we subtract the average log-pixel values. Therefore, we need to subtract 1 if we want to obtain reflectances. We can compute reflectances using

$$o_i = \exp\left(\log c_i - \frac{1}{n}\sum_{x,y} \log(c_i) - 1\right) \tag{6.34}$$

$$= \exp\left(\log L_i + \log R_i - \log L_i + 1 - 1\right) \tag{6.35}$$

$$= R_i. \tag{6.36}$$

Figure 6.12 shows the results if this type of normalization is used.

6.4 Gamut-Constraint Methods

Forsyth (1988, 1992) developed the so-called gamut-constraint methods for color constancy. He assumes that the receptor response curves are disjoint and that they are only sensitive

116 ALGORITHMS FOR COLOR CONSTANCY UNDER UNIFORM ILLUMINATION

Figure 6.12 Results obtained using a simplified version of Horn's algorithm. The gray world assumption was used for normalization.

inside a narrow-band of wavelengths. In this case, the response will be constant over the support of the receptor and one only needs to determine three scaling factors for the red, green, and blue channels. It is also assumed that there are no shadows in the image, that there is a single uniform illuminant, and that only diffuse reflections occur. Gamut-constraint methods first compute the convex hull of the gamut observed under a canonical illuminant. If a scene is observed under a different illuminant, it will have a transformed gamut. It can be transformed to the canonical gamut by determining the three scaling factors for the red, green, and blue channels.

The gamut observed under white light will usually be distributed along the gray vector. Figure 6.13 (a) shows an IT8 target that is used for color calibration of scanners. Two views of the convex hull of the gamut of this image are shown in (b) and (c). The convex hull was computed using the QHull algorithm (Barber et al. 1996). Looking at the convex hull, we can see that there are some very dark colors (black) and some very bright colors (white) in the image. Even though there are red, green, and blue patches in the image, the RGB color space is not completely filled. Only a subset of the entire gamut of the monitor is covered. Figure 6.14 shows a scene illuminated with yellow light. The color gamut of this image is highly skewed.

We now have to determine a linear map that will transform the observed gamut of colors to the canonical gamut of colors. This map will simply be a diagonal 3×3 matrix of the form

$$\mathbf{S}(s_x, s_y, s_z) = \begin{pmatrix} s_x & 0 & 0 \\ 0 & s_y & 0 \\ 0 & 0 & s_z \end{pmatrix}, \tag{6.37}$$

because we have assumed receptor response functions that are delta functions. Gamut-constraint methods initially consider a large set of possible maps. The initial set of maps is constructed using the first vertex of the convex hull of the observed gamut. For a single point, the maps allowed are simply the maps that take the vertex of the observed gamut to any point inside the gamut of the canonical illuminant. Each additional point gives us new constraints on the maps that will perform the desired transformation. We have to find a map that is consistent with the colors observed in the image. This map is found by intersecting all possible sets of maps for all image points. The set of maps for a given point is also a convex polyhedron. For each additional point, one computes the set of maps that take the point to

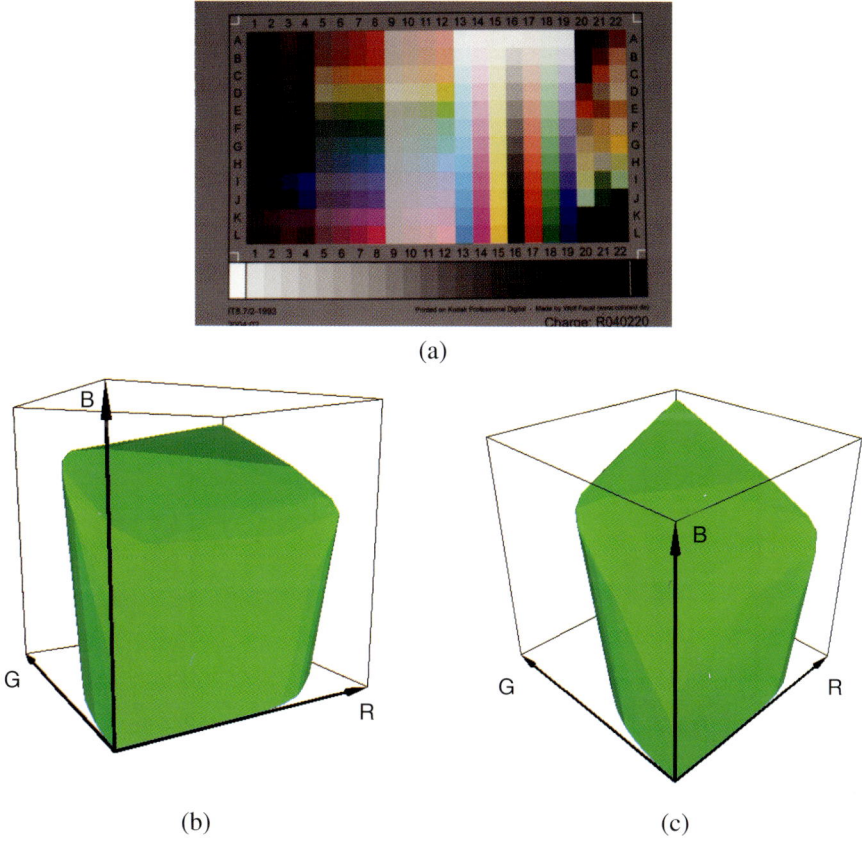

Figure 6.13 An IT8 target (made by Wolf Faust www.coloraid.de) viewed under sunlight (a). The image was taken with a Canon 10D. Two different views of the three-dimensional convex hull of the colors are shown in (b) and (c).

a point inside the canonical gamut. The convex hull of this set is then intersected with the set of maps obtained so far. Since the set of maps is a convex polyhedron, the algorithm's main operation is the intersection of convex hulls that describe the feasible maps.

Let us summarize again how the gamut-constraint method works. First, we need to determine the canonical gamut. To compute this gamut, one takes a large selection of objects with different reflectances illuminated with white light. In practice, the set of objects is chosen such that all possible reflectances occur. Then one computes the convex hull of the colors shown in the image. This convex hull will be used to describe the canonical gamut. Let $\mathcal{H}_c$ be the set of vertices of the convex hull describing the canonical gamut. Given an input image with an unknown illuminant, one computes the convex hull of the gamut of colors shown in the image. Let $\mathcal{H}_o$ be the set of vertices of the convex hull describing the observed gamut. If the input image contains many pixels, it makes sense to first compute a histogram. In this case, each color has to be processed only once, and not several times, during the computation of the convex hull.

118 ALGORITHMS FOR COLOR CONSTANCY UNDER UNIFORM ILLUMINATION

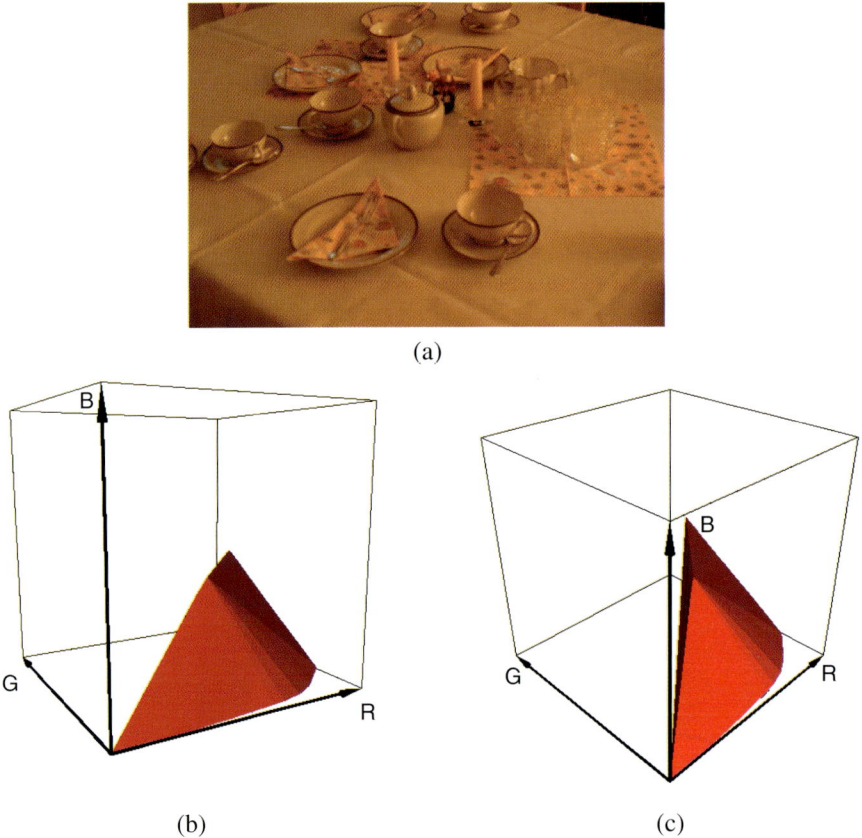

Figure 6.14 A scene illuminated with yellow light. The color gamut of this image is highly skewed.

Let $\mathbf{v} \in \mathcal{H}_o$ be a vertex of the convex hull computed from the input image. The set of feasible maps $\mathcal{M}(\mathbf{v})$ for the given vertex $\mathbf{v}$ is

$$\mathcal{M}(\mathbf{v}) = \{\mathbf{v}_c/\mathbf{v} | \mathbf{v}_c \in \mathcal{H}_c\}. \tag{6.38}$$

Here, the division is defined component-wise, i.e. if $\mathbf{c} = \mathbf{a}/\mathbf{b}$, then $c_i = a_i/b_i$ with $i \in \{r, g, b\}$. This set of feasible maps is also convex because the canonical gamut is convex and this gamut is simply scaled by the coefficients of the given vertex. Note that care must be taken to avoid a possible division by zero if some of the vertices of the convex hull of the observed gamut lie on the coordinate axis. In this case, some of the coefficients will be zero. One way to solve this problem would be to set such coefficients to a small positive value. Another way to avoid this problem altogether is to compute a histogram of the image before the convex hull of the gamut of colors is computed. A histogram segments the color space into a number of bins. The color of a bin is assumed to be located at the center of the bin. For instance, if each channel is segmented into 10 bins, then $[0.05, 0.05, 0.05]^T$ will denote black and $[0.95, 0.95, 0.95]^T$ will denote white.

For each vertex of the convex hull of the observed colors, we compute the feasible maps. We then intersect all these maps, as the actual illuminant must lie somewhere inside the intersection of these sets. Therefore, each vertex of the convex hull of the observed gamut gives us additional constraints to reduce the set of possible illuminants that may have produced the observed image. Let $\mathcal{M}_\cap$ be the computed intersection.

$$\mathcal{M}_\cap = \bigcap_{\mathbf{v} \in \mathcal{H}_o} \mathcal{M}(\mathbf{v}) \tag{6.39}$$

The intersection will also be a convex hull, as the intersection of two convex hulls is again a convex hull. Therefore, the basic operations of gamut-constraint algorithms involve computing the convex hull of a set of points and then intersecting these hulls.

If the canonical illuminant is assumed to be white light, then each vertex of $\mathcal{M}_\cap$ describes a possible illuminant. Each vertex can be viewed to be a diagonal map $\mathbf{m}$, which has the form

$$\mathbf{m} = [m_r, m_g, m_b]^T = [1, 1, 1]^T / [L_r, L_g, L_b]^T \tag{6.40}$$

where division is carried out component-wise, and $[L_r, L_g, L_b]^T$ describes the color of the illuminant. The illuminant $\mathbf{L}$ is therefore given by

$$\mathbf{L} = \left[\frac{1}{m_r}, \frac{1}{m_g}, \frac{1}{m_b} \right]^T. \tag{6.41}$$

In order to determine the coefficients of the illuminant, we need to select one point from the set of possible maps described by $\mathcal{M}_\cap$. Hopefully, this set will not be empty and we can actually choose one of the points of $\mathcal{M}_\cap$ as the illuminant. If the set is empty, then there exists no linear map that can transform the observed gamut to a subset of the canonical gamut. Forsyth (1988, 1992) suggests that the map that transforms the given gamut to a maximally large gamut be chosen. In other words, the transformed gamut will fit inside the canonical gamut and it will also be the largest gamut possible. We are only considering linear diagonal maps. These maps transform a given volume to a new volume that is multiplied by the trace of the linear diagonal map. Therefore, in order to choose the best possible map, one simply selects the map $\mathbf{m}$ that has the largest trace.

$$\mathbf{m} = \mathrm{argmax}_\mathbf{v} \{v_r v_g v_b | \mathbf{v} = [v_r, v_g, v_b]^T \in \mathcal{M}_\cap\} \tag{6.42}$$

Let $\mathbf{c} = [c_r, c_g, c_b]^T$ be the color of an input pixel. Then, the output color is computed as

$$\mathbf{o} = [m_r c_r, m_g c_g, m_b c_b]^T. \tag{6.43}$$

Let us now consider a simple example. The gamut of colors that can be produced by our display device is the RGB cube. Therefore, we will take the unit cube as our canonical gamut. Now, let us assume that the observed gamut is a subset of the canonical gamut. If we assume an illuminant of $[0.7, 0.5, 0.2]^T$, then the observed gamut will be the unit cube scaled by $s_r = 0.7$, $s_g = 0.5$, and $s_b = 0.2$ along the red, green, and blue axes respectively. Let $\mathbf{S}(s_r, s_g, s_b)$ be a scaling matrix and $\mathcal{H}_{\mathrm{cube}}$ be the set of vertices of the unit cube. Then we have

$$\mathcal{H}_o = \mathbf{S}(s_r, s_g, s_b) \mathcal{H}_{\mathrm{cube}}. \tag{6.44}$$

120 ALGORITHMS FOR COLOR CONSTANCY UNDER UNIFORM ILLUMINATION

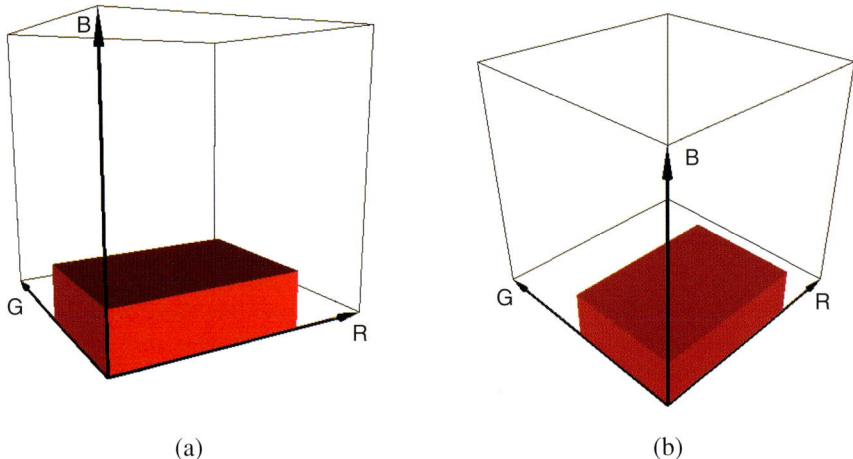

(a) (b)

Figure 6.15 Two different views of the observed gamut of colors. If we assume that the canonical gamut is the unit cube and that the illuminant is given by $[0.7, 0.5, 0.2]^T$, then the observed gamut will be the unit cube scaled by $\mathbf{S}(0.7, 0.5, 0.2)$.

This observed gamut is shown in Figure 6.15. If we compute the intersection of the feasible linear maps, we obtain

$$\mathcal{M}_\cap = \mathbf{S}\left(\frac{1}{s_r}, \frac{1}{s_g}, \frac{1}{s_b}\right) \mathcal{H}_{\text{cube}}. \tag{6.45}$$

The linear map with the largest trace will be the map created by the vertex $[1, 1, 1]^T$ of the unit cube. Therefore, the map selected from the intersection will be

$$\mathbf{S}\left(\frac{1}{s_r}, \frac{1}{s_g}, \frac{1}{s_b}\right) = \mathbf{S}\left(\frac{10}{7}, 2, 5\right) \tag{6.46}$$

which is just the map that discounts the illuminant. If we assume a color gamut that is equivalent to the unit cube and an observed color gamut that is a scaled version of the unit cube, then the gamut constraint method will perform exactly the same operation as the white patch retinex algorithm.

A drawback of the gamut-constraint method is that it may fail to find an estimate of the illuminant. This may happen if the resulting intersected convex hull $\mathcal{M}_\cap$ is the empty set. Therefore, care must be taken not to produce an empty intersection. There are several ways to address this problem. One possibility would be to iteratively compute the intersection by considering all of the vertices of the observed gamut in turn. If, as a result of the intersection, the intersected hull should become empty, the vertex is discarded and we continue with the last nonempty hull. Another possibility would be to increase the size of the two convex hulls that are about to be intersected. If the intersection should become empty, the size of both hulls is increased such that the intersection is nonempty. A simple implementation would be to scale each of the two convex hulls by a certain amount. If the intersection is still empty, we again increase the size of both hulls by a small amount.

ALGORITHMS FOR COLOR CONSTANCY UNDER UNIFORM ILLUMINATION

Figure 6.16 Results obtained using Forsyth's gamut-constraint algorithm. The algorithm assumes a single illuminant for the entire image, so we cannot use this algorithm if multiple illuminants are present. For both images, the RGB cube was assumed to be the canonical gamut.

This process is repeated until the intersection is nonempty. It would also be possible to compute the two closest points between the two convex hulls and choose the point lying half way between the two convex hulls or we could increase the size of all convex hulls by a certain percentage before computing the intersected hull in order to avoid an empty intersection. This could also be done iteratively. First we see if the intersection is indeed empty. If the intersection is empty, we increase all convex hulls by a certain percentage. If it is still empty, we increase it even further. This process continues until the intersection is nonempty. For our experiments we have used the latter method, as this method produced the best results.

Results obtained with Forsyth's gamut-constraint algorithm are shown in Figure 6.16. For both images, the canonical gamut was assumed to be the unit RGB cube. The algorithm does not perform well on either image. For the first image, this is most likely caused by the specular highlights of the glassware located on the table. Since the algorithm assumes a single illuminant, it cannot be used for images with multiple illuminants. Compared to the other algorithms, gamut-constraint algorithms are computationally rather expensive because the convex hull must be computed from the colors of the input image. Also, the convex hulls have to be intersected, which is also an expensive operation.

6.5 Color in Perspective

Forsyth's gamut-constraint algorithm assumes that only diffuse reflections occur. Finlayson (1996) noted that specular highlights may be a problem. Such highlights are usually very bright because the illuminant is reflected directly into the camera at that point. The impact of specular highlights may be reduced by first normalizing the RGB color vectors of the image in some way. What is important is the orientation of the color vector but not its length. He suggested that all RGB values be projected onto the plane located at $b = 1$. This is achieved by dividing each color channel by the value measured in the blue color channel. Let $\mathbf{c} = [c_r, c_g, c_b]^T$ be the value measured at a pixel of the input image. Then,

122 ALGORITHMS FOR COLOR CONSTANCY UNDER UNIFORM ILLUMINATION

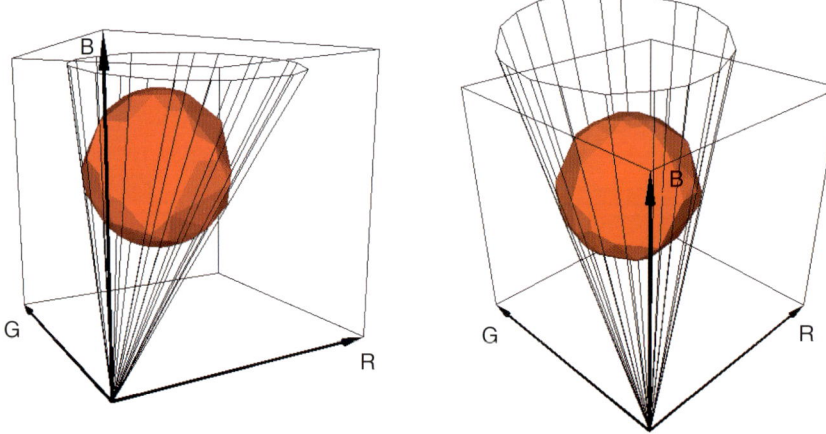

Figure 6.17 A three-dimensional color gamut projected onto the plane $b = 1$. The resulting two-dimensional gamut is simply the silhouette of the three-dimensional gamut when viewed from the origin.

the projected point on the plane at $b = 1$ is given by

$$\left[\frac{c_r}{c_b}, \frac{c_g}{c_b}, 1\right]^T. \tag{6.47}$$

Note that we have to assume $c_b \neq 0$ at this stage. The problem of avoiding $c_b = 0$ will be discussed in the subsequent text. Since this operation is equivalent to a perspective projection of all of the color points onto the viewing plane located at $b = 1$, Finlayson calls his algorithm "*Color in Perspective*." All points are now located at the plane $b = 1$ and we can drop the third coordinate. Obviously the projection could also have been performed onto the plane defined by $r = 1$ or $g = 1$ with similar results.

The projection of the convex hull of the color gamut onto a plane is the convex polygon describing the silhouette of the three-dimensional convex hull viewed from the point through which the projection was made. Such a two-dimensional projection of a color gamut onto the plane $b = 1$ is shown in Figure 6.17. The three-dimensional convex hull has become a two-dimensional convex hull on the plane $b = 1$. Instead of applying a three-dimensional map, now a two-dimensional map suffices. Figure 6.18 shows an IT8 target and its gamut of colors projected onto the plane $b = 1$. The gamut of colors of an image that was taken under a yellow illuminant is shown in Figure 6.19. The two-dimensional gamut-constraint algorithm computes a linear map that will transform the observed gamut to the gamut of colors under a canonical illuminant.

To see that indeed a two-dimensional map suffices, consider a three-dimensional map $\mathbf{S}(s_r, s_g, s_b)$, which transforms color $\mathbf{c} = [c_r, c_g, c_b]^T$ to color $\mathbf{c}' = [c'_r, c'_g, c'_b]^T$. We have

$$\mathbf{c}' = [c'_r, c'_g, c'_b]^T = \mathbf{S}(s_r, s_g, s_b)\mathbf{c} = [s_r c_r, s_g c_g, s_b c_b]^T. \tag{6.48}$$

ALGORITHMS FOR COLOR CONSTANCY UNDER UNIFORM ILLUMINATION

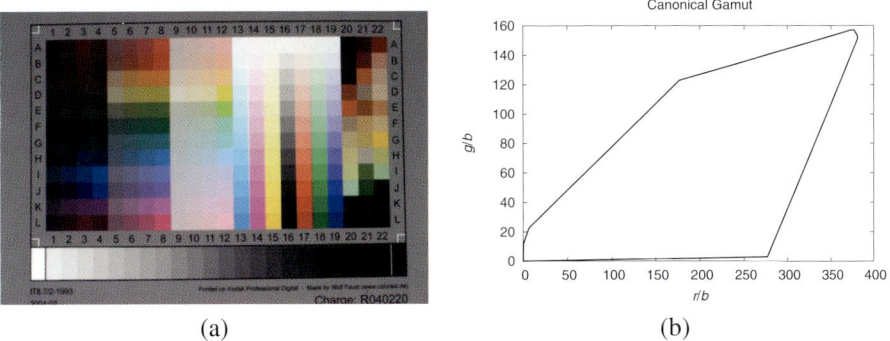

Figure 6.18 An IT8 target (made by Wolf Faust www.coloraid.de). Its gamut of colors projected onto the plane $b = 1$ is shown in (b).

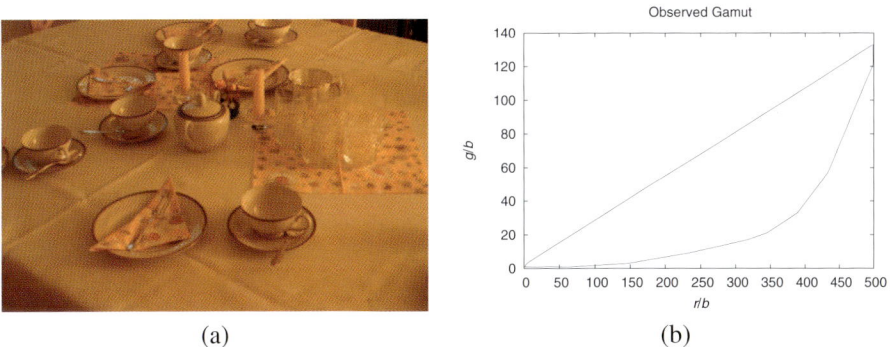

Figure 6.19 A scene illuminated with yellow light. The color gamut projected onto the plane $b = 1$ is shown in (b).

The two projections of color $\mathbf{c}$ and $\mathbf{c'}$ are $[\frac{c_r}{c_b}, \frac{c_g}{c_b}]^T$ and $[\frac{c'_r}{c'_b}, \frac{c'_g}{c'_b}]^T$, respectively. We have,

$$\left[\frac{c'_r}{c'_b}, \frac{c'_g}{c'_b}\right]^T = \left[\frac{s_r c_r}{s_b c_b}, \frac{s_g c_g}{s_b c_b}\right]^T = \mathbf{S}\left(\frac{s_r}{s_b}, \frac{s_g}{s_b}\right)\left[\frac{c_r}{c_b}, \frac{c_g}{c_b}\right]^T = \begin{pmatrix} \frac{s_r}{s_b} & 0 \\ 0 & \frac{s_g}{s_b} \end{pmatrix}\left[\frac{c_r}{c_b}, \frac{c_g}{c_b}\right]^T. \quad (6.49)$$

Thus, given a three-dimensional map $\mathbf{S}(s_r, s_g, s_b)$, the corresponding two-dimensional map will be $\mathbf{S}(\frac{s_r}{s_b}, \frac{s_g}{s_b})$. Therefore, the gamut-constraint algorithm can be applied in two dimensions instead of in three dimensions. Algorithms for computing a two-dimensional convex hull can be found in Cormen et al. (1990) and Sedgewick (1992). This has the advantage that the computation of the convex hull is much simpler in two dimensions than in three dimensions.

Note that when the three-dimensional coordinates are projected onto the plane $b = 1$, we have a problem when the blue channel is zero. If we compute the histogram, this problem can be nicely avoided. Suppose that 10 quantizations are used for each channel, giving 1000 bins in total. In this case, colors with $[r, g, 0]^T$ with $r, g \in [0, 1]$ will be mapped to the bin that corresponds to a blue intensity of 0.05. After the two-dimensional transformation is computed, colors are moved to a new location on the plane $b = 1$. Let $\mathbf{c} = [c_r, c_g, c_b]^T$ be the color of the input pixel. The location of this color on the plane $b = 1$ will be $[\frac{c_r}{c_b}, \frac{c_g}{c_b}, 1]^T$. Let $\mathbf{m} = [m_1, m_2]^T$ be the transform computed by the two-dimensional gamut-constraint algorithm. Then, the output color will be given by

$$\mathbf{o} = [m_1 \frac{c_r}{c_b}, m_2 \frac{c_g}{c_b}, 1]^T. \tag{6.50}$$

In order to display this color, we can use the original lightness of the input color to scale the output color for having the same lightness. Let L be the lightness of the input color, i.e. $L = w_r c_r + w_g c_g + w_b c_b$ with $w_r = 0.2125$, $w_g = 0.7154$, $w_b = 0.0721$, which are the coefficients for linear RGB lightness (Poynton 1997), and let $L' = w_r m_1 \frac{c_r}{c_b} + w_g m_2 \frac{c_g}{c_b} + w_b$. Then, the output color is given by

$$\mathbf{o} = \frac{L}{L'} \left[m_1 \frac{c_r}{c_b}, m_2 \frac{c_g}{c_b}, 1 \right]^T. \tag{6.51}$$

The projection could also be done onto the plane $r + g + b = 1$. However, in this case, the required transformation can no longer be described by a diagonal matrix. Results for the two-dimensional gamut-constraint algorithm are shown in Figure 6.20. The original lightness was used to rescale the computed chromaticities. Again, since a single illuminant is assumed for the entire scene, the algorithm is unable to handle input images with multiple illuminants. For each image, the possible illuminants computed by the algorithm are also shown. The illuminants are shown as CIE chromaticities.

Apart from reducing the dimensionality of the problem, Finlayson (1996) also introduced the idea of additionally constraining the set of illuminants by considering a set of possible illuminants. The three-dimensional gamut-constraint algorithm first computes the set of maps that can take the observed gamut of colors to the canonical gamut. Then, the map that produces the largest such gamut is chosen. Finlayson suggests that a set of possible illuminants be computed. Instead of simply choosing the map that produces the largest gamut, one chooses the map that will produce the largest gamut but this time also making sure that the map is also covered by the set of illuminants allowed.

The set of illuminants allowed is constructed as follows. First we choose a standard surface. Let $\mathbf{v}_s$ be the color of the standard surface viewed under the canonical illuminant. The same surface is then illuminated with a large number of illuminants. Let $\mathcal{H}_S$ be the convex hull of the observed colors of the surface. The convex hull contains all possible colors that could be observed when the illuminants are added in different amounts. If we choose a white patch as the standard surface, then the vertices of the convex hull will be just the chromaticities of the illuminants. The standard surface does not necessarily have to be white. Therefore, one computes the set of maps that take the observed color of the standard surface when viewed under the canonical illuminant to the color of the same patch when viewed using a different illuminant. This set of maps $\mathcal{M}$ is given by

$$\mathcal{M}(\mathbf{v}_s) = \{\mathbf{v}/\mathbf{v}_s | \mathbf{v} \in \mathcal{H}_S\}. \tag{6.52}$$

Since the set of illuminants $\mathcal{H}_S$ is a convex hull, this set will also be a convex hull. The set of maps is simply scaled by the inverse of the observed color of the standard patch viewed

ALGORITHMS FOR COLOR CONSTANCY UNDER UNIFORM ILLUMINATION

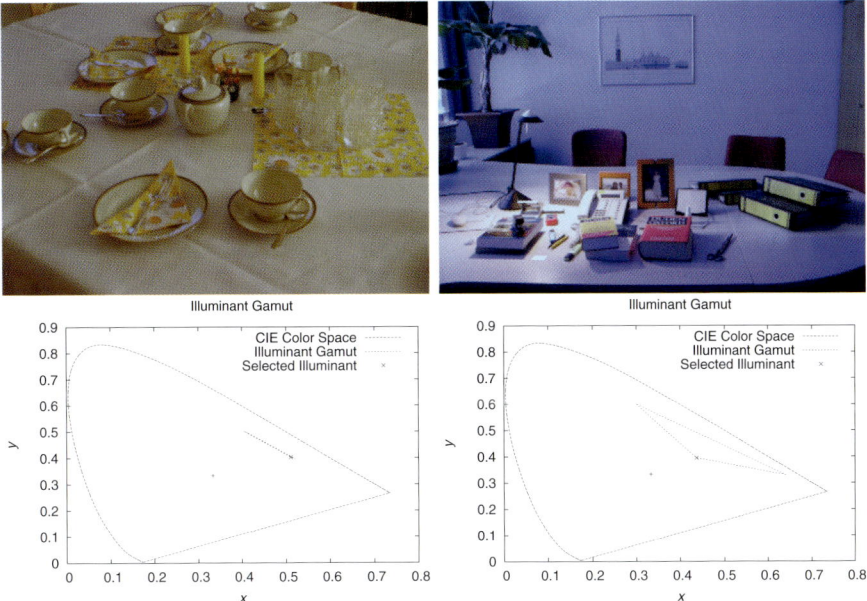

Figure 6.20 Results obtained using a two-dimensional gamut-constraint algorithm. The algorithm assumes a single illuminant for the entire image, so we cannot use this algorithm if multiple illuminants are present. The two graphs show the possible illuminants for each image. The Illuminants are displayed as CIE chromaticities.

under the canonical illuminant. The set of maps $\mathcal{M}$ essentially contains the chromaticities of all possible illuminants.

One can now use this set to additionally constrain the illuminant. The map we are looking for, which transforms the observed gamut to the canonical gamut, has to be a member of

$$\mathcal{M}_\cap = \bigcap_{\mathbf{v} \in \mathcal{H}_o} \mathcal{M}(\mathbf{v}), \tag{6.53}$$

where $\mathcal{H}_o$ is the convex hull of the observed chromaticities and $\mathcal{M}(\mathbf{v})$ is as defined earlier:

$$\mathcal{M}(\mathbf{v}) = \{\mathbf{v}_c/\mathbf{v}|\mathbf{v}_c \in \mathcal{H}_c\} \tag{6.54}$$

where $\mathcal{H}_c$ is the convex hull of the observed chromaticities under the canonical illuminant. Note that the set $\mathcal{M}_\cap$ may, in practice, be empty. An empty set indicates that no single map is consistent with the image data. Owing to noise in the data, it may be quite likely that the set is empty. We have to decide on a method to follow, should this set become empty. We can apply the same methods as described in the preceding text for the three-dimensional gamut-constraint method to avoid an empty set.

The map we are looking for, which transforms the observed gamut to the canonical gamut, also has to be a member of the set of maps that take the standard patch viewed under the given illuminant to the patch viewed under the canonical illuminant. In other words, the map we are looking for has to be a member of the set

$$\mathcal{M}^{-1}(\mathbf{v}_s) = \{1/\mathbf{v}|\mathbf{v} \in \mathcal{M}(\mathbf{v}_s)\}. \tag{6.55}$$

Again, the division is defined component-wise. Note that this set may no longer be a convex set. The set of maps allowed $\mathbf{m}_{\text{allowed}}$ is then given by

$$\mathbf{m}_{\text{allowed}} = \mathcal{M}_\cap \cap \mathcal{M}^{-1}(\mathbf{v}_s). \tag{6.56}$$

Note that this set may be empty. Since several maps may be allowed, we have to choose some method that selects the map that describes the given illuminant best. The two-dimensional gamut-constraint algorithm chooses the map $\mathbf{m}$

$$\mathbf{m} = \text{argmax}_\mathbf{v}\{v_r v_g | \mathbf{v} = [v_r, v_g]^T \in \mathbf{m}_{\text{allowed}}\} \tag{6.57}$$

as the map that takes every input pixel to a new point on the plane $b = 1$ to obtain a color-corrected output image. Barnard et al. (1997) and Funt et al. (1998) suggest that the mean map from the set of maps allowed be computed. Barnard et al. (1997) also showed how to extend the two-dimensional gamut-constraint algorithm to scenes with varying illumination by using segmentation. After we have selected one of the maps, it is applied to every image pixel. The result will be a color-corrected image. However, since we are working in chromaticity space, now we only know the correct colors of the image pixels. We can use the lightness of the corresponding pixels of the input image to create an output image that also contains the shading information of the input image using Equation 6.51.

In practice, the set of illuminants does not have to be created by using a standard patch and illuminating the patch using a variety of known light sources. Natural light sources can be approximated by a black-body radiator. We have already described the concept of a black-body radiator in Section 3.5. The radiance given off by a black-body radiator depends on its temperature. At low temperatures, the black-body radiator appears to be red, at higher temperatures yellow, and at even higher temperatures white or blue. The chromaticities of the color of a black-body radiator describe a curve from red, through yellow, white and onto blue in CIE XYZ color space. Figure 4.7 shows the curve of the black-body radiator. This curve can also be drawn as a function of x as shown in Figure 6.21. It can be approximated by the following quadratic equation

$$\hat{y} = a\hat{x}^2 + b\hat{x} + c \tag{6.58}$$

with $a = -2.7578$, $b = 2.7318$, and $c = -0.2619$ for the range $0.2 \leq \hat{x} \leq 0.7$. Figure 6.21 shows how closely this curve approximates the chromaticity curve of the black-body radiator. Since we have a quadratic equation describing the set of possible illuminants, we can compute intersections between the illuminants as estimated by the set $\mathcal{M}_\cap$. Note that the set $\mathcal{M}_\cap$ contains a set of maps that take the current illuminant to the canonical illuminant. If the canonical illuminant is assumed to be white light, then each vertex of $\mathcal{M}_\cap$ describes a possible illuminant. Each vertex can be viewed to be a diagonal map $\mathbf{m}$, which has the form

$$\mathbf{m} = [m_1, m_2]^T = [L_b, L_b]^T / [L_r, L_g]^T \tag{6.59}$$

where division is carried out component-wise, and $[L_r, L_g, L_b]^T$ describes the color of the illuminant.

$$\left[\frac{L_r}{L_b}, \frac{L_g}{L_b}, 1\right]^T = \left[\frac{1}{m_1}, \frac{1}{m_2}, 1\right]^T \tag{6.60}$$

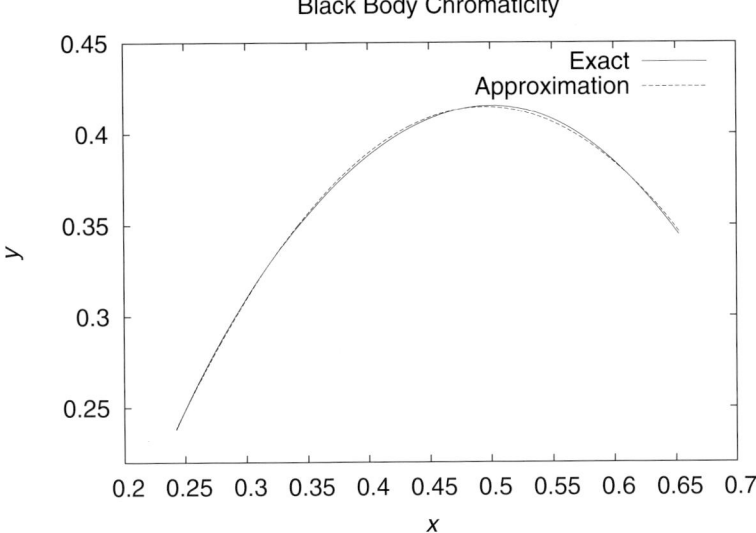

Figure 6.21 The curve of the black-body radiator in CIE XYZ space as a function of $\hat{x}$. The curve can be closely approximated by a quadratic equation.

The chromaticities of the illuminant are therefore given by

$$[\hat{x}, \hat{y}, \hat{z}] = \left[\frac{L_r}{L_r + L_g + L_b}, \frac{L_g}{L_r + L_g + L_b}, \frac{L_b}{L_r + L_g + L_b}\right] \quad (6.61)$$

$$= \left[\frac{\frac{1}{m_1}}{\frac{1}{m_1} + \frac{1}{m_2} + 1}, \frac{\frac{1}{m_2}}{\frac{1}{m_1} + \frac{1}{m_2} + 1}, \frac{1}{\frac{1}{m_1} + \frac{1}{m_2} + 1}\right]. \quad (6.62)$$

Once the chromaticities of the vertices are known, we can compute the intersection between the curve of the black-body radiator and the chromaticities of possible illuminants. The set of illuminants allowed are then given by the intersection between the curve of the black-body radiator and the chromaticities of possible illuminants. Note that the set of possible illuminants may not intersect the curve of the black-body radiator. In this case, we can choose the closest point on the curve of the black-body radiator.

Results are shown in Figure 6.22. Again, the original lightness was used to rescale the computed chromaticities. For these results, it was assumed that the illuminant can be approximated by a black-body radiator. Since the two-dimensional gamut-constraint algorithm, just like the three-dimensional gamut constraint algorithm, assumes a single uniform illuminant, it is not able to handle images with a nonuniform illumination. Also, if the additional assumption is made that the illuminant can be approximated by a black-body radiator, then the algorithm will not be able to cope with arbitrary illuminants. The algorithm will fail if we have a green illuminant, e.g. a light bulb with a green filter.

128 ALGORITHMS FOR COLOR CONSTANCY UNDER UNIFORM ILLUMINATION

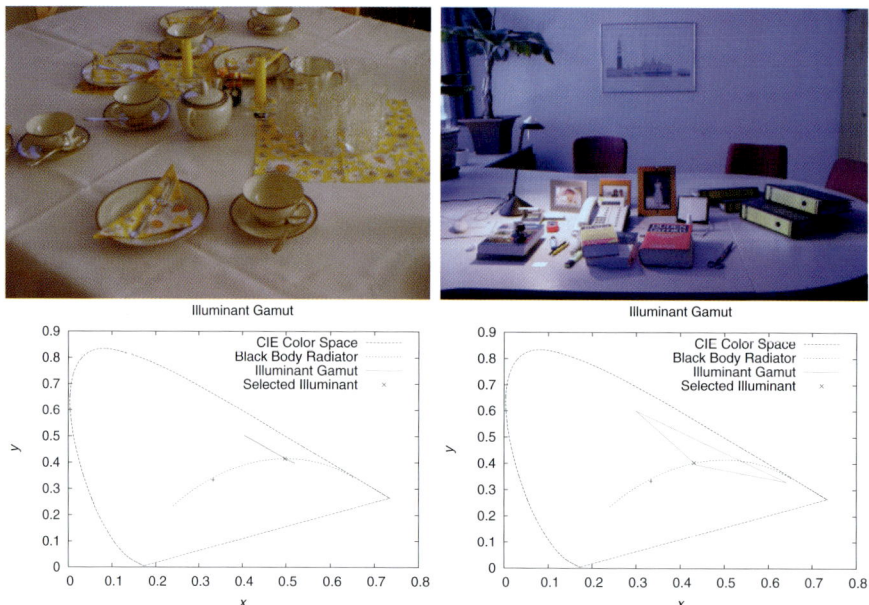

Figure 6.22 Results obtained using a two-dimensional gamut-constraint algorithm with the assumption that the illuminant can be modeled as a black-body radiator. The two graphs again show the possible illuminants as CIE chromaticities and the illuminant that was selected by the two-dimensional gamut-constraint algorithm.

6.6 Color Cluster Rotation

Paulus et al. (1998) suggested a method called *color cluster rotation*. The color values of an input image are viewed as a cloud of points inside a RGB color cube. The axes of the color cube are formed by the vectors $\mathbf{r} = [1, 0, 0]^T$ (red), $\mathbf{g} = [0, 1, 0]^T$ (green), and $\mathbf{b} = [0, 0, 1]^T$ (blue). Each color $\mathbf{c}(x, y) = [c_r(x, y), c_g(x, y), c_b(x, y)]^T$ is a point inside this three-dimensional space. Paulus et al. argue that in technical environments most objects are gray and that therefore the principal component of this cloud of points should be aligned with the gray vector $\mathbf{w} = \frac{1}{\sqrt{3}}[1, 1, 1]^T$. If it is not aligned with the gray vector, then the entire cloud is rotated such that its principal component is aligned with the gray vector.

The principal component of the cloud of points is computed as follows. First, the center of the cloud is obtained by computing the center of gravity of all the points. The center of gravity is of course equivalent to global space average color $\mathbf{a}$,

$$\mathbf{a} = \frac{1}{n} \sum_{x,y} \mathbf{c}(x, y) \tag{6.63}$$

where n is the number of pixels of the input image. The principal component is obtained by first computing the cooccurrence matrix and then finding the eigenvector that corresponds to the largest eigenvalue of this matrix. The (3 × 3) cooccurrence matrix $\mathbf{C}$ is defined as

$$\mathbf{C} = E\left[(\mathbf{c}-\mathbf{a})(\mathbf{c}-\mathbf{a})^T\right] \tag{6.64}$$

where E denotes the expected value. The eigenvectors of this matrix are computed numerically. Let $\mathbf{e}_1$ be the normalized eigenvector that corresponds to the largest eigenvalue. In order to align the vector $\mathbf{e}_1$ with the gray vector $\mathbf{w}$, one first computes the axis of rotation $\mathbf{v}$ that is orthogonal to both $\mathbf{e}_1$ and $\mathbf{w}$.

$$\mathbf{v} = \mathbf{e}_1 \times \mathbf{w} \tag{6.65}$$

The rotation angle θ is given by

$$\theta = \cos^{-1}(\mathbf{e}_1^T \mathbf{w}). \tag{6.66}$$

Output colors are computed by the following three steps. First, the cloud of points is shifted such that its center is located at the origin. Next, a rotation is performed such that the principal component is aligned with the gray vector. Finally, the cloud is shifted in the direction of the gray vector such that the intensity of local space average color is maintained. The sequence of operations is shown in Figure 6.23. Let $\mathbf{c}$ be the color of the input pixel. Then, the color of the output pixel $\mathbf{o}$ is given by

$$\mathbf{o} = \mathbf{R}(\mathbf{v}, \theta)(\mathbf{c}-\mathbf{a}) + (\mathbf{a}^T \mathbf{w})\mathbf{w} \tag{6.67}$$

where $\mathbf{R}(\mathbf{v}, \theta)$ rotates the coordinate system around $\mathbf{v}$ by angle θ. Values larger than 1 and smaller than 0 are simply clipped at the border of the coordinate system. Results obtained with this algorithm are shown in Figure 6.24. Since the algorithm assumes a color rotation that is due to a single illuminant, it is not able to handle images with multiple illuminants.

Instead of maintaining the original average intensity of the image, it is also possible to shift the color-corrected image to the center of the RGB color cube. The cloud of points can also be rescaled to fill most of the color cube. The inverse of the square root of the largest eigenvalue λ_1 is used as a scaling factor. In this case, output colors are computed as follows:

$$\mathbf{o} = \frac{1}{\sqrt{\lambda_1}} \mathbf{R}(\mathbf{v}, \theta)(\mathbf{c}-\mathbf{a}) + \frac{\sqrt{3}}{2}\mathbf{w} \tag{6.68}$$

The results of this algorithm are shown in Figure 6.25. Owing to the rescaling, the colors are now more vivid.

6.7 Comprehensive Color Normalization

Finlayson et al. (1998) have developed an algorithm called *comprehensive color normalization*. Apart from the illuminant, the lighting geometry also affects the perceived color. The comprehensive color normalization removes dependencies due to both lighting geometry and the type of illuminant. If we again look at the theory of color image formation, we see that, for a Lambertian surface, the geometry scales the sensor's response. Assuming

130 ALGORITHMS FOR COLOR CONSTANCY UNDER UNIFORM ILLUMINATION

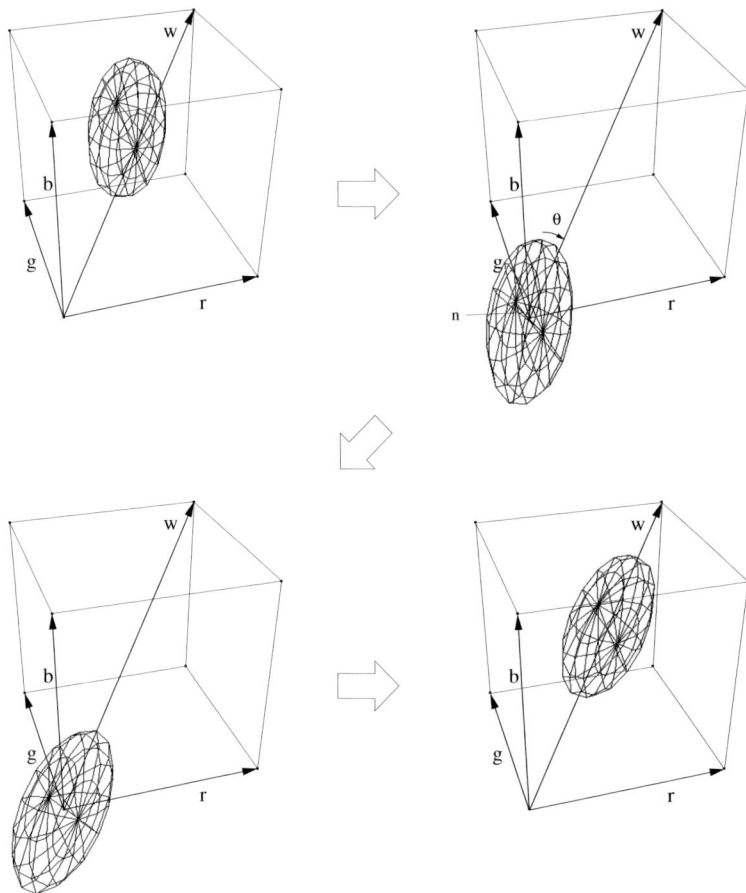

Figure 6.23 Color cluster rotation. The cluster of points is first shifted to the origin. Then, the major axis of the cluster is aligned with the gray vector. This is done by rotating the cluster around a vector that is perpendicular to the gray vector and also perpendicular to the major axis of the cluster. Finally, the cluster is shifted to maintain the original average intensity.

narrow-band sensors, the response of the sensor I_i of color channel i is given by

$$I_i(x, y) = G(x, y) R_i(x, y) L_i \qquad (6.69)$$

where $G(x, y)$ is a factor that depends on scene geometry at the corresponding object point that is shown at position (x, y), $R_i(x, y)$ is the amount of light reflected at the corresponding object position for wavelength λ_i, and L_i is the intensity of the light emitted by the light source for wavelength λ_i. The first term $G(x, y)$ only depends on the scene geometry and scales all color channels equally.

The influence of the scene geometry can be removed by normalizing all pixel values. Let $[r_1, g_1, b_1]^T$ and $[r_2, g_2, b_2]^T$ be two image pixels. If the pixels actually represent the

ALGORITHMS FOR COLOR CONSTANCY UNDER UNIFORM ILLUMINATION

Figure 6.24 Results obtained with color cluster rotation. The algorithm performs well given a uniform illumination. It is not able to handle images with multiple illuminants.

Figure 6.25 Results obtained with color cluster rotation where the cloud of points is shifted to the center of the color cube and all points are rescaled by the inverse of the square root of the largest eigenvalue.

same color but with different geometry factors, then we have

$$c_r = GR_rL_r \quad \text{and} \quad c'_r = G'R_rL_r \tag{6.70}$$

$$c_g = GR_gL_g \quad \text{and} \quad c'_g = G'R_gL_g \tag{6.71}$$

$$c_b = GR_bL_b \quad \text{and} \quad c'_b = G'R_bL_b \tag{6.72}$$

for two different scene geometry factors G and G'. We can remove the geometry factors by normalizing all components. We obtain

$$\hat{c}_r = \frac{c_r}{c_r+c_g+c_b} = \frac{GR_rL_r}{GR_rL_r+GR_gL_g+GR_bL_b} = \frac{R_rL_r}{R_rL_r+R_gL_g+R_bL_b} = \hat{c}'_r \tag{6.73}$$

$$\hat{c}_g = \frac{c_g}{c_r+c_g+c_b} = \frac{GR_gL_g}{GR_rL_r+GR_gL_g+GR_bL_b} = \frac{R_gL_g}{R_rL_r+R_gL_g+R_bL_b} = \hat{c}'_g \tag{6.74}$$

$$\hat{c}_b = \frac{c_b}{c_r+c_g+c_b} = \frac{GR_bL_b}{GR_rL_r+GR_gL_g+GR_bL_b} = \frac{R_bL_b}{R_rL_r+R_gL_g+R_bL_b} = \hat{c}'_b \tag{6.75}$$

132 ALGORITHMS FOR COLOR CONSTANCY UNDER UNIFORM ILLUMINATION

The result is independent of the geometry factors G and G'.

The influence of the illuminant can be removed by using the gray world assumption. Let $\mathbf{c}(x, y) = [c_r(x, y), c_g(x, y), c_b(x, y)]^T$ be the image color at pixel (x, y). Then, the output pixels are calculated as follows:

$$o_i = \frac{c_i}{\frac{2}{n}\sum_{x,y} c_i(x, y)} \tag{6.76}$$

where n are the number of pixels in the input image. This operation scales the color of the pixel by the inverse of twice the global space average color.

Finlayson et al. (1998) have combined both methods, normalization of pixel values and the scaling of pixel values using the gray world assumption, into a method called *comprehensive color normalization*. Both methods are applied iteratively and interleaved. In other words, first the pixel values are normalized, then the values are rescaled using the gray world assumption, followed by another normalization, and so on. This process continues until convergence. By interleaving both methods, changes that are due to the illuminant as well as changes that are due to the scene geometry are removed. Let $\mathcal{A}_{\text{norm}}$ be the algorithm that normalizes all pixels, i.e.

$$o_i(x, y) = \frac{c_i(x, y)}{\sum_{j \in \{r,g,b\}} c_j(x, y)}. \tag{6.77}$$

Let $\mathcal{A}_{\text{gray}}$ be the algorithm that performs the normalization using the gray world assumption, i.e.

$$o_i(x, y) = \frac{1}{3} \frac{c_i(x, y)}{\frac{1}{n}\sum_{x',y'} c_i(x', y')}. \tag{6.78}$$

Notice the scaling factor of $\frac{1}{3}$. Let $\mathcal{I}_0$ be the input image. Then, the comprehensive color normalization iteratively applies the two algorithms $\mathcal{A}_{\text{norm}}$ and $\mathcal{A}_{\text{gray}}$ to compute a new output image.

$$\mathcal{I}_{i+1} = \mathcal{A}_{\text{gray}}(\mathcal{A}_{\text{norm}}(\mathcal{I}_i)) \tag{6.79}$$

This process continues until the termination condition

$$\mathcal{I}_{i+1} = \mathcal{I}_i \tag{6.80}$$

is met. In practice, the algorithm is stopped after the change for each pixel falls below a suitably defined threshold. In other words, the change is so small that it not noticeable in a discretized image of 256 intensity levels for each color channel. Results for two input images are shown in Figure 6.26. Because of the normalization, all intensity information is lost. Also, comprehensive color normalization assumes a single illuminant. It does not work if multiple illuminants are present.

We now analyze why the scaling factor of $\frac{1}{3}$ is used for the gray world normalization algorithm. Let n be the number of pixels in the input image. Then, the pixels of the input image can be used to construct a $n \times 3$ matrix. In this case, the column index specifies one of the three color channels red, green, or blue and the row index specifies the pixel number. The algorithm $\mathcal{A}_{\text{norm}}$ normalizes the rows of this matrix such that the sum of the entries of a single row is one. The algorithm $\mathcal{A}_{\text{gray}}$ normalizes the columns of this matrix such that the sum of the entries of a single column is $\frac{n}{3}$. It is therefore possible that the

Figure 6.26 Comprehensive color normalization iteratively normalizes all pixels and then rescales the color channels using the gray world assumption. Therefore, intensity information is lost. Since it is based on the gray world assumption, it does not work if we have a nonuniform illumination.

Figure 6.27 Results of comprehensive color normalization when the lightness of the pixels from the input image is used to rescale the output pixels.

matrix is both in row- and column-normal forms. Finlayson et al. (1998) have shown that the process of iteratively performing both normalization algorithms always converges. The interested reader is referred to Finlayson et al. (1998) for the proof.

By normalizing all colors, the information about the original intensity is lost. The intensity can be restored by rescaling the comprehensively normalized output colors to the original lightness of the input pixels (similar to Equation 6.51). The images shown in Figure 6.26 rescaled to the intensities of the input image are shown in Figure 6.27. Also, since comprehensive color normalization uses the gray world assumption, it suffers from the same drawbacks. Comprehensive color normalization does not work if multiple illuminants are present. One possible extension to also handle multiple illuminants would be to combine comprehensive color normalization with our method that is discussed in Chapters 10 and 11. Instead of using global space average color for the comprehensive color normalization, we use local space average color. For each step of the comprehensive color normalization, we normalize the colors and then compute local space average color to perform another normalization. Since convergence happens in four to five iterations, we

134 ALGORITHMS FOR COLOR CONSTANCY UNDER UNIFORM ILLUMINATION

Figure 6.28 Results of comprehensive color normalization using local space average color instead of global space average color. The intensities of the input image were used to rescale the output pixels.

only need the same number of stages. In fact, we have implemented this algorithm. The results are shown in Figure 6.28.

6.8 Color Constancy Using a Dichromatic Reflection Model

Most of the algorithms covered so far assume that objects are matte, i.e. the amount of reflected light does not depend on the position of the observer relative to the object. Shiny objects, in contrast, reflect most of the light in a single direction. The amount of light entering the eye of the viewer not only depends on the position of the light source relative to the object but it also depends on the position of the viewer relative to the object. An ideal specular object is a mirror. If we have a perfect mirror in the image and the light source can be viewed inside the mirror, then the color of the illuminant can be obtained by analyzing the reflection. Similarly, highlights of specular objects may also be analyzed to estimate the color of the light source.

Specular objects can be modeled using the dichromatic reflection model (D'Zmura and Lennie 1986; Finlayson and Schaefer 2001; Risson 2003; Tominaga 1991; Tominaga and Wandell 1992). Basically it assumes a matte reflection in combination with a specular reflective component (Figure 6.29). The specular component causes highlights that have the color of the light source. Highlights occur at object points where the geometry is such that the light from the light source is reflected directly into the camera. A number of algorithms based on the dichromatic reflection model have been developed (Ebner and Herrmann 2005; Finlayson and Schaefer 2001; Risson 2003; Tominaga 1991; Tominaga and Wandell 1992). The dichromatic reflection model combines matte and specular reflection additively. Let $L(\lambda)$ be the illuminant for wavelength λ. Let $R_M(\lambda)$ be the object reflectance with regard to the matte reflection and let $R_S(\lambda)$ be the object reflectance with regard to the specular reflection. Then, the color **I** measured by the sensor is given by

$$\mathbf{I}(\lambda) = \int \mathbf{S}(\lambda) \, (s_M R_M(\lambda) L(\lambda) + s_S R_S(\lambda) L(\lambda)) \, d\lambda \qquad (6.81)$$

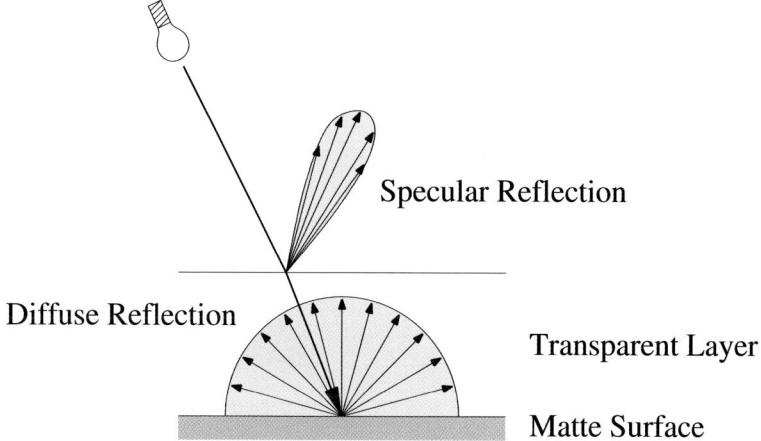

Figure 6.29 The dichromatic reflection model assumes a matte reflection in combination with a specular reflection. Part of the light is reflected at the outer surface. The remainder enters the transparent coating. The second reflection is modeled as being Lambertian. (Reproduced by permission of Pearson Education from 3D Computer Graphics Third Edition, Alan Watt, Pearson Education Limited, © Pearson Education Limited 2000.)

where $\mathbf{S}(\lambda)$ denotes the vectors with the sensors response functions, and s_M and s_S are two scale factors that depend on the object geometry.

Assuming narrow-band camera sensors that only respond to wavelength λ_i with $i \in \{r, g, b\}$, we obtain

$$I_i = s_M R_{M,i} L_i + s_S R_{S,i} L_i \tag{6.82}$$

where I_i is the response of sensor i, $R_{M,i}$ is the reflectance with respect to the matte reflection at wavelength λ_i, and $R_{S,i}$ is the reflectance with respect to the specular reflection at wavelength λ_i. If we now assume that the illuminant is constant over the entire image and that the specular reflection behaves like a perfect mirror, i.e. $R_{S,i} = 1$, then we obtain

$$I_i = s_M R_{M,i} L_i + s_S L_i = s_M c_{M,i} + s_S c_{S,i} \tag{6.83}$$

where $\mathbf{c}_M = [R_{M,r} L_r, R_{M,g} L_g, R_{M,b} L_b]^T$ is the matte color of the object point and $\mathbf{c}_S = [L_r, L_g, L_b]^T$ is the color of the illuminant. Thus, the color measured by the sensor is restricted to the linear combination of the matte color of the object point $\mathbf{c}_M$ and the color of the illuminant $\mathbf{c}_S$. The colors sensed are all restricted to a plane defined by the two vectors $\mathbf{c}_M$ and $\mathbf{c}_S$ (D'Zmura and Lennie 1986). This is illustrated in Figure 6.30. The colors sensed from a surface all lie between the diffuse color of the surface and the color of the illuminant. The highlights are very bright and have the color of the light source, whereas other points of the surface are typically darker.

When we now compute chromaticities, the three-dimensional data points are projected onto the plane $r + g + b = 1$. The linear combination results in a two-dimensional line in chromaticity space. Let $\hat{\mathbf{c}}_M$ be the chromaticity of the matte color and let $\hat{\mathbf{c}}_S$ be the chromaticity of the specular color. Then, the dichromatic line on the plane $r + g + b = 1$

136 ALGORITHMS FOR COLOR CONSTANCY UNDER UNIFORM ILLUMINATION

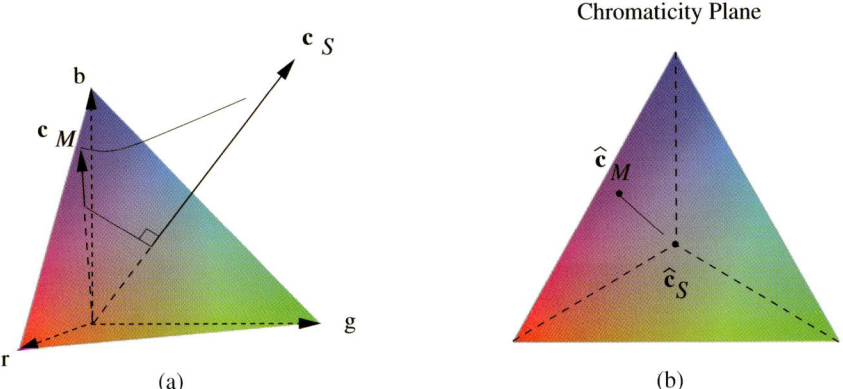

Figure 6.30 The chromaticity plane is centered on the color of the illuminant. Colors sensed from a surface are located inside a plane spanned by the diffuse color of the object and the color of the illuminant. The graph in (b) shows the projection onto the chromaticity plane.

is described by

$$\begin{bmatrix} \hat{x} \\ \hat{y} \end{bmatrix} = s\hat{\mathbf{c}}_M + (1-s)\hat{\mathbf{c}}_S. \tag{6.84}$$

where s is a scaling factor. For each uniquely colored surface of the scene, we will obtain one line in chromaticity space. Figure 6.31 shows the dichromatic lines for two surfaces. Since we have assumed a unique illuminant for the entire scene, all lines must have one point in common, which is the color of the illuminant. In order to find the color of the illuminant, we compute the dichromatic line for each surface and then intersect the lines of all surfaces. The lines can be computed by performing linear regression on the chromaticity data for each surface.

Finlayson and Schaefer (2001) note that algorithms based on the dichromatic reflection model perform well only under idealized conditions and that the estimated illuminant does not turn out to be that accurate. Small amounts of noise may change the point of intersection of two dichromatic lines quite drastically. According to Finlayson and Schaefer, the method works well for highly saturated surfaces under laboratory conditions but does not work well for real images. In order to make the method more robust, Finlayson and Schaefer suggest that the intersection of the dichromatic lines with the curve of the black-body radiator be computed. In other words, they assume that the illuminant is not arbitrary, but can be approximated by a black-body radiator. Finlayson and Schaefer assume that the scene is presegmented. For each segment, they compute the dichromatic line and intersect it with the curve of the black-body radiator. In theory, this algorithm can work if there is only a single surface in the image. If the dichromatic line does not intersect the curve of the black-body radiator, one could choose the closest point. If there are two intersections, some heuristics can be used to rule out one of the intersections.

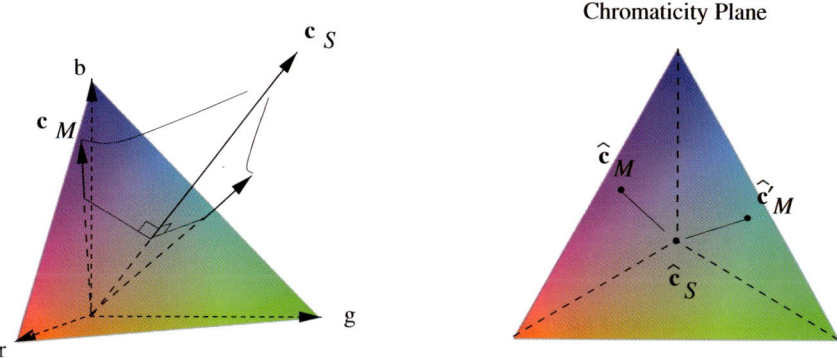

Figure 6.31 Two different surfaces have a different diffuse component; however, the specular component is the same. The color of the illuminant is located at the intersection of the dichromatic lines.

Risson (2003) takes the algorithm of Finlayson and Schaefer a step further by also addressing the segmentation problem. Risson proposed that the illuminant be determined from a digital color image by segmentation and filtering. The algorithm consists of the following steps as shown in Figure 6.32. If necessary, the input image is smoothed using a median or a Gaussian filter. Then, the image is segmented into regions of homogeneous color. All regions that are not compliant with the dichromatic reflection model are removed. Examples of noncompliant regions are achromatic regions, regions that belong to the sky or are part of shadows. Achromatic regions can be removed by computing the saturation for each region and removing regions whose saturation is below a certain threshold. Also, in order to apply the dichromatic reflection model, the region has to have at least two pixels. In order to be on the safe side, all regions that have less than a certain number of pixels are removed. For the remaining regions, the dichromatic line is computed. Risson suggests that a linear regression on the x- and y-coordinates in CIE XYZ color space may be performed.

Alternatively, we can also compute the covariance matrix for the pixel colors of the region in XYZ color space and then determine the eigenvector that corresponds to the largest eigenvalue. Let c_j with $j \in \{1, ..., n\}$ be the pixel colors of an image region that contains n pixels. Let $\mathbf{xy}_j = [x, y]^T$ be the xy-coordinates of the pixel colors c_j in XYZ color space. Then the (2×2) covariance matrix $\mathbf{C}$ is given by

$$\mathbf{C} = E\left[(\mathbf{xy} - \mathbf{m})(\mathbf{xy} - \mathbf{m})^T\right] \qquad (6.85)$$

where E denotes the expected value and $\mathbf{m}$ is the mean of the xy-coordinates. The eigenvectors of this matrix can be computed directly since the matrix is only a 2×2 matrix. Once we know the eigenvalues, we can check if the point cloud from the region indeed resembles a line. For instance, we can remove regions whose smaller eigenvalue is larger

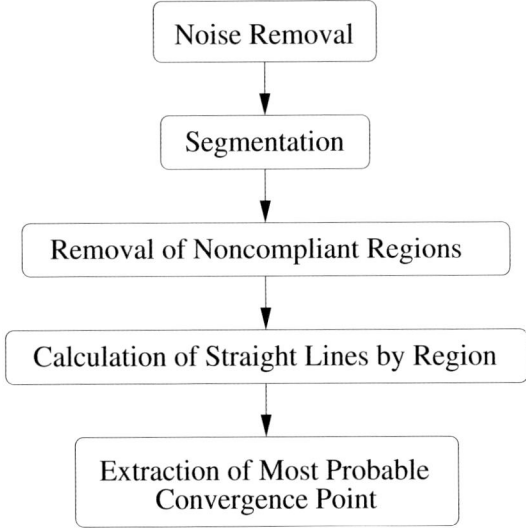

Figure 6.32 Algorithm of Risson (2003). First, noise is removed by prefiltering the image. Next, the image is segmented. Regions that do not satisfy the assumptions of the dichromatic reflection model are removed. The dichromatic line is computed for the remaining regions. Finally, the most probable point of intersection is computed.

than a threshold. We could also remove regions whose eigenvalues are very close together. We only have a line if one eigenvalue is much larger than the other. Let $\mathbf{e}_j$ be the normalized eigenvector of region j that corresponds to the largest eigenvalue. The dichromatic line $\mathcal{L}_j$ of region j is then given by

$$\mathcal{L}_j = \{\mathbf{m}_j + s\mathbf{e}_j | \text{with } s \in \mathbb{R}\}. \tag{6.86}$$

The illuminant is located at the position where all the lines $\mathcal{L}_j$ intersect.

Owing to small inaccuracies, the lines do not intersect in a single point in XYZ color space. Thus, we have to choose a method to find this intersection. One possibility would be to compute the intersection for all possible combinations between two lines. This gives us a set of intersections (Herrmann 2004)

$$\{\mathbf{p}_i = (x_i, y_i) | \text{with } i \in \{1, \ldots, \tfrac{1}{2}n(n-1)\}\}. \tag{6.87}$$

The actual intersection could be estimated by computing the average of the intersection points. In this case, the position of the illuminant is given by

$$\mathbf{p} = \frac{2}{n(n-1)} \sum_i \mathbf{p}_i. \tag{6.88}$$

An alternative way would be to take the median of the x- and y-coordinates. This would remove outliers. In this case, the position of the illuminant is given by

$$\mathbf{p} = [\text{Median}\{x_i\}, \text{Median}\{y_i\}] \tag{6.89}$$

ALGORITHMS FOR COLOR CONSTANCY UNDER UNIFORM ILLUMINATION

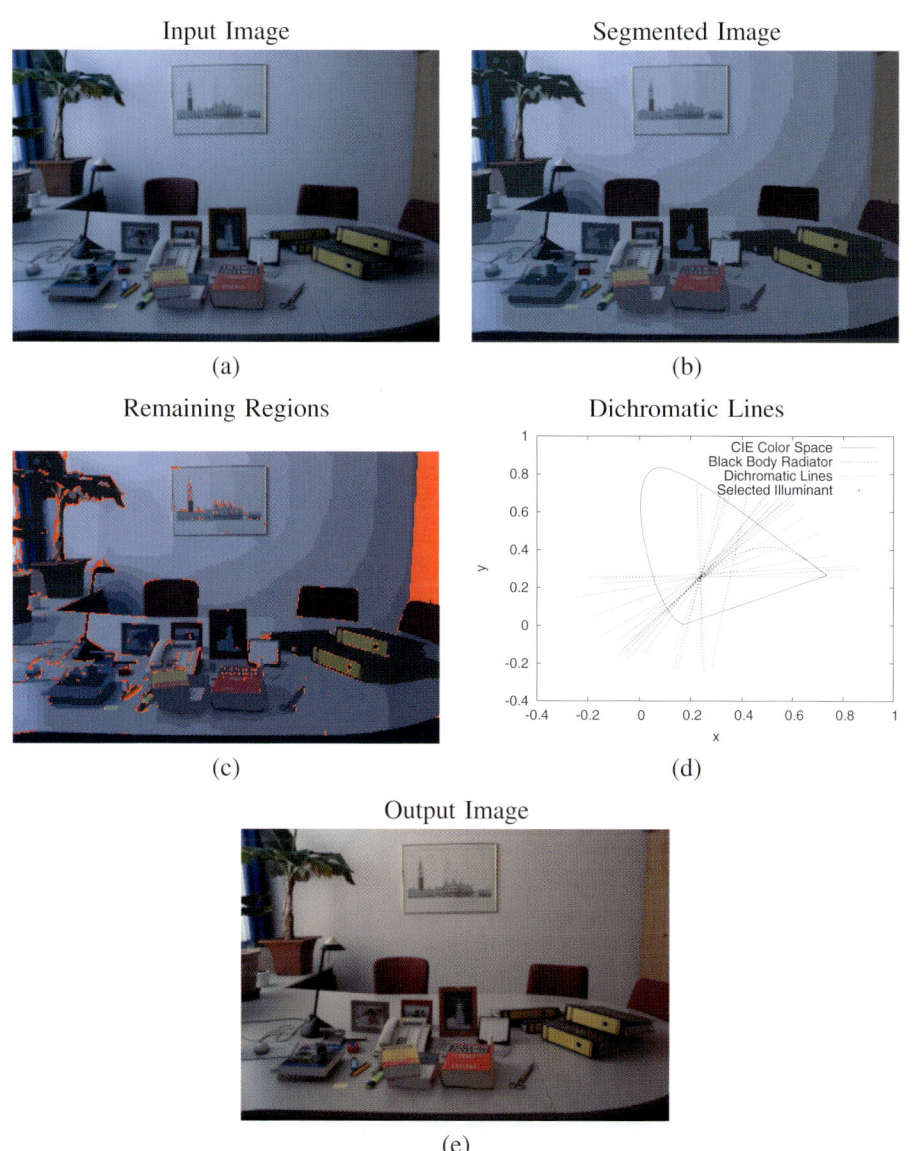

Figure 6.33 Individual steps of the algorithm described by Risson (2003). Image (a) is the input image. Removal of noise has been omitted here. Image (b) shows the segmented image. Image (c) shows the remaining regions. The removed regions are marked in red. For each region, the dichromatic line is computed. Only a subset of all lines is shown (d). The point of intersection is marked by a vertical line and a horizontal line. Image (e) is the output image.

140 ALGORITHMS FOR COLOR CONSTANCY UNDER UNIFORM ILLUMINATION

where Median denotes the computation of the median. Figure 6.33 shows the steps of the algorithm for a sample image.

Risson (2003) also suggests that the most probable intersection point with the curve of the black-body radiator be found. This assumes that the illuminant can indeed be approximated by a black-body radiator. If this is not the case, i.e. for a green illuminant, then this method should not be used. If we make the assumption that the illuminant must be located on the curve of the black-body radiator, then we can compute, for each dichromatic line, the intersection with the curve of the black-body radiator. The result will be a set of (x, y) coordinates describing the intersection points. Discarding the y-coordinate, we can either compute the average or take the median x-coordinate to arrive at the most probable intersection point. The corresponding y-coordinate can be computed from the approximation given in Section 6.5. Alternatively, we can also compute a histogram for the x-coordinates and then determine the intersection point from the bucket that has the maximum count. Again, the y-coordinate can be computed using the approximation given earlier. Instead of averaging the x-coordinates or taking the median, we could also find the corresponding temperature of the black-body radiator for each intersection. After computing the set of temperatures, we could again take the average temperature or select the median temperature of all intersections. Once the temperature of the illuminant is estimated, we can then compute the corresponding chromaticities.

In our experiments, the best results are obtained when no constraints are made regarding the type of illuminant (Herrmann 2004). Therefore, we do not use the constraint that the illuminant has to lie on the curve of the black-body radiator. We have used the median x- and y-coordinate of all possible dichromatic line intersections in order to estimate the color of the illuminant. The use of the median has the advantage that outliers are removed. This method produced the best results (Ebner and Herrmann 2005). After estimating the position of the illuminant, we compute the RGB coordinates of the illuminant and then divide each channel by the corresponding illuminant coordinate. The lightness is adjusted such that the output lightness is identical to the lightness of the input image. Since only a single uniform illuminant is assumed, the algorithm cannot be used if multiple illuminants are present. Results for Risson's algorithm (Risson 2003) are shown in Figure 6.34. For the first image, the illuminant was correctly found to be reddish. However, the value that was determined for the illuminant raises the blue channel to a great extent.

Figure 6.34 Results for the algorithm of Risson (2003). The algorithm cannot be used if multiple different illuminants are present.

ALGORITHMS FOR COLOR CONSTANCY UNDER UNIFORM ILLUMINATION

Before we will evaluate all algorithms that have been discussed in this chapter, we will turn to algorithms that do not assume a uniform illuminant in the next chapter. All algorithms that have been described in this chapter as well as in the next chapter will be evaluated in detail in Chapter 13 and Chapter 14.

7

Algorithms for Color Constancy under Nonuniform Illumination

All the algorithms discussed in this chapter assume that the illuminant is nonuniform across the image. The intensity measured by the sensor can be modeled as

$$I_i(x, y) = G(x, y) R_i(x, y) L_i(x, y) \tag{7.1}$$

where $G(x, y)$ is a factor that depends on the scene geometry, $R_i(x, y)$ is the reflectance for wavelength λ_i, and $L_i(x, y)$ is the irradiance at position (x, y) for wavelength λ_i. Note that in contrast to the algorithms described in Chapter 6, the algorithms that are discussed in this chapter do not assume that the illuminant is uniform across the entire image. Here, the irradiance depends on the coordinates (x, y).

7.1 The Retinex Theory of Color Vision

Land (1964, 1974, 1983, 1986b), an influential personality in color constancy research, has developed the retinex theory of color perception. According to the retinex theory, three different receptors are used inside the retina that primarily respond to long, middle, and short wavelengths. These receptors measure the energy for different parts of the visible spectrum. Each set of receptors acts as a unit to process the measured energy. At that time, Land did not know whether the processing took place only inside the retina or also in the visual cortex. Land called this retinal-cerebral system, which processes one part of the visible spectrum, the *retinex*. Together with Land and McCann (1971), Land developed a computational theory of color constancy.

In their experiments, Land and McCann identified edges as being important in the determination of the reflectances of an object (Land and McCann 1971). They demonstrated this using two sheets of paper. One reflected 40% of the incident light and the other reflected 80% of the incident light. Two fluorescent tubes were used to illuminate the two sheets of papers. One was placed in front of the two papers and the other one was placed on the side of it. The second tube was positioned such that the amount of reflected light was

Color Constancy M. Ebner
© 2007 John Wiley & Sons, Ltd

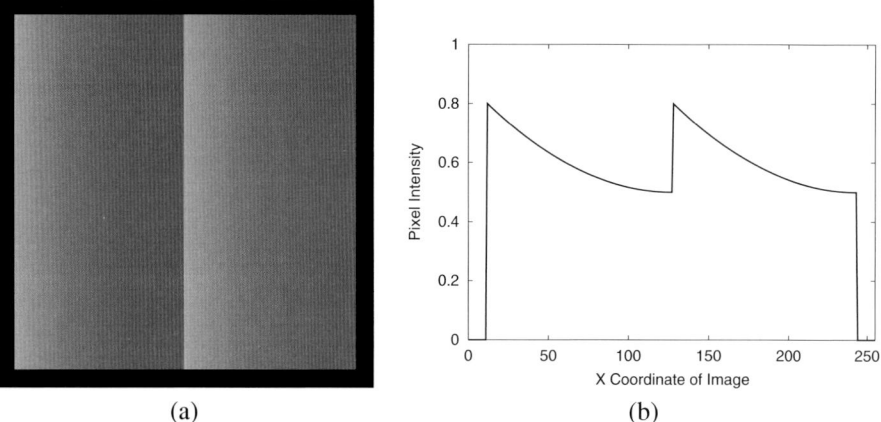

(a) (b)

Figure 7.1 Land placed two sheets of papers with reflectances of 40% (a) and 80% (b) next to each other. The illumination was chosen such that the amount of light reflected from the center of one sheet of paper was equivalent to the light reflected from the center of the other sheet of paper. The left sheet appears darker than the sheet on the right. However, if a pencil is placed over the edge where both sheets join, then both sheets seem to have the same reflectance. The image shown here was generated artificially. The graph on the right shows the pixel intensities for a horizontal line of the image.

equivalent for the two centers of the sheets of papers. The resulting image is shown in Figure 7.1. The stimulus is shown on the left, the pixel intensities along a horizontal line of this image are shown on the right. The sheet of paper on the left looks darker than the one on the right. It may be necessary to move close to the center of the stimulus to see the effect. However, if an object, such as a pencil or a long strip of paper, is used to cover the edge between the two sheets, then both sheets of paper seem to have the same reflectance. Try covering the edge with some object. Now both areas appear to have the same brightness.

From this experiment, Land and McCann drew the conclusion that edges are an important source of information to achieve color constancy. Imagine again the Mondrian image illuminated by some light source. The illumination can be nonuniform. Suppose that we are equipped with two detectors with which we can measure the reflected light from any point on the Mondrian image. Let us first point each detector toward a random point of the same rectangle. If a nonuniform illumination is used, then the measurements will differ slightly. However, if the two detectors are moved closer together, then the measurements will be approximately equal. Land and McCann suggest calculating the ratio of the two measurements. If the two detectors measure the light reflected from two nearby points on the same rectangle, then the ratio will be approximately unity. It will be unity, because for two adjacent points, the illuminant as well as the reflectance of the rectangle will be equivalent. Let us now use the two detectors to measure the light reflected by two adjacent rectangles. One detector is placed on a random point of the first rectangle and the other one is placed on a random point on the second rectangle. As the two detectors are moved toward a single point on the border of the two rectangles, the ratio will approach

COLOR CONSTANCY UNDER NONUNIFORM ILLUMINATION

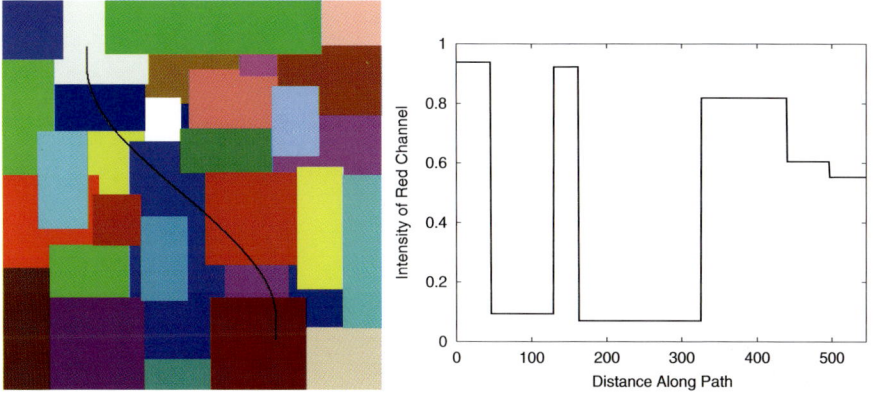

Figure 7.2 A random path between two rectangles of the color Mondrian. The graph on the right shows the intensity of the red channel along this path.

the ratio of the two reflectances of the rectangles. Since the illuminant is equivalent for both rectangles for this point, it will fall out of the equation. What remains is the ratio of the reflectances.

In order to calculate the ratio of reflectances between any two rectangles of the Mondrian image, Land and McCann suggest choosing a random path from a point on the first rectangle to a point on the second rectangle. Figure 7.2 shows a Mondrian image with a random path from one point of the Mondrian to another. The intensity of the red channel is also shown for each rectangle along the path. In the following we will only consider a single color band. Land's retinex theory is applied to all color bands independently. The intensities of the red channel for the seven rectangles are

$$0.938, 0.094, 0.923, 0.070, 0.819, 0.605, \text{ and } 0.553. \tag{7.2}$$

First, the ratio of luminances between the first two rectangles along the path is calculated. We calculate $\frac{0.938}{0.094} = 9.98$. When the path crosses the boundary to another rectangle, the ratio is multiplied with the ratio of luminances between the current and the next rectangle along the path. For the second and third rectangle along the path we obtain $\frac{0.094}{0.923} = 0.102$. Both ratios can be multiplied to obtain the ratio between the first and the third rectangle, $\frac{0.938}{0.094} \cdot \frac{0.094}{0.923} = \frac{0.938}{0.923} = 1.02$. We can continue this process until we have reached the last rectangle. We obtain

$$\frac{0.938}{0.094} \cdot \frac{0.094}{0.923} \cdot \frac{0.923}{0.070} \cdot \frac{0.070}{0.819} \cdot \frac{0.819}{0.605} \cdot \frac{0.605}{0.553} = \frac{0.938}{0.553} = 1.70. \tag{7.3}$$

Obviously, this ratio is equivalent to the ratio between the first and the last rectangle along the path. Since all the intermediate luminances cancel, the resulting ratio does not depend on the actual path taken. It only depends on the position of the starting point and on the position of the ending point.

Let us now consider the case of varying illumination. A color gradient is applied to the original Mondrian image such that the color at the starting point of the path on the first

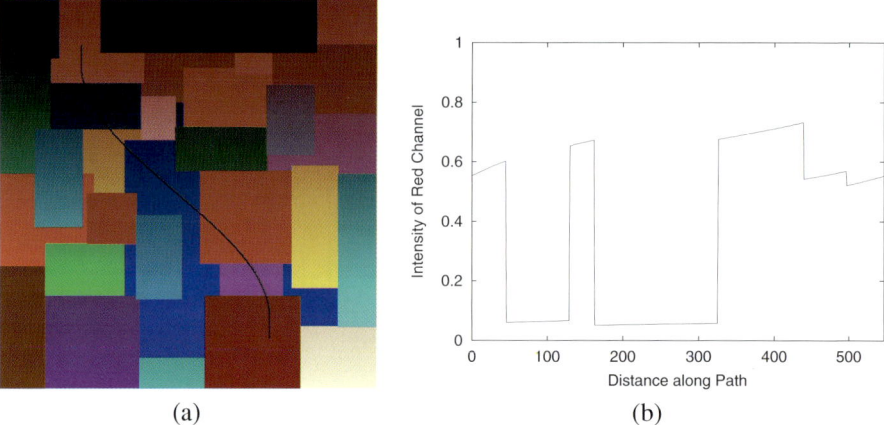

(a) (b)

Figure 7.3 Mondrian image with color gradient. The color gradient was chosen such that the color of the first point of the path is equivalent to the color of the second point of the path.

rectangle is equivalent to the color at the last point of the path. The Mondrian image with this color gradient is shown in Figure 7.3. Again, the intensities of the red channel along the path are shown in (b). As can be seen from the graph on (b), the intensity of the red channel at the beginning of the path is equivalent to the intensity of the red channel at the last point of the path. The intensity of the red channel at the starting point is 0.553. It increases to 0.602 at the point where the path crosses the border between the first and the second rectangle along the path. The intensity of the red channel is 0.060 at that point for the second rectangle. It increases to 0.066 at the point where the path crosses the border between the second and the third rectangles along the path. If we calculate the ratios of these intensities for all the border points and then multiply the ratios, we obtain

$$\frac{0.602}{0.060} \cdot \frac{0.066}{0.653} \cdot \frac{0.673}{0.051} \cdot \frac{0.058}{0.675} \cdot \frac{0.733}{0.542} \cdot \frac{0.569}{0.521} = 1.70. \quad (7.4)$$

This is the same ratio that we obtained for the original color Mondrian. So even though the first and the last points of the path have the same color, we have arrived at a value that specifies relative color between the two rectangles. The reason this works is that the illuminant at the border between two rectangles where the path moves from one rectangle to the next is almost equivalent for the two adjacent points. The illuminant cancels out of the ratio. This method can be used to calculate the relative reflectances of any two points on the Mondrian by connecting them through a random path. Land and McCann (1971) used this fact to develop an algorithm for color constancy on the basis of these ratios.

So far, we only have ratios of reflectances. To compute the reflectance of a rectangle, some normalization is required. Land and McCann suggest choosing an area of high reflectance as a reference patch. This reference patch is assumed to have a reflectance of 1.0. Suppose that a random rectangle is selected, along with a random path from the rectangle to the reference patch. Then the sequential product of the ratios along the path is equivalent to the ratio of the reflectance of the selected rectangle to the reflectance of the reference patch, i.e. 1.0. In other words, we have the reflectance of the selected rectangle.

The remaining question is how to select this reference patch. One method would be to compute the sequential products of the ratios for randomly selected points of the input image. Next, the patch with the highest sequential product is chosen. This patch becomes the reference patch from which all random paths start. Land and McCann reject this possibility on the grounds that this solution is biologically not plausible. They settled on a method that does not require a scanning step that selects the area of highest reflectance.

To compute the reflectances for the points of an image, Land and McCann suggest following multiple paths through the input image. Initially, the starting point is assumed to have a reflectance of 1.0. The sequential product is calculated along the path by forming the ratio between the luminance of the next patch relative to the luminance of the current patch. Whenever the sequential product becomes larger than 1.0, the sequential product is reset to 1.0. This procedure is shown for a short path in Figure 7.4. The rectangles at the top of Figure 7.4 symbolize the rectangles of the Mondrian. The reflectance is specified inside each rectangle in percent. Pairs of detectors are spread along the path. Each pair of detectors measures one ratio. The sequential product is computed along this path. Pairs of detectors that are located inside a rectangle compute a ratio close to 1.0.

Detectors that straddle a boundary between two rectangles compute a ratio that is different from 1.0. The computed sequential product is shown at the bottom of Figure 7.4. This sequential product becomes larger than 1.0 at the detector that is located between the

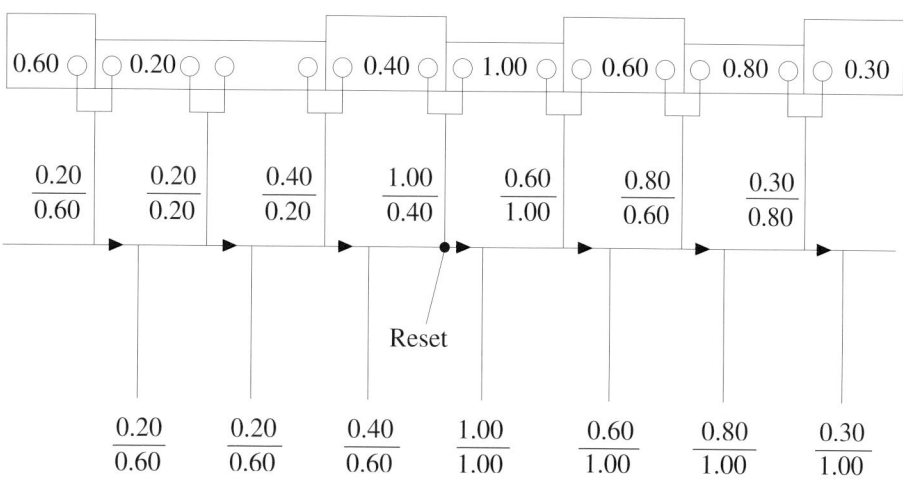

Figure 7.4 Computation of the sequential product along a path on the Mondrian. The rectangles of the Mondrian are symbolized at the top. The reflectance is shown in percent inside each rectangle. Pairs of detectors are spread along the path. These pairs of detectors are used to compute the ratio of the measured luminance. A ratio of 1.0 is assumed for the beginning of the path. Whenever this ratio becomes larger than 1.0, it is reset to 1.0. After the rectangle with a reflectance of 1.0 is passed, the sequential product outputs the reflectance of the rectangle (Reproduced from Land EH and McCann JJ 1971 Lightness and retinex theory. Journal of the Optical Society of America, 61(1), 1-11 by permission from The Optical Society of America.)

two areas with reflectance 0.4 and 1.0. At this point, the sequential product is reset to 1.0 and the computation continues. The estimated reflectances of the seven rectangles are

$$1.00, 0.33, 0.67, 1.00, 0.60, 0.80, \text{ and } 0.30. \tag{7.5}$$

The reflectances at the beginning of the path do not correspond to the actual reflectances of the rectangles. After the path passes the rectangle that has a reflectance of 1.0, the remaining reflectances are estimated correctly. Suppose that we had taken the path in the other direction. In this case, the rectangle with a correct reflectance of 0.30 would have falsely been assumed to have a reflectance of 1.0. Resets would now occur between the first and the second rectangles (counted from right to left) as well as between the third and the fourth rectangles. The estimated reflectances in this case are

$$1.00, 1.00, 0.75, 1.00, 0.40, 0.20, 0.60 \tag{7.6}$$

(also from right to left).

The reflectances are correct only after the patch with the maximum reflectance has been passed. In order to obtain the correct estimate of the reflectance for all the rectangles, Land and McCann assume that the paths are very long. Also, the estimated reflectances from multiple paths are averaged. For the example in the preceding text (with two paths, one forward and one backward) one would obtain

$$0.80, 0.27, 0.53, 1.00, 0.68, 0.90, 0.65 \tag{7.7}$$

(from left to right). The result becomes more accurate when many long paths are averaged.

For an electronic realization of their system, Land and McCann suggest using logarithmic receptors. If logarithmic receptors are used, then the sequential product turns into a sum. Let $\log c_{i,j}$ be the measured data from the j-th receptor of color band i with $i \in \{r, g, b\}$ along the path. Then the ratio between two adjacent receptors can be computed using a simple subtraction.

$$\log c_{i,j+1} - \log c_{i,j} = \log \frac{c_{i,j+1}}{c_{i,j}} \tag{7.8}$$

Also, the sequential product of the ratios can be computed by a simple summation of the logarithmic ratios.

$$\sum_{j=0}^{n} \left(\log c_{i,j+1} - \log c_{i,j} \right) = \sum_{j=0}^{n} \log \frac{c_{i,j+1}}{c_{i,j}} \tag{7.9}$$

If the sequential product gets larger than zero it is reset to zero. Exponentiation can be used to obtain the reflectances along the path.

A schematic drawing of such a system is shown in Figure 7.5. The logarithmic receptors are shown in (a). Positive and negative voltages are used to compute the ratio between adjacent descriptors using summation. The result is added to the existing sequential sum of the logarithms. If the voltage becomes positive it is simply clamped to zero. The schematic diagram for one receptor pair is shown in (b) of Figure 7.5. Land and McCann actually built a working prototype to demonstrate their method. A camera was used to view an image with white, gray, and black segments. Photocells were placed at the boundary of the segments such that one photocell measures the light reflected from one segment and

COLOR CONSTANCY UNDER NONUNIFORM ILLUMINATION

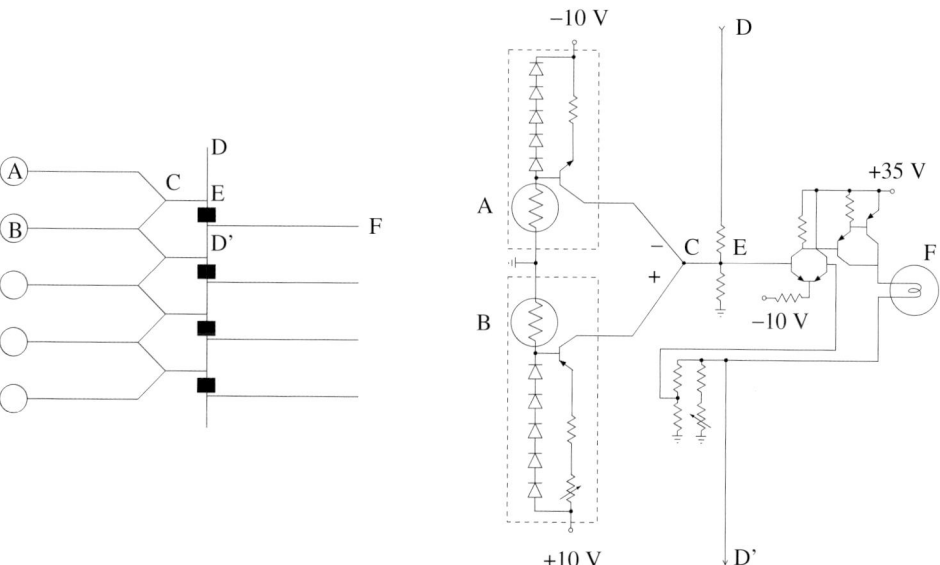

Figure 7.5 Schematic diagram of Land and McCann's prototype (a). Logarithmic receptors are spaced equidistant along a path. The output of two adjacent receptors A and B is added at point C. Because the first term is added with a negative voltage, the ratio of reflectances between A and B is computed at C. At point E the existing sequential product D is added to the result from C. The black square denotes that the sequential product is set to zero if it becomes larger than zero. The reflectances are read out at point F. A circuit for two adjacent receptors A and B is shown in (b). A light bulb (F) is used to visualize the reflectances (Reproduced from Land EH and McCann JJ 1971 Lightness and retinex theory. Journal of the Optical Society of America, 61(1), 1-11 by permission from The Optical Society of America).

the other photocell measures the light reflected from the next segment. The output was visualized using several light bulbs. Each light bulb corresponds to the reflectance of one segment. The bulbs were driven such that they had an anti-logarithmic response.

To handle nonuniform illumination, a threshold must be applied to each ratio along the path. If the logarithm of the ratio is larger than the threshold, it is reset to zero. In this case, the output color in Land's retinex theory is calculated as follows (Land 1983; Land et al. 1983). Let $\Theta'(x)$ be a step function.

$$\Theta(x) = \begin{cases} x & \text{if } x < 0 \\ 0 & \text{otherwise} \end{cases} \qquad (7.10)$$

All values smaller than 0 are left unchanged. All others are set to zero. Let us now compute the relative reflectances along a random path from a point $\mathbf{x}$ to a random point $\mathbf{x}'$. Let

$$\mathbf{x} \rightsquigarrow \mathbf{x}' = \{\mathbf{x}_0, \ldots, \mathbf{x}_n\} \qquad (7.11)$$

with $\mathbf{x} = \mathbf{x}_0$ and $\mathbf{x}_n = \mathbf{x}'$ be the random path from $\mathbf{x}$ to $\mathbf{x}'$. Then the relative reflectances between $\mathbf{x}$ and $\mathbf{x}'$ for color band $i \in \{r, g, b\}$ are given by

$$R_{i,\mathbf{x}\leadsto\mathbf{x}'} = \sum_{j=0}^{n-1} \Theta\left(\log\left(\frac{c_i(\mathbf{x}_{j+1})}{c_i(\mathbf{x}_j)}\right)\right). \tag{7.12}$$

That is, the thresholded logarithms of the ratios are summed up along a path from $\mathbf{x}$ to $\mathbf{x}'$. The reflectances are estimated by averaging the result from several different paths. Let $\mathcal{P}$ be the set of n random points, then the output o_i of color channel i is computed as

$$o_i(\mathbf{x}) = \frac{1}{n} \sum_{\mathbf{x}' \in \mathcal{P}} R_{i,\mathbf{x}'\leadsto\mathbf{x}}. \tag{7.13}$$

Essentially, this operation computes the average reflectance ratio between the reflectance of the current point and the average reflectance of random samples surrounding the point (Brainard and Wandell 1992; Hurlbert 1986). If the threshold operation is used and the paths are assumed to pass through all of the image pixels then the retinex algorithm is simply normalization with respect to the response at the largest location (Brainard and Wandell 1992).

Let us assume for a moment that the illumination is uniform, then the threshold operation is not needed. If we take the current point as a reference and consider multiple paths from the surrounding points that pass through the current point, we obtain

$$o_i(\mathbf{x}) = \frac{1}{n} \sum_{\mathbf{x}' \in \mathcal{P}} \log\left(\frac{c_i(\mathbf{x})}{c_i(\mathbf{x}')}\right) \tag{7.14}$$

$$= \frac{1}{n} \sum_{\mathbf{x}' \in \mathcal{P}} \left(\log(c_i(\mathbf{x})) - \log(c_i(\mathbf{x}'))\right) \tag{7.15}$$

$$= \log(c_i(\mathbf{x})) - \overline{\log}(c_i(\mathbf{x})) \tag{7.16}$$

where $\overline{\log}(c_i(\mathbf{x})) = \frac{1}{n} \sum_{\mathbf{x}' \in \mathcal{P}} \log(c_i(\mathbf{x}'))$ is the average log reflectance of the surrounding pixels or simply the geometric mean of the surrounding pixels. In other words, the retinex algorithm computes

$$o_i(\mathbf{x}) = \log\left(\frac{c_i(\mathbf{x})}{\left(\prod_{\mathbf{x}' \in \mathcal{P}} c_i(\mathbf{x}')\right)^{\frac{1}{n}}}\right). \tag{7.17}$$

If we assume for a moment that the illuminant were uniform across the image, i.e. $c_i(\mathbf{x}) = R_i(\mathbf{x})L_i$, then we obtain

$$o_i(\mathbf{x}) = \log c_i(\mathbf{x}) - \frac{1}{n} \sum_{\mathbf{x}' \in \mathcal{P}} \log(c_i(\mathbf{x}')) \tag{7.18}$$

$$= \log R_i(\mathbf{x}) + \log L_i - \log L_i - \frac{1}{n} \sum_{\mathbf{x}' \in \mathcal{P}} \log(R_i(\mathbf{x}')) \tag{7.19}$$

$$= \log R_i(\mathbf{x}) - \frac{1}{n} \sum_{\mathbf{x}' \in \mathcal{P}} \log(R_i(\mathbf{x}')). \tag{7.20}$$

Let us now assume that the reflectances are uniformly distributed over the range [0, 1]. Suppose that n reflectances are sampled along the path $\mathcal{P}$. Then we can rewrite the second term as

$$\frac{1}{n}\sum_{i=1}^{n}\log\left(\frac{i}{n}\right) = \frac{1}{n}\log\prod_{i=1}^{n}\frac{i}{n} = \frac{1}{n}\log\frac{n!}{n^n} = \log\frac{n!^{\frac{1}{n}}}{n}. \quad (7.21)$$

If we have a sufficiently large number of reflectances, i.e. $n \to \infty$, and using Stirling's formula (Weisstein 1999a),

$$\lim_{n\to\infty}\frac{(n!)^{\frac{1}{n}}}{n} = \frac{1}{e} \quad (7.22)$$

we obtain for the output color o_i

$$o_i(\mathbf{x}) = \log R_i(\mathbf{x}) + 1 \quad (7.23)$$

which is independent of the illuminant.

Land (1974) and McCann et al. (1976) developed a computer model for a 24 × 20 array of receptors. This array was used to process input from a simplified Mondrian. Random paths were set up inside this array by choosing random starting points and directions. The measured intensities along the path are processed in the same manner as described in the preceding text for the prototype, i.e. a sequential product using two adjacent intensities along the path is computed. Whenever the sequential product becomes larger than 1, it is reset to 1. The paths travel in a straight line along the initial direction. When the path hits the boundary of the array it is either reflected in a random direction or it travels along the boundary. For each position of the array, output is computed by averaging the results obtained from all paths that pass through this position of the array. McCann et al. (1976) showed that this computer model is able to predict human color perception without using a reflectance standard. They used a color Mondrian as a sample input. For each receptor they measured the radiance using filters that approximate the response curves of the cones of the retina. A human observer matched color chips from the Munsell book of colors as described in Section 2.5. The reflectance, which was estimated by the model, correlated well with the scaled integrated reflectances measured for the Munsell chips. Code for the retinex algorithm in MATLAB is given by Funt et al. (2004). See also Ciurea and Funt (2004) for dependence of the output of the retinex algorithm on its parameters.

Land performed the following experiment to confirm that the computed output of his algorithm corresponds to human color perception (Land 1983). He placed a purple circular piece of paper at the center of the Mondrian (Figure 7.6). Three projectors were used to illuminate the Mondrian. One projector was equipped with a long-wave filter, the second was equipped with a middle-wave filter and the third with a short-wave filter. A neutral-density filter was used to attenuate the light everywhere except for the circular piece of paper. This neutral-density filter was simply a slide with a hole in it. The three projectors were carefully aligned such that the projected hole of the neutral-density filter was in alignment with the circular piece of paper. Land then computed the output color using his retinex theory. According to the retinex theory the computed color of the circular piece of paper is white. This is in agreement with the color perception of a human observer. To a human observer, the piece of paper also looked white when the light outside the paper was attenuated. Land then removed the slides from the projectors. Now the piece

Figure 7.6 A purple circular piece of paper is placed on top of the color Mondrian. This piece of paper looks purple to a human observer if the whole Mondrian is brightly illuminated (a). However, if the surrounding illumination is attenuated while at the same time the illumination of the circular piece of paper is kept unchanged, then the piece of paper looks white to a human observer (b). Note that in order to illustrate the results of this experiment the circular area was actually drawn with white to produce the image shown in (b).

of paper looked purple. Land then used his retinex theory to compute the output color. According to his retinex theory the output color is purple, which is again in line with human observation.

From this experiment, we see that the surround does have an impact on color perception. If only a single patch is illuminated brightly, then this patch is assumed to be white as no other color information is available. However, if the surround is also illuminated then the viewed scene becomes sufficiently complex such that color constant descriptors can now be computed.

A drawback of this formulation of Land's retinex theory is the use of a threshold (Marr 1982). A threshold is used to distinguish between a smooth transition due to a varying illumination and a color edge, which is due to a change of reflectance. In practice, it is quite difficult to select a suitable threshold. If the threshold operation is omitted, then the algorithm assumes a single uniform illuminant. In this case, all changes of the measured intensity are treated as a change of reflectance.

Land (1983) reports that the viewed image can be surprisingly far out of focus without an impact on color perception. However, in this case a sharp color edge turns into a smooth color gradient. It therefore seems unlikely that human color perception uses some type of threshold. Also, Marr (1982) notes that one is able to see a rainbow. If the retinex algorithm were used by the visual system, one would assume that the rainbow were thresholded out because a rainbow is a smooth transition of all colors of the spectrum.

COLOR CONSTANCY UNDER NONUNIFORM ILLUMINATION

As an alternative formulation of the retinex theory, Land (1986a) proposed to take the log of the ratio of the lightness for a small central field of the region of interest to the average lightness over an extended field. Let c_i be the color at a particular point in the image and let a_i be the space average lightness at the same point. In this case, the output lightness is calculated as

$$o_i = \log\left(\frac{c_i}{a_i}\right). \tag{7.24}$$

In other words, Land subtracts the logarithm of the average lightness from the logarithm of the lightness at the current pixel

$$o_i = \log(c_i) - \log(a_i). \tag{7.25}$$

Land suggests computation of local space average lightness using a sensitivity pattern that is shown in Figure 7.7. The density of the dots varies as $\frac{1}{r^2}$ where r is the radius measured from the center of the pattern. In effect, this operation subtracts a blurred version of the input image from the original input image. This operation can also be formulated as (Moore et al. 1991)

$$o_i(x, y) = \log c_i(x, y) - \log\left(c_i \otimes \frac{1}{r^2}\right) \quad r \neq 0 \tag{7.26}$$

where $\otimes$ denotes convolution (Gonzalez and Woods 1992; Horn 1986; Jähne 2002; Jain et al. 1995; Parker 1997; Press et al. 1992; Radig 1993; Shapiro and Stockman 2001).

Although no experimental results are given for real input images, Land illustrates the method on a white square spinning on a black background. Land's algorithm is able to explain the occurrence of so called *Mach bands*, which appear to an observer when looking at the spinning disk. Note that the output computed by Land's algorithm may be positive as well as negative. Therefore, a suitable transformation must be applied to bring the output values to the range [0, 1].

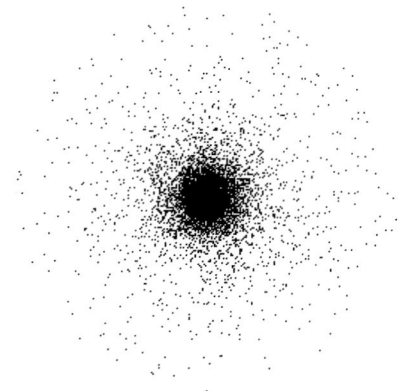

Figure 7.7 Sensitivity pattern used by Land to compute the average intensity at a given point. The density of the dots varies with $1/r^2$ where r is the radius from the center.

7.2 Computation of Lightness and Color

Image irradiance is proportional to scene radiance. We assume that we are viewing a flat surface illuminated by a single light source. In this case, scene radiance is proportional to the product of the irradiance falling on the surface and the reflectance of the surface. Let us consider light of a single wavelength first. If we assume orthographic projection that is a direct correspondence between image and surface coordinates, we have

$$c(x, y) = R(x, y)L(x, y) \tag{7.27}$$

where $c(x, y)$ is the image intensity at point (x, y) for a single band of the image, $R(x, y)$ is the reflectance at point (x, y) on the surface and $L(x, y)$ is the irradiance falling on point (x, y) on the surface. The algorithm developed by Horn (1974, 1986) consists of several stages. In the subsequent text, the output generated by stage i with $i \in \{1, ..., 5\}$ will be called o_i. Horn separates reflectance and irradiance into two components by taking the logarithm of the sensor's response. This turns the product on the right-hand side into a sum. We obtain

$$o_1(x, y) = \log c(x, y) = \log R(x, y) + \log L(x, y) \tag{7.28}$$

as the output of the first stage. It is assumed that the irradiance changes only slightly over the images, while sharp discontinuities are present where the reflectance changes. That is, we have sharp edges in the image, where the color changes, and smooth edges where the irradiance changes. The algorithm works by enhancing changes in reflectances using an edge detector. A rotational symmetric operator, the Laplacian, is used for edge detection, because edges should be detected irrespective of their orientation. The Laplacian is the lowest order linear combination of derivatives that is rotationally symmetric (Horn 1974). The output of the second stage is given by

$$o_2(x, y) = \nabla^2 (\log R(x, y) + \log L(x, y)) = \nabla^2 \log R(x, y) + \nabla^2 \log L(x, y) \tag{7.29}$$

with

$$\nabla^2 = \frac{\partial^2}{\partial x^2} + \frac{\partial^2}{\partial y^2}. \tag{7.30}$$

The first component $\nabla^2 \log R(x, y)$ is zero almost everywhere. It is nonzero only at positions where the reflectance changes. Each change in reflectance will produce a doublet edge with two closely spaced impulses with opposite sign. A positive impulse is located on the darker side of the edge and a negative impulse is located on the brighter side of the edge. The larger the change in reflectance, the larger the impulse gets. It is proportional to the difference in the logarithm of the reflectance across the edge. The second component $\nabla^2 \log L(x, y)$ will be finite because $L(x, y)$ varies smoothly over the surface. If $\log L(x, y)$ varies linearly, then $\nabla^2 \log L(x, y)$ will be zero.

A threshold operator is used to separate these two components. Let $\Theta(x)$ be a threshold operator

$$\Theta(x) = \begin{cases} x & \text{if } |x| > \theta \\ 0 & \text{if } |x| \leq \theta \end{cases} \tag{7.31}$$

where θ determines the threshold. All values larger than θ pass this operator. If we apply this threshold operator to the output of the Laplacian, we obtain

$$o_3(x, y) = \Theta(\nabla^2 \log R(x, y) + \nabla^2 \log L(x, y)) = \nabla^2 \log R(x, y) \tag{7.32}$$

with a suitably defined threshold. After the component, which is due to the illuminant, has been removed, we have a function that only depends on the reflectance. The output, after thresholding, is the second derivative of the logarithm of the reflectances,

$$o_3(x, y) = \nabla^2 \log R(x, y) = \nabla^2 R'(x, y) \tag{7.33}$$

with $R' = \log R$. This is called *Poisson's equation* (Bronstein et al. 2001; Kuypers 1990; Nolting 1992). The solution of this equation is one of the fundamental problems in electrodynamics. It can be solved using Green's function (Horn 1986). The solution to Poisson's equation is given by

$$R'(x, y) = \int\int o_3(\xi, \eta) g(x - \xi, y - \eta) d\xi d\eta \tag{7.34}$$

where $g(x, y; \xi, \eta)$ is Green's function. If we substitute $o_3(\xi, \eta)$, we obtain

$$R'(x, y) = \int\int \left(\frac{\partial^2}{\partial x^2} + \frac{\partial^2}{\partial y^2}\right) R'(\xi, \eta) g(x - \xi, y - \eta) d\xi d\eta \tag{7.35}$$

$$= \int\int R'(\xi, \eta) \left(\frac{\partial^2}{\partial x^2} + \frac{\partial^2}{\partial y^2}\right) g(x - \xi, y - \eta) d\xi d\eta. \tag{7.36}$$

If we now choose $g(r) = \frac{1}{4\pi} \log(r^2)$ with $r^2 = x^2 + y^2$, we get

$$R'(x, y) = \int\int R'(\xi, \eta) \frac{1}{4\pi} \left(\frac{\partial^2}{\partial x^2} + \frac{\partial^2}{\partial y^2}\right) \log\left((x - \xi)^2 + (y - \eta)^2\right) d\xi d\eta. \tag{7.37}$$

The expression $\frac{1}{4\pi} \nabla^2 \log(x^2 + y^2)$ is equivalent to the delta function $\delta(x, y)$. This gives us

$$R'(x, y) = \int\int R'(\xi, \eta) \delta(x - \xi, y - \eta) d\xi d\eta = R'(x, y) \tag{7.38}$$

which is the desired result. Therefore, we need to convolve the output of the Laplacian with the Kernel $\frac{1}{4\pi} \log(r^2)$ to calculate the reflectances.

$$o_4(x, y) = \frac{1}{4\pi} \int\int \left(\frac{\partial^2}{\partial x^2} + \frac{\partial^2}{\partial y^2}\right) R'(\xi, \eta) \log\left((x - \xi)^2 + (y - \eta)^2\right) d\xi d\eta \tag{7.39}$$

After that, we exponentiate and obtain an estimate of the reflectances.

$$o_5(x, y) = \exp(o_4(x, y)) \tag{7.40}$$

Unfortunately, it is not possible to completely recover the original reflectances. For instance, scaling and a constant factor cannot be recovered. Any solution to the equation $\nabla f(x, y) = 0$ can be added to the input without changing the result of the algorithm. One has to define suitable boundary conditions. Because of this, Horn calls the result lightness

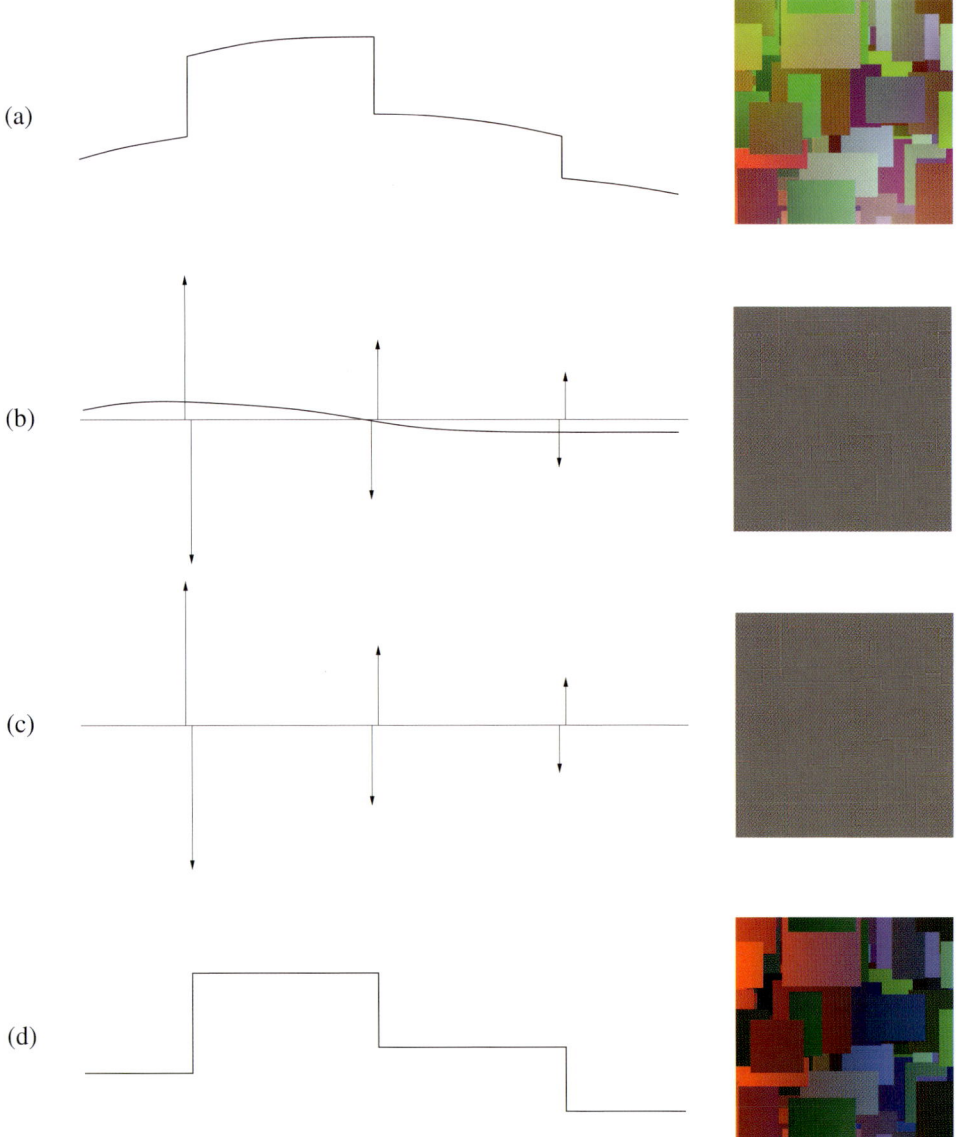

Figure 7.8 (a) Logarithm of input brightness, (b) edge detection using a Laplacian, (c) thresholding operation, and (d) integration. The impulses shown in (b) and (c) are caused by a change in reflectance. The thresholding operation removes small values that are caused by a change in the lighting conditions. The one-dimensional case is shown on the left (Reproduced by permission of The MIT Press. Horn BKP 1986 Robot Vision, The MIT Press, Cambridge, MA.) The two-dimensional case is shown on the right.

COLOR CONSTANCY UNDER NONUNIFORM ILLUMINATION 157

instead of reflectance. Horn normalizes the result by assuming that there is one point in the image that reflects all of the light, i.e. there is a white patch. Therefore, he adds a constant such that the brightest value computed in layer 4 is zero. This is equivalent to a normalization by multiplying the output of layer 5 by a constant such that the maximum output of layer 5 is one.

To summarize the steps of the algorithm, we first take the logarithm of the signal measured by the sensors, perform edge detection using a Laplacian, apply a thresholding operation, integrate twice, and then exponentiate the result. Finally, a normalization step is applied, which transforms the result to the desired output range. Figure 7.8 shows the individual steps of Horn's algorithm for the one-dimensional case. The two-dimensional case is shown on the right. The first and last steps have been omitted.

So far, we have only considered the continuous case. However, we always work with discretized images. Therefore, we must also discretize the operators we use. Care must be taken at the boundary of the image. This important point was made by Blake (1985). Horn (1974) originally assumed that the viewed scene is embedded inside a constant background. If this assumption does not hold, then the Laplacian must be adjusted at the border. Similarly, the integration step must also be adjusted at the border. Figure 7.9 shows two discretized versions of the Laplacian and the adjustments which are necessary at the border of the image. Differences between the center and surrounding values are only computed

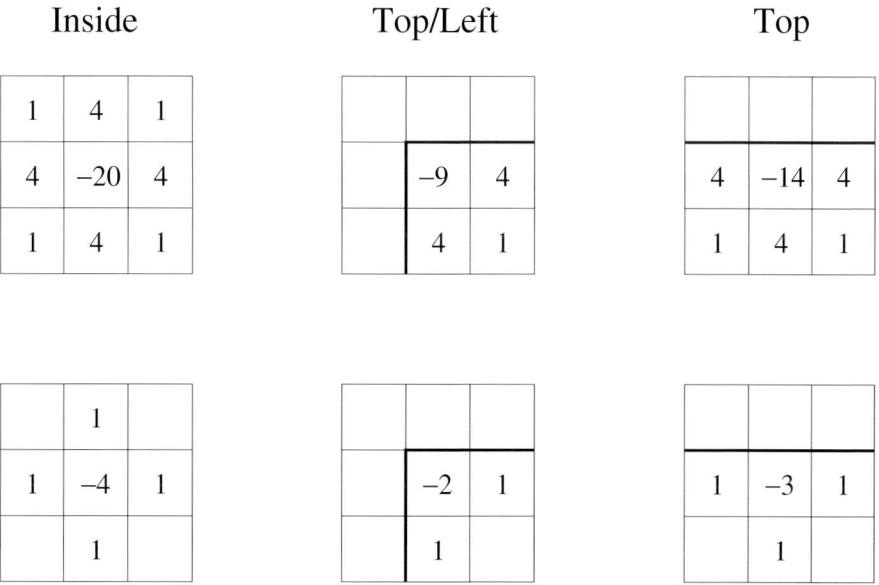

Figure 7.9 Two discretized versions of the Laplace operator. Care must be taken at the border of the image. The operators shown in the first column are the standard form of the Laplace operator. The second column shows the Laplace operator to be applied at the top/left corner of the image. The two rightmost operators are used at the top border of the image.

if the particular pixel values are available. If the pixel values are not available then no contribution is made to the value computed by the Laplacian.

In the continuous case, the Laplacian produces infinite peaks at positions in the image where the reflectance changes. This makes it relatively easy to select a threshold. In the discrete case, one has no choice but to choose some threshold. We must separate the change due to a change in reflectance from the change due to a change of the illuminant. We have assumed that the illuminant produces smooth changes in image intensity. If this change is linear, the Laplacian will produce an output of zero.

Let Δ be the smallest change in intensity that can be resolved by the sensor. In the discrete case, the output of the Laplacian will be zero, provided that the change in image intensity from one pixel to the next is Δ (See (a) and (b) of Figure 7.10). If the change in image intensity is even smaller than that, the change will materialize as a step function (See (c) and (d) of Figure 7.10). The output of the Laplacian is produced in stage o_2. If we apply the Laplacian shown on the top of Figure 7.9, then we have $|o_2(x, y)| \leq 6\Delta$ for a horizontal or a vertical step (c) and $|o_2(x, y)| \leq 9\Delta$ for a diagonal step (d). Thus,

$$|o_2(x, y)| \leq 10\Delta \tag{7.41}$$

is a suitable threshold to separate a change due to a change of the illuminant from a change that is due to a change in reflectance for the discrete case. If we use the Laplacian shown

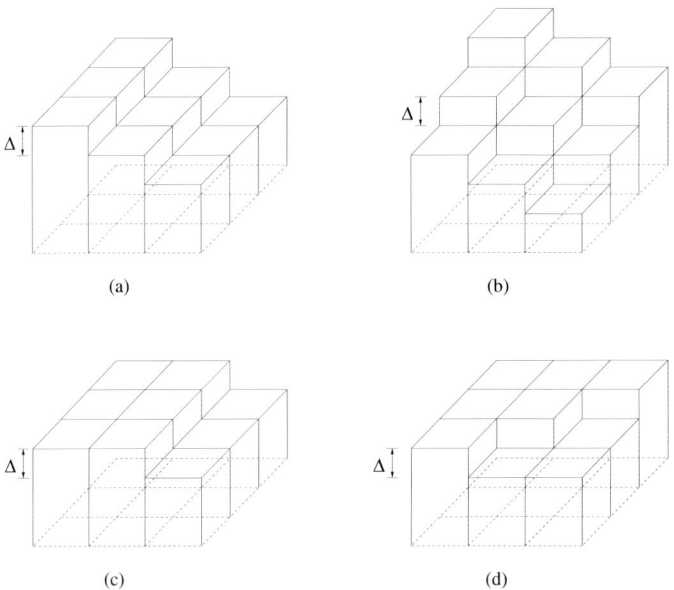

Figure 7.10 (a) Linear horizontal change in image brightness, (b) linear diagonal change in image brightness, (c) horizontal step due to a small change in image brightness, (d) diagonal step due to a small change in image brightness. The smallest change in image brightness resolved by the sensor is Δ.

on the bottom of Figure 7.9 we obtain $|o_2(x, y)| \leq 1\Delta$ for a horizontal or a vertical step (c) and $|o_2(x, y)| \leq 2\Delta$ for a diagonal step (d). In this case,

$$|o_2(x, y)| \leq 3\Delta \qquad (7.42)$$

would be an appropriate threshold.

Taking the logarithm of the input values makes it difficult to select a suitable threshold. If we take the logarithm, then the threshold depends on the intensity of the pixel. Let c be the intensity of the pixel and Δ a small change in intensity. If we assume the Laplacian shown on the top of Figure 7.9, then we obtain $6\log(c + \Delta) - 6\log(c) = 6\log(1 + \frac{\Delta}{c})$ for a horizontal or a vertical step (c) and $8\log(c + \Delta) - 8\log(c) = 8\log(1 + \frac{\Delta}{c})$ for a diagonal step (d). So we need to select an adaptive threshold that depends on the intensity of the image pixel. A uniform threshold will not work, because the Laplacian will produce a larger output for small intensity changes at low intensities compared to small intensity changes at high intensities. Similar to the linear case,

$$|o_2(x, y)| \leq 9\log\left(1 + \frac{\Delta}{c}\right) \qquad (7.43)$$

is a suitable threshold to separate a change due to a change of the illuminant from a change that is due to a change in reflectance. Similarly,

$$|o_2(x, y)| \leq 3\log\left(1 + \frac{\Delta}{c}\right) \qquad (7.44)$$

is a suitable threshold for the Laplacian shown on the bottom of Figure 7.9.

Blake (1985) suggested another variant on how to perform the thresholding operation. He suggested splitting the application of the Laplace operator followed by a thresholding into three operations. First, the gradient of the input image is computed. Then, a thresholding operation is applied. Finally, another derivative operator is applied to compute the second derivative. This has the advantage that the threshold operation is always applied to two adjacent image pixels. In contrast, Horn's method applies the threshold operation to the output of the Laplace operator. As we have just seen, we need a threshold that is a little larger than 2Δ to eliminate diagonal changes of the illuminant. However, if we use a threshold larger than 2Δ, we are no longer able to distinguish between a change in reflectance of size 2Δ. If the threshold is computed after the first derivative, then we can still distinguish the two cases. Another advantage of applying the threshold operation after taking the first derivative is that the Laplacian is actually invertible.

After thresholding, one needs to undo the Laplacian. In the continuous case, we convolved the output with the function $\frac{1}{4\pi}\log r^2$. In the discrete case, we have to solve the discretized version of

$$o_3(x, y) = \left(\frac{\partial^2}{\partial x^2} + \frac{\partial^2}{\partial y^2}\right) R'(x, y) \qquad (7.45)$$

which is equivalent to

$$o_3(x, y) = -20R'(x, y) + 4\left(R'(x + 1, y) + R'(x, y + 1) + R'(x - 1, y) + R'(x, y - 1)\right)$$
$$+ (R'(x + 1, y + 1) + R'(x - 1, y + 1) + R'(x - 1, y - 1) + R'(x + 1, y - 1)) \qquad (7.46)$$

for the discretized Laplacian shown on the top of Figure 7.9. This version of the Laplacian was used by Horn (1986). One can solve this equation using Jacobi's method (Bronstein et al. 2001). If we assume that the values $R'(x, y)$ are known for $x' \neq x$ and $y' \neq y$, we can solve for $R'(x, y)$.

$$R'(x, y) = \frac{1}{5} \left(R'(x+1, y) + R'(x, y+1) + R'(x-1, y) + R'(x, y-1) \right)$$
$$+ \frac{1}{20} \left(R'(x+1, y+1) + R'(x-1, y+1) + R'(x-1, y-1) + R'(x+1, y-1) \right)$$
$$- \frac{1}{20} o_3(x, y) \quad (7.47)$$

This equation is applied iteratively. The output is computed in layer 4. Let o_4^{n+1} be the output of the n-th iteration, then we obtain

$$o_4^{n+1}(x, y) = \frac{1}{5} \left(o_4^n(x+1, y) + o_4^n(x, y+1) + o_4^n(x-1, y) + o_4^n(x, y-1) \right)$$
$$+ \frac{1}{20} \left(o_4^n(x+1, y+1) + o_4^n(x-1, y+1) + o_4^n(x-1, y+1) + o_4^n(x-1, y-1) \right)$$
$$- \frac{1}{20} o_3(x, y). \quad (7.48)$$

For the smaller version of the Laplacian (shown on the bottom of Figure 7.9), the equations are even simpler. In this case, we obtain

$$o_4^{n+1}(x, y) = \frac{1}{4} \left(o_4^n(x+1, y) + o_4^n(x, y+1) + o_4^n(x-1, y) + o_4^n(x, y-1) \right) - \frac{1}{4} o_3(x, y). \quad (7.49)$$

If this method is implemented on a sequential computer, the iteration is stopped when the difference between two successive estimations of the reflectance drops below a threshold ϵ.

$$\forall_{x, y} |o_4^{n+1}(x, y) - o_4^n(x, y)| < \epsilon \quad (7.50)$$

In a parallel hardware implementation, the algorithm can be run indefinitely and the resulting estimate could be used for further processing. The complete algorithm for the discrete case is shown in Figure 7.11. First, edges are enhanced with a Laplacian. This produces spikes at positions where the reflectance changes. Next, a thresholding operation is applied. This removes any small changes that might be due to a change of the illuminant. Finally, one integrates over the spikes iteratively.

Using Jacobi's method to compute the inverse of the Laplacian is rather slow. Faster convergence may be achieved using successive over-relaxation (SOR) (Bronstein et al. 2001; Demmel 1996). The iterative solver can also be written in the Gauss–Seidel formulation where already computed results are reused.

$$o_4^{n+1}(x, y) = \frac{1}{4} \left(o_4^n(x+1, y) + o_4^n(x, y+1) + o_4^{n+1}(x-1, y) + o_4^{n+1}(x, y-1) - o_3(x, y) \right) \quad (7.51)$$

Note that the first two terms on the right-hand side of the preceding equation are from iteration n and the last two terms are from the current iteration. The Gauss-Seidel method

COLOR CONSTANCY UNDER NONUNIFORM ILLUMINATION

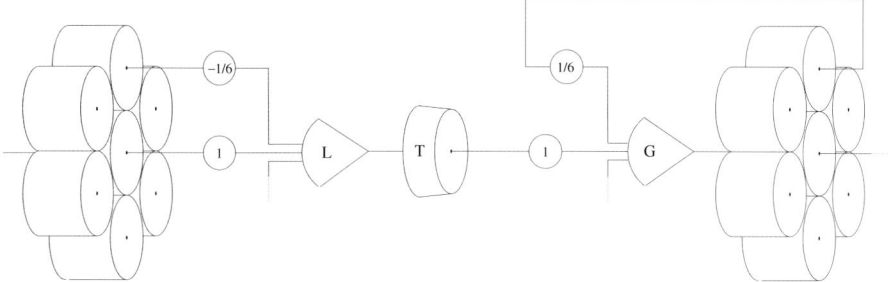

Figure 7.11 Parallel implementation of Horn's algorithm for color constancy. The individual steps are edge enhancement using a Laplacian and a threshold operation, followed by integration (Reprinted from Horn BKP 1974 Determining lightness from an image. Computer Graphics and Image Processing, 3, 277-299. Copyright 1974 with permission from Elsevier.)

can also be written in its corrective form

$$o_4^{n+1}(x, y) = o_4^n(x, y) + \frac{1}{4}\Big(o_4^n(x + 1, y) + o_4^n(x, y + 1) \tag{7.52}$$

$$+ o_4^{n+1}(x - 1, y) + o_4^{n+1}(x, y - 1) - o_3(x, y)\Big) - o_4^n(x, y)$$

$$= o_4^n(x, y) + \text{Corrective-Term}(x, y) \tag{7.53}$$

with

$$\text{Corrective-Term}(x, y) = \frac{1}{4}\Big(o_4^n(x + 1, y) + o_4^n(x, y + 1) \tag{7.54}$$

$$+ o_4^{n+1}(x - 1, y) + o_4^{n+1}(x, y - 1) - o_3(x, y)\Big) - o_4^n(x, y).$$

The corrective term basically tells us the direction we need to move toward in order to get a better solution. The idea behind SOR is to move farther in the direction of the corrective term. The size of the corrective term can be varied by introducing a factor w.

$$o_4^{n+1}(x, y) = o_4^n(x, y) + w\text{Corrective-Term}(x, y) \tag{7.55}$$

$$= (1 - w)o_4^n(x, y) - \frac{w}{4}o_3(x, y) \tag{7.56}$$

$$\frac{w}{4}\Big(o_4^n(x + 1, y) + o_4^n(x, y + 1) + o_4^{n+1}(x - 1, y) + o_4^{n+1}(x, y - 1)\Big)$$

Convergence is achieved for $w \in (0, 2)$. Use of $w < 1$ corresponds to under-relaxation, use of $w > 1$ is known as *over-relaxation*.

Terzopoulos (1986) has shown how to speed up the computation using multi-grid relaxation methods. He uses a hierarchy of multi-resolution grids where data propagates upward as well as downward through the hierarchy. The use of a multi-resolution pyramid allows information to propagate faster over larger distances.

The Poisson equation can also be solved using the discrete sinc transform (Zenger and Bader 2004). The discrete sine transform of a two-dimensional function $f(x, y)$ defined on a $n \times n$ grid is given by (Frigo and Johnson 2004; Jähne 2002; Zenger and Bader 2004)

$$f(x,y) = 2\sum_{k=1}^{n}\sum_{l=1}^{n} \tilde{f}(k,l) \sin\frac{xk\pi}{n+1} \sin\frac{yl\pi}{n+1}. \tag{7.57}$$

where $\tilde{f}$ are the coefficients in the frequency domain. Since we are looking for the solution to

$$o(x,y) = \left(\frac{\partial^2}{\partial x^2} + \frac{\partial^2}{\partial y^2}\right) R'(x,y) \tag{7.58}$$

$$= R'(x-1,y) + R'(x+1,y) + R'(x,y-1) + R'(x,y+1)$$
$$-4R'(x,y), \tag{7.59}$$

we now substitute the discrete sine transform of both $o(x,y)$ and $R'(x,y)$ and see what we obtain.

$$\sum_{k=1}^{n}\sum_{l=1}^{n} \tilde{o}(k,l) \sin\frac{xk\pi}{n+1} \sin\frac{yl\pi}{n+1} =$$

$$\sum_{k=1}^{n}\sum_{l=1}^{n} \tilde{R}'(k,l) \sin\frac{(x-1)k\pi}{n+1} \sin\frac{yl\pi}{n+1} + \sum_{k=1}^{n}\sum_{l=1}^{n} \tilde{R}'(k,l) \sin\frac{(x+1)k\pi}{n+1} \sin\frac{yl\pi}{n+1}$$

$$-4 \sum_{k=1}^{n}\sum_{l=1}^{n} \tilde{R}'(k,l) \sin\frac{xk\pi}{n+1} \sin\frac{yl\pi}{n+1}$$

$$+ \sum_{k=1}^{n}\sum_{l=1}^{n} \tilde{R}'(k,l) \sin\frac{xk\pi}{n+1} \sin\frac{(y-1)l\pi}{n+1} + \sum_{k=1}^{n}\sum_{l=1}^{n} \tilde{R}'(k,l) \sin\frac{xk\pi}{n+1} \sin\frac{(y+1)l\pi}{n+1}$$
$$\tag{7.60}$$

Using the trigonometric identities $\sin(A+B) = \sin A \cos B + \cos A \sin B$, and $\sin(A-B) = \sin A \cos B - \cos A \sin B$, one obtains

$$\sum_{k=1}^{n}\sum_{l=1}^{n} \tilde{o}(k,l) \sin\frac{xk\pi}{n+1} \sin\frac{yl\pi}{n+1} =$$

$$2 \sum_{k=1}^{n}\sum_{l=1}^{n} \tilde{R}'(k,l) \sin\frac{xk\pi}{n+1} \cos\frac{k\pi}{n+1} \sin\frac{yl\pi}{n+1}$$

$$-4 \sum_{k=1}^{n}\sum_{l=1}^{n} \tilde{R}'(k,l) \sin\frac{xk\pi}{n+1} \sin\frac{yl\pi}{n+1}$$

$$2 \sum_{k=1}^{n}\sum_{l=1}^{n} \tilde{R}'(k,l) \sin\frac{xk\pi}{n+1} \sin\frac{yl\pi}{n+1} \cos\frac{l\pi}{n+1}. \tag{7.61}$$

COLOR CONSTANCY UNDER NONUNIFORM ILLUMINATION

This equation is fulfilled if

$$\tilde{o}(k, l) = \left(2\cos\frac{k\pi}{n+1} + 2\cos\frac{l\pi}{n+1} - 4\right)\tilde{R}'. \qquad (7.62)$$

Thus, the Poisson equation may be solved by applying the discrete sine transform, multiplying the result by

$$\frac{1}{2\cos\frac{k\pi}{n+1} + 2\cos\frac{l\pi}{n+1} - 4} \qquad (7.63)$$

and then transforming the result back to the spatial domain by applying another sine transform.

Horn (1986) shows how to implement the inverse operation efficiently in hardware using a network of resistors. The integration is done in parallel. The output voltage measured at each grid point is the desired result. Each grid point is connected to its four immediate neighbors by a resistor of resistance R. It is also connected to the four diagonal neighbors by resistors of resistance $4R$. We will see in a moment that this network of resistors performs exactly the desired computation. Using Kirchhoff's law (Orear 1982), we can calculate the current at each grid point. As shown in Figure 7.12 we only consider a single node of this network. Let the voltage at the center be $\bar{V}$. Let the voltage at the n neighboring points be V_i. Then the current in the i-th resistor is $I_i = (V_i - \bar{V})/R_i$, where R_i is the resistance of resistor i. The current at each grid point sums to zero.

$$\sum_{i=1}^{n} I_i = 0 \qquad (7.64)$$

$$\sum_{i=1}^{n} \left(\frac{V_i - \bar{V}}{R_i}\right) = 0 \qquad (7.65)$$

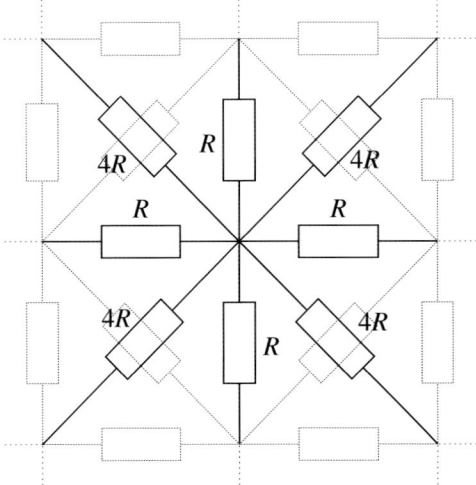

Figure 7.12 A network of resistors can be used to perform the last step of the algorithm, i.e. the integration. Each point of the grid is connected to its immediate neighbors by a resistor of resistance R. It is also connected to its diagonal neighbors by a resistor of resistance $4R$.

If we also inject a current I at the center point, then we have,

$$\sum_{i=1}^{n}\left(\frac{V_i - \bar{V}}{R_i}\right) + I = 0. \tag{7.66}$$

If we now solve for $\bar{V}$, we obtain,

$$\bar{V} = \frac{1}{\sum_{i=1}^{n}\frac{1}{R_i}} \left(\sum_{i=1}^{n}\frac{V_i}{R_i} + I\right). \tag{7.67}$$

Each node is connected to eight neighbors, i.e. $n = 8$. If the i-th resistor connects to one of the immediate neighbors, we have $R_i = R$. If the i-th resistor connects to one of the diagonal neighbors, we have $R_i = 4R$. Therefore, we obtain

$$\sum_{i=1}^{n}\frac{1}{R_i} = \left(4\frac{1}{R} + 4\frac{1}{4R}\right) = \frac{5}{R} \tag{7.68}$$

for the sum of inverse resistances. Let the injected current I be equal to $-\frac{1}{4R}o_3(x, y)$. Then the voltage at node (x, y) is given by

$$\bar{V}(x, y) = \frac{R}{5}\left(\frac{V(x+1, y)}{R} + \frac{V(x, y+1)}{R} + \frac{V(x-1, y)}{R} + \frac{V(x, y-1)}{R}\right)$$
$$+ \frac{R}{5}\left(\frac{V(x+1, y+1)}{4R} + \frac{V(x-1, y+1)}{4R} + \frac{V(x-1, y-1)}{4R} + \frac{V(x+1, y-1)}{4R}\right)$$
$$- \frac{R}{5}\frac{1}{4R}o_3(x, y), \tag{7.69}$$

which is exactly the expression that performs the required integration as described earlier.

$$\bar{V}(x, y) = \frac{1}{5}\left(V(x+1, y) + V(x, y+1) + V(x-1, y) + V(x, y-1)\right)$$
$$+ \frac{1}{20}\left(V(x+1, y+1) + V(x-1, y+1) + V(x-1, y-1) + V(x+1, y-1)\right)$$
$$- \frac{1}{20}o_3(x, y) \tag{7.70}$$

If we compare Horn's algorithm to the original formulation of Land's retinex theory, we see that Horn's algorithm is just an alternative two-dimensional formulation of Land's retinex theory. Formal connections between the color constancy algorithms of Land, Horn, and Blake were established by Hurlbert (1986). Land suggested choosing random paths for each input pixel. He used logarithmic receptors to measure intensity for the three wavelengths red, green, and blue (RGB). Differences between adjacent receptors are computed and the result is summed along the path. Only values that are larger than a threshold are summed up. We now see that the application of the Laplacian corresponds to the differences between adjacent receptors. The threshold operation only keeps values larger than the threshold. Spatial averaging replaces the sum along the path. Results for Horn's algorithm with a threshold value of zero are shown in Figure 7.13.

COLOR CONSTANCY UNDER NONUNIFORM ILLUMINATION 165

Figure 7.13 Output images for Horn's algorithm. The threshold was set to zero.

Figure 7.14 The first image is the input image. The second image shows the output of the Laplacian. The third image shows the thresholded Laplacian. Additionally, 1% of the pixels has been set to zero. The integrated output image is shown on the right. Horn's algorithm relies on the exact separation between changes of the reflectance and changes of the illuminant.

Like Land's retinex theory, Horn's algorithm suffers from the same problems. The use of a threshold is not easy in practice. One explicitly has to define a threshold that will distinguish a change of reflectance from a change of the illuminant. If the threshold is too low, changes of the illuminant will be taken as changes of the reflectance. If the threshold is too high, then changes of the reflectance will be removed accidentally. However, Horn's algorithm relies on the ability to clearly separate a change of reflectance from a change of the illuminant. In Figure 7.14 we show what happens if some additional pixels are set to zero during thresholding. Here, we have assumed that 1% of the pixels are accidentally set to zero. Now the areas of the Mondrian are no longer sharply delineated. There are holes in the boundary and the spatial averaging no longer works. Since an exact separation between the two types of changes is impossible for natural images Horn's algorithm can only be used if a single uniform illumination is given. In this case Horn's algorithm is a variant of the white patch retinex algorithm. We have already discussed the simplified version of Horn's algorithm for a uniform illuminant in Section 6.3.

For the Mondrian images, it was easy to select a threshold. A change of reflectance could be clearly separated from a change of the illuminant. This does not hold for real input images. Figure 7.15 shows the output images obtained when the threshold value is set to six. Some reflectance changes are accidentally removed. After applying the inverse

Figure 7.15 Output images for Horn's algorithm. The threshold was set to six.

Figure 7.16 Output images for Blake's algorithm. The threshold was set to three.

Laplacian, the result no longer resembles the input image. Results for Blake's version, which computes the first derivative, applies the threshold operation and then computes the second derivative instead of thresholding the output of the Laplacian, and are shown in Figure 7.16. By thresholding the first derivative, the Laplacian can still be inverted. The result is much more pleasing.

7.3 Hardware Implementation of Land's Retinex Theory

Moore et al. (1991) have implemented a version of Land's retinex algorithm in hardware using analog CMOS VLSI resistive grids (see also Moore and Allman (1991)). The resistive grids are used to compute a blurred version of the input image. Moore et al. propose to take the logarithm of the input intensities for each channel. Next, a blurred image of the logarithm of the input intensities is computed. Then, the blurred image is subtracted from the logarithm of the input intensities $\log(c_i)$, i.e.

$$o'_i(x, y) = \log(c_i(x, y)) - \log(c_i) \otimes e^{-\frac{|r|}{\sigma}} \qquad (7.71)$$

where o'_i with $i \in \{r, g, b\}$ are the color constant descriptors, $\otimes$ denotes convolution, r is the distance from the current pixel and σ defines the extent of the convolution. This algorithm is quite similar to the algorithm proposed by Land (see Section 7.1) except that a different method is used to compute a blurred surrounding. Note that the output is positive if the logarithm of the current intensity is above the intensity of the blurred image and negative if it is below the intensity of the blurred image. To bring the output values back into the range [0, 1], Moore et al. (1991) first compute the minimum and maximum values over all color channels. They subtract the minimum value from the output color o' and then rescale the result to the full range, i.e. they compute output colors o_i

$$o_i(x, y) = \frac{o'_i(x, y) - \min}{\max - \min} \quad (7.72)$$

where

$$\max = \max\{o'_i(x, y)| \text{ for all } (x, y) \text{ and } i \in \{r, g, b\} \} \quad (7.73)$$

$$\min = \min\{o'_i(x, y)| \text{ for all } (x, y) \text{ and } i \in \{r, g, b\} \}. \quad (7.74)$$

Figure 7.17 shows the resistive grid implementation. The resistive grids were built using subthreshold analog CMOS VLSI. Each grid contained approximately 60 000 transistors. Images of size 50 × 50 could be processed. Larger images had to be down-sampled before processing. In their hardware implementation, Moore et al. did not take the logarithm of the video signal before processing. They only took the logarithm when they simulated the algorithm in software. Results for this algorithm are shown in Figure 7.18.

A drawback of this algorithm, as noted by Moore et al. (1991), is color induction across edges. Moore et al. give an example of a uniformly colored region next to a black region. If one has a large uniform area of a particular color next to a dark region, then this color is subtracted from the dark region where it will induce the complementary color. If we have a green region next to a black region then the blurred image will be green. If this green blurred image is subtracted from the black region, this region will appear to be colored with red and blue, i.e. the complementary colors of green. Also, if one has a large uniform region in the image, this region will appear gray after the algorithm has been applied. In the

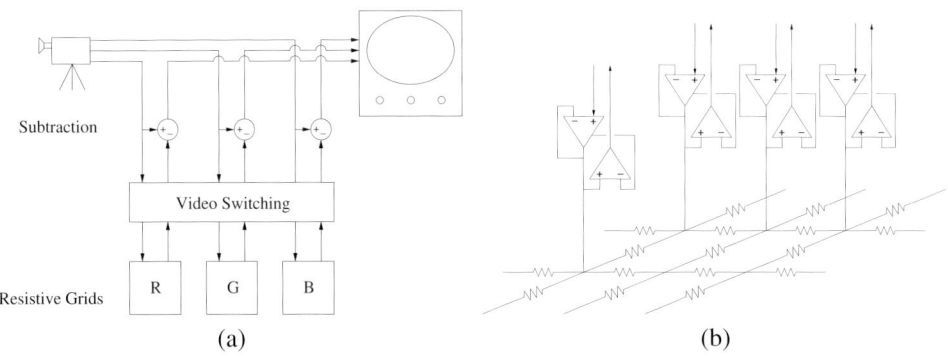

Figure 7.17 Moore's hardware implementation of Land's algorithm (Reproduced by permission of IEEE. Moore A and Allman J and Goodman RM 1991 A real-time neural system for color constancy. IEEE Transactions on Neural Networks, IEEE, 2(2), 237-247, March). The output of the camera is smoothed using three separate resistive grids (a). A single resistive grid is shown in (b). The smoothed image is then subtracted from the original image.

Figure 7.18 Results for the algorithm of Moore et al.

center of the region, the color will be the same as for the blurred image. Thus, the method will compute an output of zero, which will appear gray after the final transformation has been applied.

Moore et al. (1991) have also developed an extension to Land's retinex algorithm in order to reduce the effect of color induction. The modified algorithm computes the magnitude of the first spatial derivative. This derivative is also smoothed with a resistive grid. A blurred version of the input image is computed as before. The smoothed version of the derivative is used to modulate the blurred version of the input image before it is

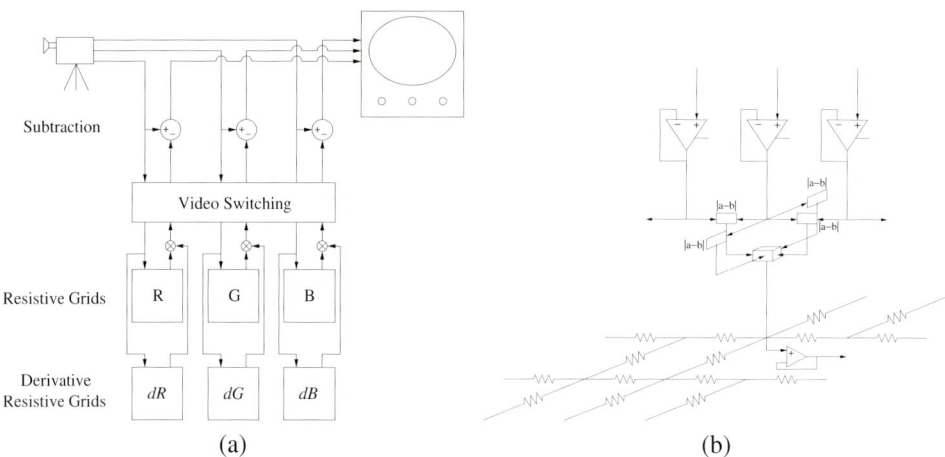

Figure 7.19 Moore's extended algorithm (Reproduced by permission of IEEE. Moore A and Allman J and Goodman RM 1991 A real-time neural system for color constancy. IEEE Transactions on Neural Networks, IEEE, 2(2), 237-247, March). Apart from the input image, the magnitude of the local spatial derivative is also smoothed for each color channel on a separate resistive grid (a). The resistive grids that smooth the derivatives are denoted with dR, dG, and dB. The smoothed derivatives are used to modulate the output from the resistive grids that compute the blurred image before the data is subtracted from the original image. The resistive grid that is used to compute the smoothed derivatives is shown in (b).

COLOR CONSTANCY UNDER NONUNIFORM ILLUMINATION

Figure 7.20 Results for the extended algorithm of Moore et al.

subtracted from the input image.

$$o'_i(x, y) = \log(c_i(x, y)) - \left(\log(c_i) \otimes e^{-\frac{|r|}{\sigma}}\right)\left(|\partial \log c(x, y)| \otimes e^{-\frac{|r|}{\sigma}}\right) \tag{7.75}$$

The hardware implementation of this algorithm is shown in Figure 7.19. Two resistive grids are used for each color channel. The first resistive grid computes a blurred version of the input image. The second resistive grid receives the average absolute value of the first spatial derivative of the original image as an input. The second resistive grid is shown in (b) of Figure 7.19. The average of the absolute difference between neighboring pixel values (denoted by $|a - b|$) is fed into the resistive grid where it is averaged. The output of the second resistive grid is used to modulate the output of the first resistive grid before it is subtracted from the input image. With this algorithm, Moore et al. try to reduce the effect of color induction. In areas with many edges, the algorithm works exactly as the previous algorithm. In large uniform areas, however, the blurred version of the absolute value of the first derivative is almost zero. Here, the colors of the input image are passed right through to the output colors. In such regions, the algorithm does not perform any color correction. Results for the extended retinex algorithm are shown in Figure 7.20.

7.4 Color Correction on Multiple Scales

Rahman et al. (1999) extended the algorithm of Moore et al. They use a Gaussian to compute the blurred image and perform color correction on multiple scales. This method is not only used for color constancy but also for dynamic range compression. Output color o_i is computed as

$$o_i = \sum_{j=1}^{n} w_j \left(\log c_i(x, y) - \log\left(c_i(x, y) \otimes G_j(x, y)\right)\right) \tag{7.76}$$

where $\otimes$ denotes convolution, w_j are weights and n is the number of scales over which color correction is performed. The weights w_j can be used to emphasize a correction step for a particular scale. If no preference is given, then w_j is set to $\frac{1}{n}$. All functions G_j with $j \in \{1, ..., n\}$ are chosen such that

$$k_j \int\int G_j(x, y)dxdy = 1 \tag{7.77}$$

for some constant k_j. Typically, a Gaussian $G_j = e^{-\frac{r^2}{\sigma_j^2}}$ with $r = \sqrt{x^2 + y^2}$ is used. The parameter σ_j defines the extent over which local space average color will be computed.

This formulation is similar to Land's alternative version of his retinex theory where each term of the sum computes

$$o = \log(c) - \log(a_{\sigma_j}), \tag{7.78}$$

and a_{σ_j} is the local space average color for some parameter σ_j. Rahman et al. (1999) also discuss the possibility of first taking the logarithm of the input pixel and then convolving the image with a Gaussian. In this case, the convolution does not compute a weighted sum but a weighted product. Let c and c' be two pixel colors. If we first average the values and then take the logarithm, then we obtain $\log\left(\frac{c+c'}{2}\right)$. However, if we take the logarithm first and then average the values, we obtain $\log\left(\sqrt{cc'}\right)$. Therefore, if we first convolve the image with a Gaussian and then take the logarithm, we compute local space average color using the arithmetic mean. If we first take the logarithm and then convolve the image with a Gaussian, we compute local space average color using the geometric mean (Bronstein et al. 2001).

Finally, for display purposes, the output o must be transformed to the range [0, 1] or [0, 255]. Rahman et al. (1999) apply an identical offset and gain function to all pixels. In addition, Rahman et al. propose to multiply the output o_i by the following factor

$$\ln\left(\frac{kc_i(x, y)}{\sum_{i \in \{r,g,b,\}} c_i(x, y)}\right) \tag{7.79}$$

where k is a constant. The constant k is used to control the perceived color saturation. Figure 7.21 shows the results for the algorithm described in Rahman et al. (1999).

7.5 Homomorphic Filtering

Stockham, Jr. (1972) suggested in 1972 to perform image processing in the context of a visual model of the human visual system. Stockham worked with single band images. Faugeras (1979) extended the idea to color image processing. In particular, he suggested the use of homomorphic filtering for color image enhancement. In homomorphic filtering (see also Parker (1997)) the input image is transformed to a space where the desired operation is easier to perform. The slowly varying component of a color channel can be attributed to a slowly varying illuminant whereas the rapidly varying component of a color channel is likely be caused by a change in reflectance. In trying to obtain a color constant image, one simply has to remove the low-frequency component of the image. This can readily be done in frequency space.

Thus, in order to remove illumination effects we first need to transform the input image to frequency space using a Fourier transform. The one-dimensional continuous Fourier transform is defined as (Bronstein et al. 2001)

$$f(\omega) = \int f(x) e^{-2\pi i \omega x} dx. \tag{7.80}$$

This transform can be considered a coordinate transformation where spatial coordinates are transformed into frequencies. Any given curve can be described by adding sine and

COLOR CONSTANCY UNDER NONUNIFORM ILLUMINATION

Figure 7.21 Results for the algorithm developed by Rahman et al. The first row shows the results without using the gain factor. The images shown in the second row show the results when the gain factor is used with a value of $k = 5$. In both cases, three Gaussians were applied with $\sigma_j \in [1.0, 0.16, 0.03]$.

cosine waves, possibly infinitely many, with different amplitudes. The Fourier transform calculates these amplitudes $f(\omega)$. The inverse of the Fourier transform is given by

$$f(x) = \frac{1}{2\pi} \int f(\omega) e^{2\pi i \omega x} d\omega. \tag{7.81}$$

Using the Fourier transform, we can transform a given image to frequency space and back. Since we are working with two-dimensional discrete images, a two-dimensional discrete Fourier transformation is used. Let n_x and n_y be the width and the height of the image; then the two-dimensional discrete Fourier transformation is given by

$$c(\omega, \nu) = \frac{1}{\sqrt{n_x n_y}} \sum_{x=0}^{n_x} \sum_{y=0}^{n_y} c(x, y) e^{-2\pi i \omega x} e^{-2\pi i \nu y} \tag{7.82}$$

The inverse operation is defined similarly.

A straightforward implementation of the Fourier transform requires $O(N^2)$ operations, where $n = n_x \times n_y$ is the number of pixels of the input image. A number of optimizations can be applied to reduce the time needed to perform this transformation. First, the transformation is separable. Therefore, the transformation can be performed independently for the two coordinates. Second, if the image dimensions are a power of two then the Fourier transformation can be computed recursively, which gives a considerable speedup. This is known as the fast Fourier transformation. The fast Fourier transformation requires $O(N \log N)$ operations. It is described in detail in many textbooks on image processing, e.g. (Gonzalez and Woods 1992; Jähne 2002; Parker 1997; Press et al. 1992; Shapiro and Stockman 2001).

Use of homomorphic filtering for color constancy is basically a variant of Land's retinex theory. In Land's retinex theory or in the variants described by Horn (1974, 1986) or Blake (1985), low-frequency components are removed using a threshold function. Homomorphic filtering was initially used for gray-scale images. The description by Parker (1997) is also given for gray-scale images. However, the method may also be used for color images taken by a camera equipped with narrow-band sensors. In this case, the algorithm is applied independently to the three color bands RGB. First, the logarithm of the input intensity is computed for all image pixels. The intensity measured by an image sensor $I(x, y)$ is proportional to the product between the intensity of the light source $L(x, y)$ and the reflectance $R(x, y)$ at the corresponding object point. We obtain

$$\log c(x, y) = \log I(x, y) \approx \log L(x, y) + \log R(x, y) \qquad (7.83)$$

if the relationship between pixel intensities $c(x, y)$ and the intensities measured by the sensor is linear. It is assumed that the illuminant varies slowly over the image. Changes in reflectance are assumed to be discontinuous. Illumination effects can be removed by suppressing the slowly varying components of the image.

In Fourier space, this operation can be performed easily. Figure 7.22 shows the magnitude of the components in Fourier space for a sample image. The origin was shifted such that the low-frequency components are located in the center of the image. High-frequency components are located away from the center. Low-frequency components should be attenuated while high frequency components should be emphasized. This can be done by applying a high emphasis filter. Such filters are shown in (a) in Figure 7.23. A high emphasis filter specifies a gain factor for each of the Fourier coefficients depending on the radius from the center of the transformed image. Low-frequency components are reduced by using a low gain factor while high-frequency components are emphasized using a high gain factor. After the high emphasis filter is applied, the inverse of the Fourier transform is computed, i.e. a transformation from frequency space to spatial coordinates is made. Next, the application of the logarithm is undone by exponentiating all pixels computed. This reverses the

Figure 7.22 The input image is shown in (a). The image in (b) shows the magnitude of the coefficients in frequency space. The values are shown on a double logarithmic scale.

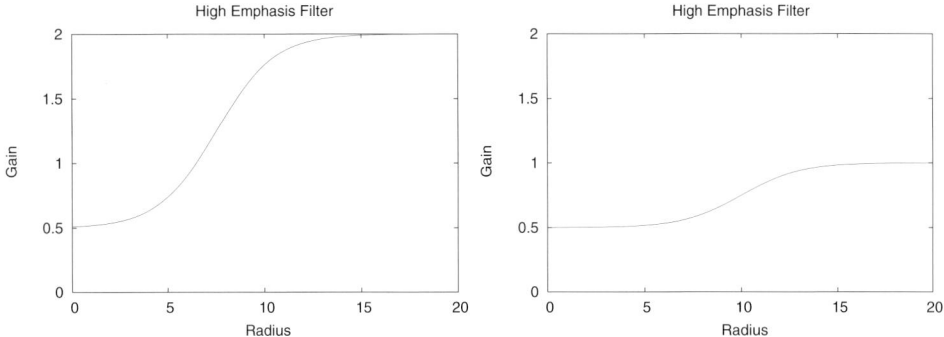

Figure 7.23 Two high emphasis filters. A filter that is similar to the one shown in (a) was used by Parker (1997). This particular filter works well for Parker's sample image. In practice, we found that the filter shown in (b) produces better results.

application of the logarithm that was taken initially to separate the product between the reflectance and the intensity of the illuminant into a sum.

Application of the high emphasis filter followed by an inverse Fourier transform may lead to exceptionally large values for some image pixels. Parker (1997) suggests to clip these outliers. In addition, he rescales the image band to the full intensity range. The exact operation that performs this rescaling can be found on the CD-ROM that accompanies Parker's book (Parker 1997). He sorts all pixel values of the image and then takes the intensity located at the third percentile as the maximum intensity. Intensities larger than this value are clipped to the maximum. Also, intensities lower than zero are clipped to zero.

Parker (1997) uses a high emphasis filter similar to that in of Figure 7.23(a) in order to enhance a gray-scale image. This high emphasis filter reduces low-frequency components to half their original amplitude. The amplitude of high-frequency components is doubled. This particular filter works well for Parker's sample image. The performance on other images is not as good. The filter shown in Figure 7.23(b) works better in practice. However, the result is still far from satisfactory. The output images for the two types of filters are shown in Figure 7.24. The computed output images are not very pleasing. Even though the image shown on the right looks better than the one on the left we can notice several problems with this algorithm. Slowly varying components are removed from the image; however, the background is still blue. Also there is a noticeable color difference between the color at the center of a uniformly colored area and the color at the edge of the area. The objects seem to be surrounded by a halo.

We have just seen how homomorphic filtering may be applied independently to the three different color channels. Faugeras (1979) originally suggested performing image processing in a perceptually uniform coordinate system. In this framework, the images should be transformed first into the coordinate system that is used by the human visual system. Once the image processing is done, the processed coordinates are transformed back to the original space. Figure 7.25 shows the individual steps of this framework. On the left side, the processing starts with the original tristimulus values R, G, B of three color primaries.

Figure 7.24 Output images produced by homomorphic filtering. The images in the top row were computed using the first high emphasis filter. The image in the bottom row were computed using the second high emphasis filter.

A matrix $\mathbf{T}_C$ is applied to the tristimulus data. The result will be the response of the cone receptors denoted by L, M, S that respond to light in the long, middle, and short part of the spectrum respectively. Since the response of the receptors is proportional to the logarithm of the intensity, the logarithm is applied at the next stage. After the logarithm is applied, the data is transformed by $\mathbf{T}_P$ such that one of the coordinate axes is the achromatic axis and the other two axes are the red–green and yellow–blue axes respectively.

$$\mathbf{T}_P = \begin{pmatrix} 0.612 & 0.369 & 0.019 \\ 1 & -1 & 0 \\ 1 & 0 & -1 \end{pmatrix} \qquad (7.84)$$

The information of the achromatic axis and the two color axes are then processed independently. After processing, the original transformation is undone. First, the transformation into the perceptual space is undone using $\mathbf{T}_P^{-1}$. Then the exponential function is applied and, finally, the data is transformed back to the tristimulus values by applying the matrix $\mathbf{T}_C^{-1}$. Among the possible applications discussed by Faugeras are brightness enhancement and color balance. He also demonstrated that image encoding is done best using the perceptual color space of the human visual system.

Instead of performing the homomorphic filtering operation independently on the three color bands, one can transform the image into the coordinate system used by the human visual system and then filter the image in this space. After filtering, the transformation into the coordinate system used by the human visual system can be undone. Note that this method works only if we assume that we have receptors that are similar to delta

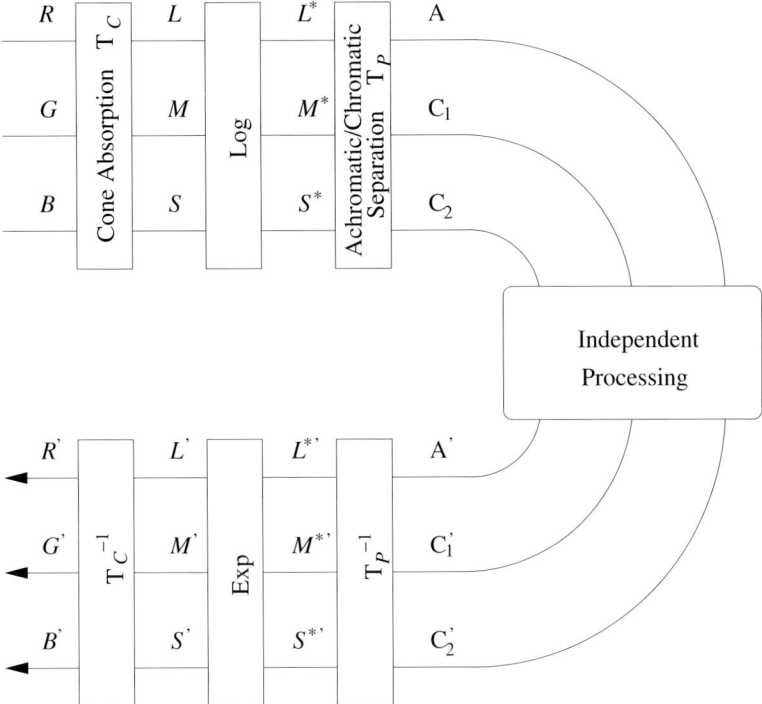

Figure 7.25 Digital image processing within the framework of what is known about the human visual system (Reproduced by permission of IEEE. Faugeras OD 1979 Digital color image processing within the framework of a human visual model. IEEE Transactions on Acoustics, Speech and Signal Processing, ASSP-27(4) 380-393.)

functions, i.e. the receptors respond to a single wavelength. Only then will the product between reflectance and the radiance be turned into a sum, once the logarithm is applied to the individual channels. If the sensors are not narrow band, the model may only be applied to a certain extent. For this to work, a change of illuminant has to be interpreted as a multiplicative scaling of the data measured by the sensors. Figure 7.26 shows the result when the low frequencies of the two color channels are attenuated in a perceptual uniform color space as proposed by Faugeras (1979) using the filter shown in Figure 7.27.

7.6 Intrinsic Images

Finlayson and Hordley (2001a,b) have developed a method for color constancy at a pixel. They make two assumptions. One that the camera's sensors are sufficiently narrow band and that the illuminant can be approximated by a black-body radiator. If the sensor's response functions are not sufficiently narrow band, a sharpening technique can be used (Barnard et al. 2001; Finlayson and Funt 1996; Finlayson et al. 1994a,b). The power spectrum of a black-body radiator was described in Section 3.5. The radiance of a black-body radiator at a temperature T, measured in Kelvin, at wavelength λ is given by (Haken and Wolf 1990;

Figure 7.26 Output images produced by attenuating low frequencies of the two color channels using the filter shown in Figure 7.27.

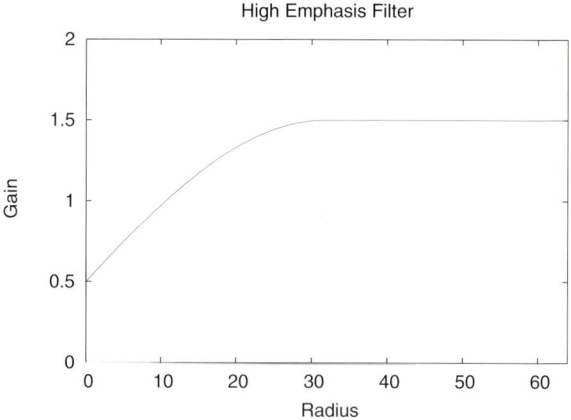

Figure 7.27 High emphasis filter similar to the one used by Faugeras (1979) to filter the two color channels in a perceptually uniform color space.

Jähne 2002)

$$L(\lambda, T) = \frac{2hc^2}{\lambda^5} \frac{1}{(e^{\frac{hc}{k_B T \lambda}} - 1)} \tag{7.85}$$

where $h = 6.626176 \cdot 10^{-34} Js$ is Planck's constant, $k_B = 1.3806 \cdot 10^{-23}$ J/K is Boltzmann's constant, and $c = 2.9979 \cdot 10^8$ m/s is the speed of light. Many natural light sources can be approximated by a black-body radiator. According to Finlayson and Hordley, many fluorescent light sources can also be approximated by a black-body radiator. Fluorescent lights tend to have highly localized emission spikes. However, one is not interested in the power spectrum of the illuminant. What is important is how the illuminant combines with the sensor's absorbance characteristic to finally form RGB values. The difference

COLOR CONSTANCY UNDER NONUNIFORM ILLUMINATION 177

between perceived colors measured in $L^*a^*b^*$ space using the actual fluorescent illuminant and perceived colors using an approximation was found to be acceptable.

The equation of the black-body radiator can be simplified by noting that the temperature T used to approximate many light sources is usually no larger than 10 000 K. Also, the visible spectrum ranges from 400 nm to 700 nm. Therefore, we have

$$e^{\frac{hc}{k_B T \lambda}} \gg 1 \qquad (7.86)$$

for $\lambda < 700$ nm and $T < 10\,000$ K. The equation is now simplified to

$$L(\lambda, T) = c_1 \lambda^{-5} e^{-\frac{c_2}{T\lambda}} \qquad (7.87)$$

with constants $c_1 = 2hc^2$ and $c_2 = \frac{hc}{k_B}$. Different intensities of the black-body radiator can be modeled by introducing another constant k.

$$L(\lambda, T) = k c_1 \lambda^{-5} e^{-\frac{c_2}{T\lambda}} \qquad (7.88)$$

If we have narrow-band sensors where each sensor responds to a single wavelength λ_i with $i \in \{r, g, b\}$, then the intensity I_i measured by the sensor is simply the product of a geometry factor G, the illuminant $L_i(x, y)$ at the corresponding object location (x, y) and the reflectance R at this location.

$$I_i(x, y) = G(x, y) R_i(x, y) L_i(x, y) \qquad (7.89)$$

If we assume that the illuminant can be approximated by a black-body radiator and that pixel colors are linearly related to the data measured by the sensor, i.e. $c_i(x, y) = I_i(x, y)$, we obtain

$$c_i = G R_i k c_1 \lambda_i^{-5} e^{-\frac{c_2}{T\lambda_i}}. \qquad (7.90)$$

Note that, from now on, we omit the coordinate index (x, y). If we now apply the logarithm to both sides, the product of irradiance and reflectance is split into a sum.

$$\log(c_i) = \log(kG) + \log(c_1 \lambda_i^{-5} R_i) - \frac{c_2}{T\lambda_i} \qquad (7.91)$$

The first term only depends on the power of the illuminant and the scene geometry but is independent of the color of the illuminant and the reflectance. The second term only depends on the wavelength to which the sensor responds and on the reflectance of the object. The last term only depends on the color of the illuminant. The first term can be removed by computing differences between the data measured for two different color channels. If we compute the differences between the red and green channel, which we denote by ρ_{rg}, as well as between the blue and green channel, which we denote by ρ_{bg}, we obtain

$$\rho_{rg} = \log(c_r) - \log(c_g) = \log(c_1 \lambda_r^{-5} R_r) - \frac{c_2}{T\lambda_r} - \log(c_1 \lambda_g^{-5} R_g) + \frac{c_2}{T\lambda_g} \qquad (7.92)$$

$$\rho_{bg} = \log(c_b) - \log(c_g) = \log(c_1 \lambda_b^{-5} R_b) - \frac{c_2}{T\lambda_b} - \log(c_1 \lambda_g^{-5} R_g) + \frac{c_2}{T\lambda_g}. \qquad (7.93)$$

Finlayson and Hordley (2001a,b) now make the following simplification. Let $R'_i = c_1 \lambda_i^{-5} R_i$ and let $E_i = -\frac{c_2}{\lambda_i}$. We obtain

$$\rho_{rg} = \log\left(\frac{R'_r}{R'_g}\right) + \frac{1}{T}(E_r - E_g) \tag{7.94}$$

$$\rho_{bg} = \log\left(\frac{R'_b}{R'_g}\right) + \frac{1}{T}(E_b - E_g). \tag{7.95}$$

The two equations can be viewed as a two-dimensional vector where the first coordinate is equal to ρ_{rg} and the second is equal to ρ_{bg}. If we now think of the temperature T as a parameter, we see that the two equations define a line where $\left[\log\left(\frac{R'_r}{R'_g}\right), \log\left(\frac{R'_b}{R'_g}\right)\right]$ is a point on the line and other points on the line are reached by adding $[E_r - E_g, E_b - E_g]$ with varying amounts. The constants E_i only depend on the wavelength to which the sensor responds. For a different reflectance, we obtain a translated line which has the same orientation because $[E_r - E_g, E_b - E_g]$ is independent of reflectance.

This is illustrated in Figure 7.28. Seven different surfaces (red, green, blue, yellow, cyan, magenta, and white) were illuminated by a black-body radiator of different temperatures. A camera that responds only to wavelengths $\lambda_r = 450$ nm, $\lambda_g = 540$ nm, and $\lambda_b = 610$ nm is assumed. The temperature of the black-body radiator was varied from 1000 to 10 000 K in steps of 20%. We see that the log-chromaticity differences form a line for each surface. The direction of the lines only depends on the type of sensors used in the camera.

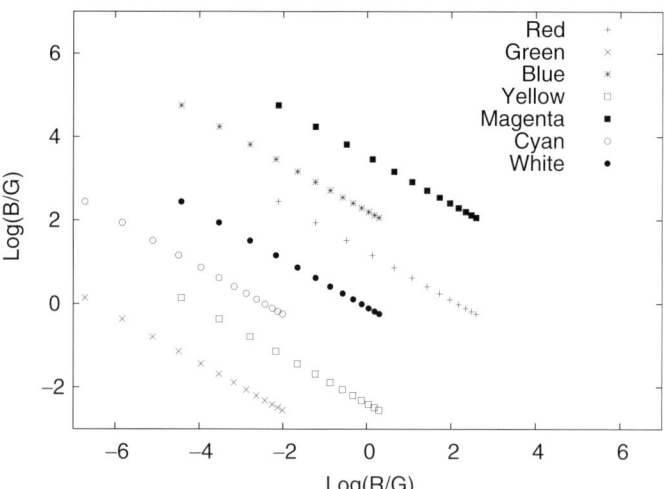

Figure 7.28 Log-chromaticity differences for simulated measurements of seven different surfaces (red, green, blue, yellow, cyan, magenta, and white) viewed under a black-body illuminant. For each surface, the temperature of the black-body radiator was varied from 1000 to 10 000 K in steps of 20%. The log-chromaticity differences of a single surface are neatly lined up.

COLOR CONSTANCY UNDER NONUNIFORM ILLUMINATION

If a gamma correction has been applied, the result will be similar (Finlayson et al. 2002). In this case, the pixel colors will be given by

$$c_i = I_i^\gamma \tag{7.96}$$

for some gamma value γ. After we take the logarithm of the pixel color, the gamma value will simply be a scaling factor.

$$\log(c_i) = \gamma \log(I_i) \tag{7.97}$$

Thus, even if some unknown gamma correction has been applied, the two-dimensional vectors $[\rho_{rg}, \rho_{bg}]$ for a given reflectance viewed under black-body illumination will form a line in log-chromaticity difference space.

The temperature dependence can be removed by projecting the data points in a direction orthogonal to the line. The vector orthogonal to the line is given by $[E_b - E_g, -(E_r - E_g)]^T$. The projected coordinate is then given by

$$\begin{bmatrix} E_b - E_g \\ -(E_r - E_g) \end{bmatrix} \begin{bmatrix} \rho_{rg} \\ \rho_{bg} \end{bmatrix} = (E_b - E_g) \log\left(\frac{R'_r}{R'_g}\right) - (E_r - E_g) \log\left(\frac{R'_b}{R'_g}\right) \tag{7.98}$$

which is independent of the illuminant because E_i is constant for the given sensor and the R'_i is a function of the reflectances R_i and the wavelengths to which the sensor responds.

Note, however, that once we project the data points in a direction orthogonal to the line, reflectances that differ by a multiplier $[e^{\frac{s}{\lambda_r}}, e^{\frac{s}{\lambda_g}}, e^{\frac{s}{\lambda_b}}]^T$ with $s \in \mathbb{R}$ can no longer be distinguished. This multiplier would move the origin in a direction parallel to the lines defined by the black-body radiator. Figure 7.29 shows a color circle. All points have the same intensity in RGB space. When we compute log-chromaticity difference coordinates for points sampled along the circle, we obtain the graph shown in Figure 7.29 (b). The

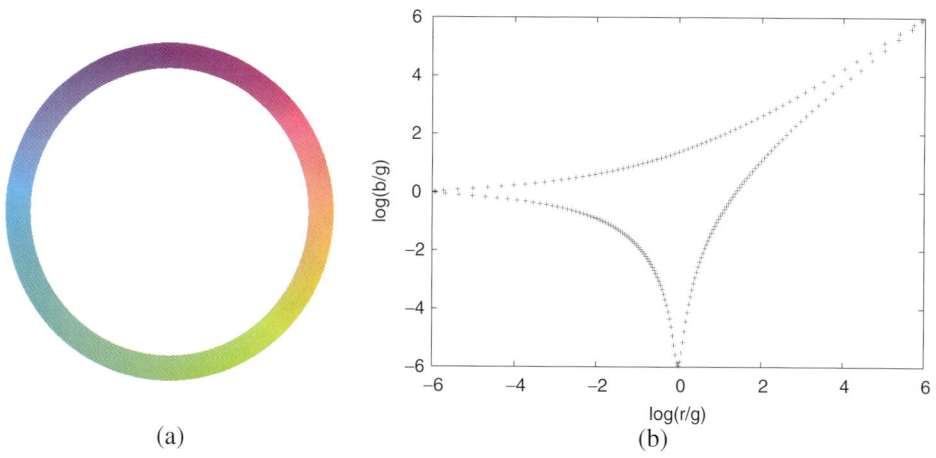

Figure 7.29 A circle of colors in RGB color space is shown in (a). When we transform data points that are sampled along the circle to log-chromaticity difference space, we obtain the graph in (b).

data points still form a distorted circle in log-chromaticity difference space. Thus, if we now project the data points along the invariant direction, some information will be lost. Suppose that we were to project all data points on the vertical axis. Now we can no longer distinguish between blue and red or between cyan and yellow.

The direction in which to project the log-chromaticity differences is unique for each camera. One possible way to compute this direction is to take a sequence of images of a calibration target, such as the Macbeth color checker, which consists of different colored patches. Let us assume that we have n different patches and m images. This will give us m data points $\rho = [\rho_{rg}, \rho_{bg}]$ for each patch. The data points of each patch all line up approximately. Let $\rho_{j,k}$ be the data point from the j-th patch and k-th image. For each patch, we can move the line to the origin of the coordinate system by subtracting the mean. Let $\tilde{\rho}$ be the new coordinates with the mean subtracted.

$$\tilde{\rho}_{j,k} = \rho_{j,k} - \frac{1}{m}\sum_{k=1}^{m}\rho_{j,k} \qquad (7.99)$$

We can now do a covariance analysis on the data and find the direction in which the spread is largest. Let $\mathbf{M}$ be a $2 \times nm$ size matrix of the data points.

$$\mathbf{M} = (\tilde{\rho}_{1,1}, \ldots, \tilde{\rho}_{1,m}, \ldots \tilde{\rho}_{n,1}, \ldots, \tilde{\rho}_{n,m}) \qquad (7.100)$$

The (2×2) covariance matrix $\mathbf{C}$ is then given by

$$\mathbf{C} = E\left[\mathbf{M}\mathbf{M}^T\right]. \qquad (7.101)$$

We compute the two eigenvectors of this symmetric matrix. The eigenvector that corresponds to the largest eigenvalue tells us the direction in which the points line up. Once this direction is known for the data points measured from the calibration target, one can project any data point measured by the camera in a direction orthogonal to this eigenvector. Let $\mathbf{e}$ be the eigenvector that corresponds to the largest eigenvalue and let $\mathbf{e}_\perp$ be the direction orthogonal to the eigenvector $\mathbf{e}$. The invariant data points o are then given by

$$o = \rho \mathbf{e}_\perp. \qquad (7.102)$$

Note that each color is now described by a one-dimensional scalar. For display purposes or object recognition based on color, the scalar can be transformed to the range $[0, 1]$.

In order to arrive at the invariant one-dimensional color descriptor, we have assumed that the illuminant can be approximated by a black-body radiator. Indeed, many natural light sources such as daylight or incandescent light can be approximated by a black-body radiator. Quite frequently, however, a scene is not only illuminated by a single light source but by several different light sources. Finlayson and Hordley (2001a,b) argue that the method also works for a supposition of two light sources that can be approximated by a black-body radiator. Let $L(\lambda, T_1) = c_1 \lambda^{-5} e^{-\frac{c_2}{T_1 \lambda}}$ be the radiance given off by the first light source and let $L(\lambda, T_2) = c_1 \lambda^{-5} e^{-\frac{c_2}{T_2 \lambda}}$ be the radiance given off by the second light source. The combined radiance $L(\lambda)$ at an object position is then given by

$$L(\lambda) = s_1 L(\lambda, T_1) + s_2 L(\lambda, T_2) \qquad (7.103)$$

for some suitable scaling factors s_1 and s_2. In Section 4.4, we have seen that the chromaticities of the black-body radiator in XYZ space form a convex curve. Therefore, the

COLOR CONSTANCY UNDER NONUNIFORM ILLUMINATION

supposition of two black-body illuminants of different chromaticities cannot be approximated by a single black-body illuminant

$$s_3 c_1 \lambda^{-5} e^{-\frac{c_2}{T_3 \lambda}} \neq s_1 c_1 \lambda^{-5} e^{-\frac{c_2}{T_1 \lambda}} + s_2 c_1 \lambda^{-5} e^{-\frac{c_2}{T_2 \lambda}} \quad (7.104)$$

for all choices of constants s_3 and T_3. Finlayson and Hordley argue that for temperatures between 2800 K and 10 000 K, which is the range of typical light sources, the curve is only weakly convex. Finlayson and Hordley check this by generating 190 weighted combinations of different black-body illuminations and computing the chromaticities. All chromaticities lie either on or close to the curve described by the black-body radiator. Note that, for the method to work successfully, we must be able to describe all light sources that illuminate the scene by a black-body radiator. If we have any light sources, such as green or purple lights that cannot be approximated by a black-body radiator, then the method will not produce the desired output.

Finlayson and Drew (2001) show that a similar method can also be applied to a 4-sensor camera. Since four measurements are available for each image point, the measurements project onto a plane that is orthogonal to the direction caused by the black-body illuminant. Matte points that are similarly colored are projected onto the same point on the plane. Specularities are extended surrounding the projected specular point. Under the assumption that the illuminant can be approximated by a black-body radiator this point is unique for each camera. By moving outward from the specular point they show that an image can be made invariant to shading, shadows, and specularities.

Choosing one of the sensors and computing log-chromaticity differences between the chosen channel and the two other channels has the drawback that the response of the chosen channel may be very low, leading to noisy results. Another question is, which channel should be chosen? Finlayson and Drew (2001) suggest dividing by the geometric mean of the three channels to remove the dependence on the shading information G and the dependence on the intensity k. Thus instead of computing

$$\rho_{rg} = \log(c_r) - \log(c_g) \quad (7.105)$$

$$\rho_{bg} = \log(c_b) - \log(c_g) \quad (7.106)$$

one computes

$$\rho_i = \log(c_i) - \log(c_M) \quad (7.107)$$

with $c_M = \sqrt[3]{c_r c_g c_b}$. This leads to

$$\rho_i = \log(c_i) - \frac{1}{3} \sum_{j \in \{r,g,b\}} \log(c_i) \quad (7.108)$$

$$= \log(c_1 \lambda_i^{-5} R_i) - \left(\frac{c_2}{T \lambda_i}\right) - \frac{1}{3} \log\left(\prod_{j \in \{r,g,b\}} c_1 \lambda_j^{-5} R_j\right) + \frac{1}{3} \sum_{j \in \{r,g,b\}} \frac{c_2}{T \lambda_j} \quad (7.109)$$

$$= \log(c_1 \lambda_i^{-5} R_i) - \frac{1}{3} \log\left(\prod_{j \in \{r,g,b\}} c_1 \lambda_j^{-5} R_j\right) - \frac{c_2}{T \lambda_i} + \frac{1}{3} \sum_{j \in \{r,g,b\}} \frac{c_2}{T \lambda_j} \quad (7.110)$$

$$= \log\left(\frac{R_i'}{R_M'}\right) - \frac{1}{T} (E_i - E_M) \quad (7.111)$$

with $R'_i = c_1 \lambda_i^{-5} R_i$, $R'_M = \sqrt[3]{\prod_{j \in \{r,g,b\}} R'_j}$, $E_i = -\frac{c_2}{\lambda_i}$, and $E_M = -\frac{c_2}{3} \sum_{j \in \{r,g,b\}} \frac{1}{\lambda_j}$. Again, we obtain a line parameterized by the temperature T.

Since the vector $\rho = [\rho_r, \rho_g, \rho_b]$ is orthogonal to the vector $\mathbf{u} = \frac{1}{\sqrt{3}}[1, 1, 1]^T$,

$$\rho \mathbf{u} = 0 \tag{7.112}$$

all points ρ are located on the two-dimensional plane defined by $\mathbf{u}$. Two-dimensional coordinates would therefore be sufficient to specify any point of the geometric mean two-dimensional chromaticity space. Finlayson et al. (2004) therefore define the following coordinate system for the geometric mean chromaticity space. Let $\{\chi_1, \chi_2\}$ be the two basis vectors. The two vectors can be found by considering the projector $\mathcal{P}_\mathbf{u} = \mathbf{I} - \mathbf{u}\mathbf{u}^T$ that takes three-dimensional coordinates onto the plane

$$\mathcal{P}_\mathbf{u} = \mathbf{I} - \mathbf{u}\mathbf{u}^T = \begin{pmatrix} \frac{2}{3} & -\frac{1}{3} & -\frac{1}{3} \\ -\frac{1}{3} & \frac{2}{3} & -\frac{1}{3} \\ -\frac{1}{3} & -\frac{1}{3} & \frac{2}{3} \end{pmatrix} = \mathbf{U}\mathbf{U}^T \tag{7.113}$$

with

$$\mathbf{U} = \begin{pmatrix} \sqrt{\frac{2}{3}} & 0 \\ -\sqrt{\frac{1}{6}} & -\sqrt{\frac{1}{2}} \\ -\sqrt{\frac{1}{6}} & \sqrt{\frac{1}{2}} \end{pmatrix}. \tag{7.114}$$

The geometric mean two-dimensional chromaticity space is therefore defined as

$$\chi = \begin{bmatrix} \chi_1 \\ \chi_2 \end{bmatrix}^T = \mathbf{U}^T \rho. \tag{7.115}$$

Note that the matrix $\mathbf{U}$ simply rotates vectors that are already in the plane defined by $\mathbf{u}$ to a standard coordinate system. Figure 7.30 shows the data points of a color circle transformed

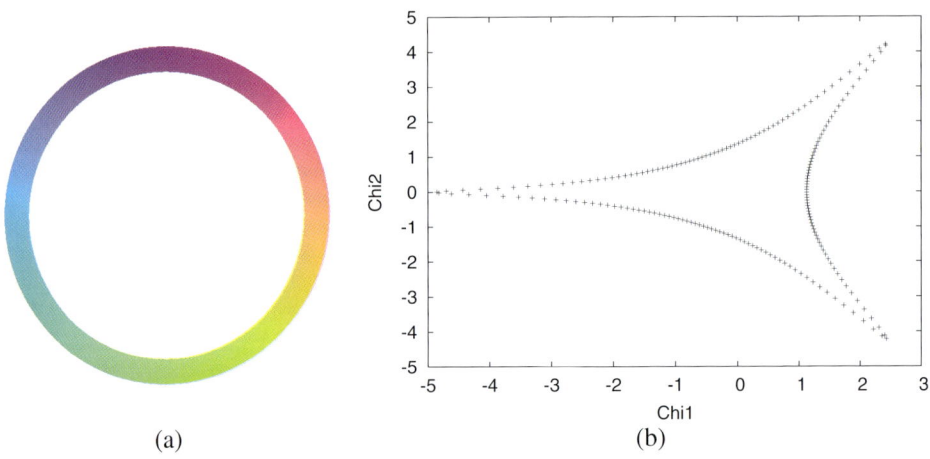

(a)   (b)

Figure 7.30 A circle of colors in RGB color space is shown in (a). When we transform data points sampled along the circle to geometric mean two-dimensional chromaticity space, we obtain the graph in (b).

to geometric mean two-dimensional chromaticity space. Thus, it is clear that if we project along a line in geometric mean two-dimensional chromaticity space, we will lose some information about the color of the objects. Using geometric mean chromaticities, we can now transform the input image of a calibrated camera to an intrinsic image. Intrinsic images contain only one intrinsic characteristic of the scene being viewed (Tappen et al. 2002). Here, the intrinsic images only depend on the reflectance of the object points. The two-dimensional coordinates $[\chi_1, \chi_2]^T$ have to be projected onto a line that is orthogonal to the vector $\mathbf{e} = [E_r - E_M, E_g - E_M, E_b - E_M]^T$. For a given camera, this vector can be found experimentally by imaging a calibration image under several different illuminants. Figure 7.31 shows a section of a color Mondrian. Several images of this Mondrian were taken during the course of the day. Sometimes the Mondrian was illuminated by direct sunlight, sometimes it was illuminated by a cloudy sky. Some images were also taken during sunset. The graph in (b) shows the χ-chromaticities of the patches for the entire image set. One data point is shown for each patch of each image. Each data points represents the average color of the patch. Pixels at the border of the patch were excluded from the average. We clearly see that the χ-chromaticities of each patch form lines. As described in the preceding text, we can determine the orientation of these lines. For the given camera, a Canon 10D, the orientation of the invariant direction was found to be at an angle of $150.75°$.

Finlayson et al. (2004) describe how to compute intrinsic images for an uncalibrated camera from a single image. The projection has to be done onto a line inside the geometric mean two-dimensional chromaticity space. The question is, which line is the correct one. Figure 7.32 shows the projection of the data points for two different lines. For the image in (a), the line is oriented exactly orthogonal to the invariant direction. For the image in (b), the line is turned slightly away. Finlayson et al. argue that the correct orientation, we are looking for, is the orientation where the resulting image has minimum entropy. They suggest the following method on how to find the correct direction for an image from an uncalibrated camera. Let $\{[\cos\theta, \sin\theta]| \text{ with } \theta \in [1, \ldots, 180]\}$ be the set of lines for which

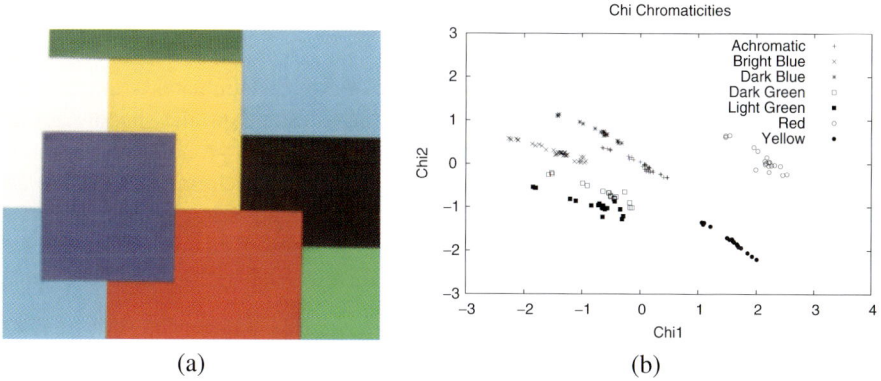

Figure 7.31 A section from a color Mondrian is shown in (a). The Mondrian contains the colored patches: black, white, bright blue, dark blue, yellow, bright green, dank green, and red. Several images of the Mondrian were taken during the course of the day. The graph in (b) shows the χ-chromaticities of each patch for all images of the sequence.

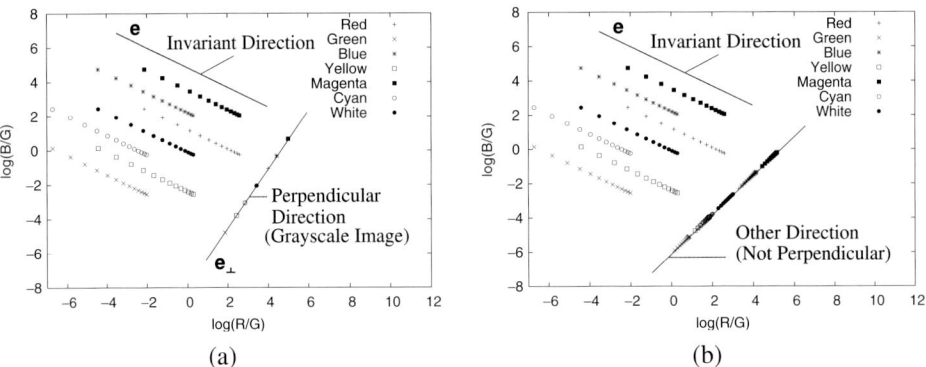

Figure 7.32 Projection obtained when one projects along the invariant direction (a). If we project onto any other line, then the projected data points will be spread along the line (b).

we compute the entropy. For each line, the invariant data points g are given by

$$g = \chi_1 \cos\theta + \chi_2 \sin\theta. \qquad (7.116)$$

A gray-scale image is formed from the projected data points $g(x, y)$ that are transformed to the range $[0, 1]$. Next, a histogram is computed for the image. From the histogram, we can compute the probability $p(g)$ that the gray value g occurs in the image. Finally, the entropy E is computed as

$$E = -\sum_g p(g) \ln p(g). \qquad (7.117)$$

The correct direction θ for the line will be the one where the entropy is minimal, i.e.

$$\theta_{\min} = \operatorname{argmin}_{\theta'} E(\theta'). \qquad (7.118)$$

Figure 7.33 shows how the method works. The image in (a) is the input image. It contains a strong shadow of a person. Thus, there are two illuminants - direct sunlight illuminating the surrounding area and indirect lighting illuminating the shadow area. The graph in (b) shows the entropy for all possible angles. It has a pronounced minimum at $\theta_{\min} = 57.4°$. We obtain the intrinsic image if we project the χ-chromaticities onto a line oriented at $57.4°$. The value obtained by entropy minimization is in close agreement with the value that was obtained by the calibration procedure described in the preceding text. According to the calibration procedure, the invariant direction is located at an angle of $150.75°$, which corresponds to a projection direction of $60.75°$.

The intrinsic image only depends on the reflectance of the corresponding object points. Since the intrinsic image (d) is only a function of reflectance, the image is free from shadows. The two images to the left and right of the intrinsic image show the gray-scale image for $\theta_{\min} - 20$ and $\theta_{\min} + 20$. By minimizing the entropy, we choose the angle that leads to the simplest possible image. For the image shown, it leads to the intrinsic image where the shadow has been removed. For other types of images, minimizing the entropy may not lead to the desired result.

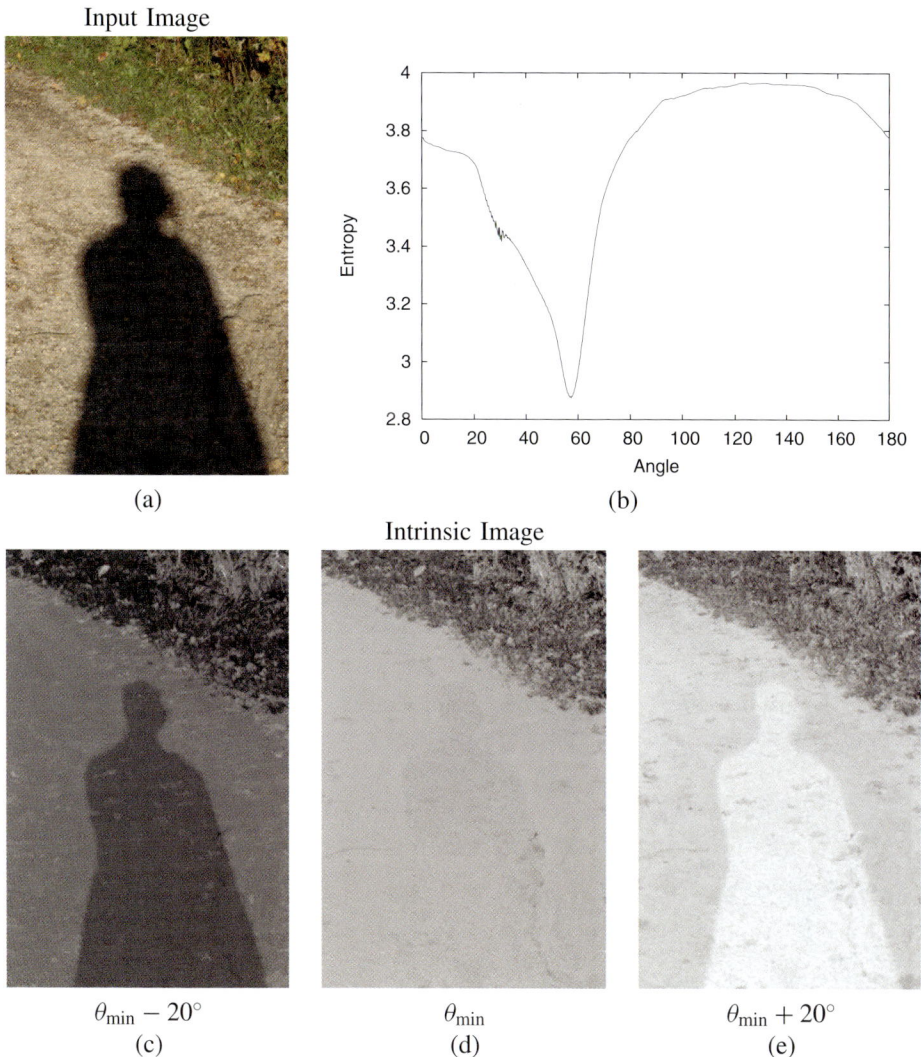

Figure 7.33 Below the input image, three gray-scale images are shown. Image (d) is the intrinsic image. The χ-chromaticities were projected onto a line with orientation $\theta_{min} = 57.8$. At this angle, the entropy of the gray-scale image becomes minimal. The two images in (c) and (e) were computed for an angle of $\theta_{min} - 20°$ and $\theta_{min} + 20°$. The plot shown in (b) shows the entropy for all possible angles.

The intrinsic image that is based only on reflectance can be used for object recognition. This has the added advantage that we only have to process a single band image instead of three bands. However, it would also be interesting to see if we can move from an intrinsic image back to a full color image. As we have seen in the preceding text, some color information is lost. Therefore, it is not possible to fully recover a color image. Drew et al. (2003) suggested the following method to go back to a color image. The projected coordinates do not have to be interpreted as a gray-scale image. The coordinates along the projection line are still two-dimensional.

$$\chi_\theta = \begin{bmatrix} \chi_1 \\ \chi_2 \end{bmatrix} = g \begin{bmatrix} \cos\theta \\ \sin\theta \end{bmatrix} = g\mathbf{e}_\perp \tag{7.119}$$

We can compute the corresponding three-dimensional coordinates in geometric mean chromaticity space by multiplying the projected coordinates by $\mathbf{U}$.

$$\tilde{\rho} = \mathbf{U}\chi_\theta \tag{7.120}$$

The influence of the illuminant moves the data points in a direction perpendicular to the projection line. Drew et al. (2003) suggest adding a little illumination in order to obtain a color image from the one-dimensional data points. All points along the line are shifted along the invariant direction such that the median of the brightest 1% of the image pixels has the 2D chromaticity of the original image. For each point of the image, we check if it belongs to the brightest 1% of the image pixels. We then compute the χ-chromaticities and project these points onto the invariant direction $\mathbf{e} = [\sin\theta, -\cos\theta]^T$.

$$\chi_{\text{Illuminant}} = \chi\mathbf{e} = \chi_1 \sin\theta - \chi_2 \cos\theta \tag{7.121}$$

We then compute the median value of this data. Let χ_{Shift} be this median value. An illuminated image is obtained by moving all projected χ-chromaticity data points by χ_{Shift} along the direction $\mathbf{e}$.

$$\tilde{\rho} = \mathbf{U}(\chi_\theta + \chi_{\text{Shift}}\mathbf{e}) \tag{7.122}$$

Figure 7.34 illustrates this process for our sample image. Since we originally applied the logarithm to compute geometric mean chromaticities we now exponentiate to compute

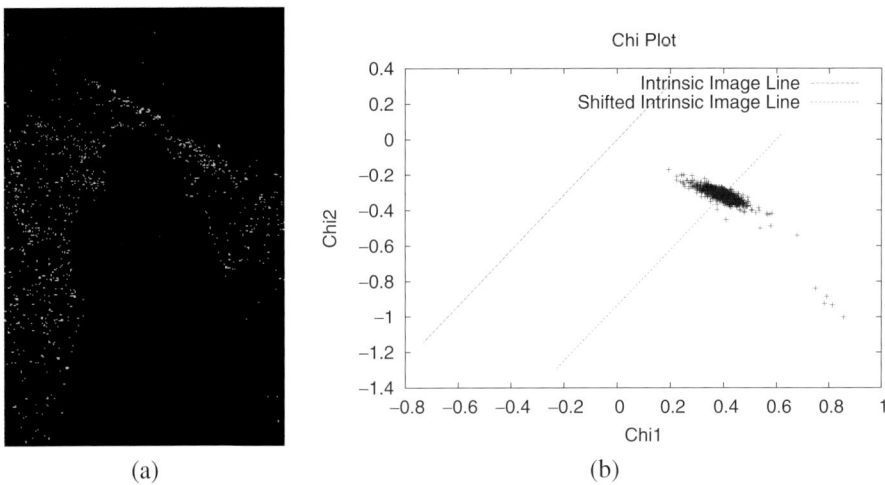

Figure 7.34 The image in (a) shows 1% of the brightest image pixels. In (b), the intrinsic image line and χ-chromaticities of the points are shown. The intrinsic image line is shifted along the invariant direction. The length of the shift was computed such that the median of the brightest 1% of the image pixels has the two-dimensional chromaticity of the original image.

output colors.

$$\mathbf{o} = [e^{\rho_r}, e^{\rho_g}, e^{\rho_b}] \tag{7.123}$$

From this we can compute the normalized RGB chromaticities. Figure 7.35 shows the resulting image. The RGB intrinsic image only depends on the reflectances of the object points. We can obtain an image with shading information by scaling the colors of the RGB intrinsic image such that the lightness of the pixels is identical to the lightness of the input pixels. This image is shown in Figure 7.35(b).

As we have noted in the preceding text, some of the information is lost when projecting the data points along the invariant direction. The method worked well on the sample image because it contained mostly green and brown colors. However, for other types of images with many different colors, it may not be clear into which direction the color shift should be made. A decision has to be made on which colors to restore. Thus, instead of looking at the colors of the brightest 1% of the image pixels, we can also check if most of the colors are located to the left of the vector $\mathbf{e}_\perp$ or to the right. If most of the colors are located to the left, then we should perform the color shift in this direction to maintain the overall look of the image. In this case, we first determine the direction in which most of the colors are located and then determine the median offset of the colors in this direction. Figure 7.36 shows an image of a color Mondrian. Below the color Mondrian, the RGB intrinsic images are shown. For the image in (b), the color shift was performed by considering only the points located in the positive invariant direction, whereas for the image in (c), the color shift was performed by considering only the points located in the negative invariant direction. For the image in (b), the reds, greens, browns, and yellows are retained. For the image in (c) the blues, cyan, and magenta are retained. Figure 7.37 shows the results for our two test images.

(a)

(b)

Figure 7.35 Intrinsic RGB image. The image was created by moving all projected data points along the invariant direction. The image in (b) shows the same data where each pixel was rescaled to have the original lightness of the input image.

188 COLOR CONSTANCY UNDER NONUNIFORM ILLUMINATION

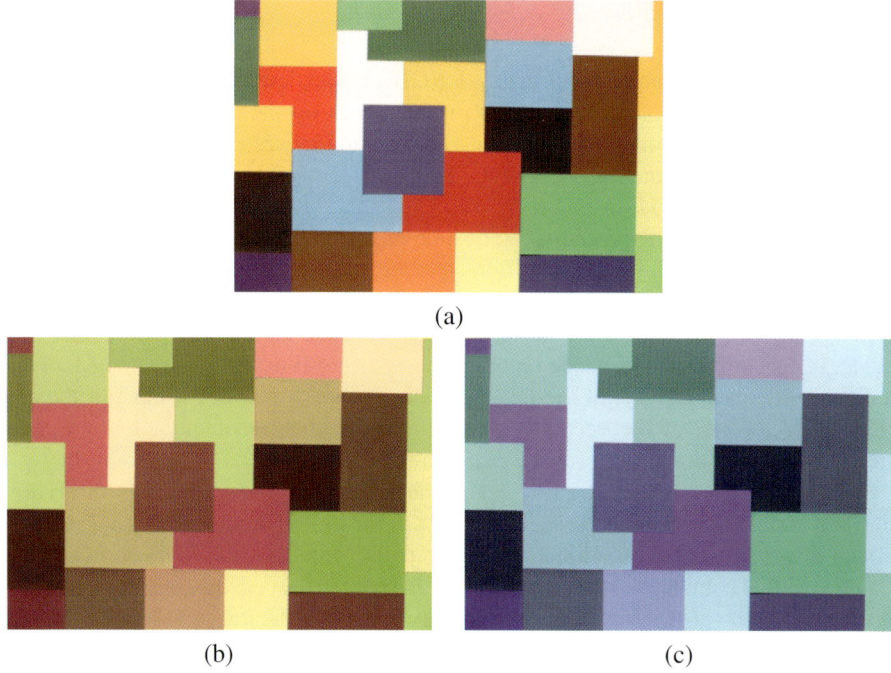

Figure 7.36 RGB intrinsic images for a color Mondrian. The input image is shown in (a). The image in (b) was created by applying the extra light shift in the positive direction whereas the image in (c) was created by applying the extra light shift in the negative direction.

Figure 7.37 Results for our sample images. Each pixel was rescaled to have the original lightness of the input image.

7.7 Reflectance Images from Image Sequences

Weiss (2001) describes how a reflectance image may be obtained from an image sequence taken with a stationary camera. Obviously, if the camera is stationary and no moving objects

COLOR CONSTANCY UNDER NONUNIFORM ILLUMINATION 189

are in the image, then the reflectance is constant for each image pixel. Only the illuminant may vary over time. Weiss suggests first taking the logarithm of the input pixels that separates reflectance and illumination components into a sum. Assuming that the measured pixel values c_i are proportional to the product of the reflectance and the illumination, i.e.

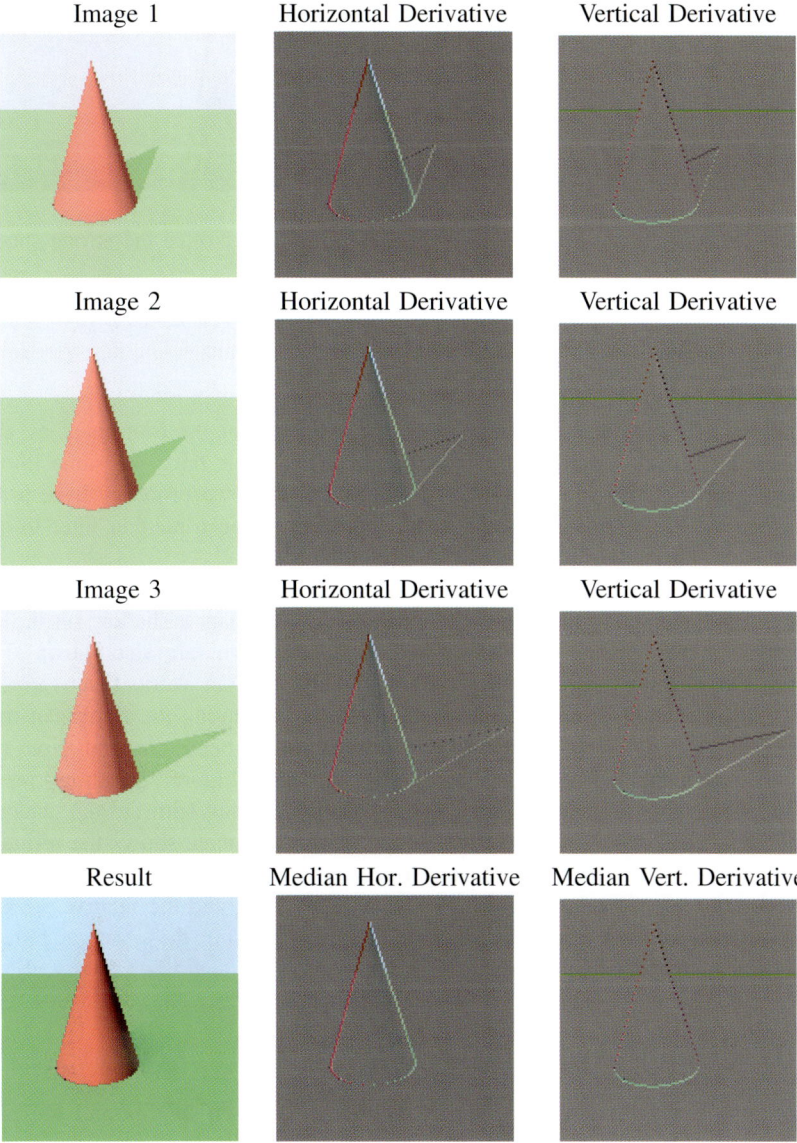

Figure 7.38 Results for three synthetic input images. A cylinder is illuminated from the side. For each input image, horizontal and vertical derivatives are shown. Median horizontal and vertical derivatives are shown below. In the lower left, the re-integrated image is shown.

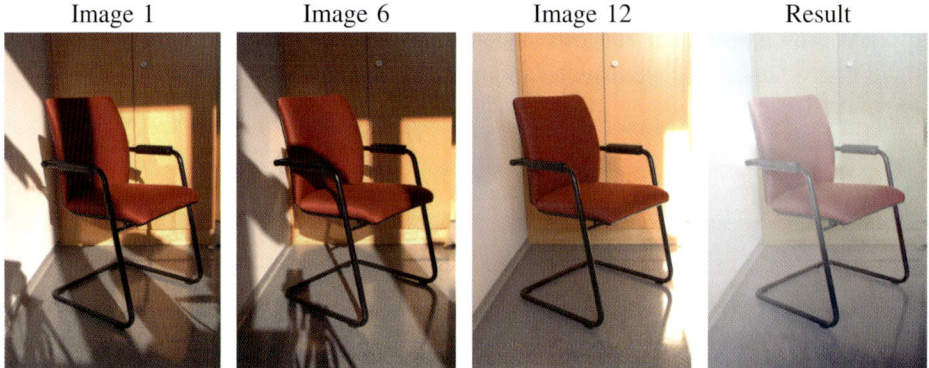

Figure 7.39 Results for a sequence of 12 images of a chair. Notice the strong shadow on the chair. The resulting image is shown on the right.

$c_i = R_i L_i$, we obtain
$$\log(c_i) = \log R_i + \log L_i. \qquad (7.124)$$
If the illuminant is constant, we can remove the influence of the illuminant by computing the first derivative. This works if two neighboring pixels have the same illuminant. If two neighboring pixels have a different illuminant, then we probably have a shadow edge between the two pixels. Weiss suggests simply applying a median filter to the first derivatives to eliminate the influence of the illuminant.

Figure 7.38 illustrates the method on a synthetic image. Three frames of a cylinder are shown. The cylinder is illuminated from the side and casts a shadow on the ground below. For each frame, the horizontal and vertical derivatives are also shown. The last row shows the median horizontal derivatives and the median vertical derivatives. The image in the lower right shows the image obtained by computing the Laplacian from the median derivatives and then re-integrating. The shadow has been removed. Note that the cylinder still appears round because of shading information. Figure 7.39 shows results for a short sequence of real images. The sequence shows a chair with strong shadows. As the sun moves, the shadows move in the image. The final image shows the re-integrated image. The shadows have been removed. Notice the reflections of the chair in the floor. Such reflections are only a function of the reflecting surface, and the relative positions of the viewed object and the observer. Since neither the chair nor the observer moved, the reflection is also contained in the intrinsic image.

7.8 Additional Algorithms

The algorithms for color constancy that have been discussed in this and the previous chapter should give a broad overview of the many different types of algorithms that have been developed to achieve color constancy. The set of algorithms, however, is far from being complete. Many algorithms have not been discussed in detail. The reader should take note that this does not necessarily say anything about the quality of the algorithms that have not been discussed. Because of time constraints a selection had to be made. Among these algorithms that were not discussed in depth are color by correlation (Finlayson et al. 1997), color by correlation in a three-dimensional color space (Barnard et al. 2000) and a two-dimensional gamut-constraint algorithm for scenes with varying illumination (Barnard et al. 1997). Brill (1978) suggested achieving color constancy based on the ratio of volumes of parallel-epipeds in tristimulus space. Brainard and Freeman (1997) suggested a color constancy algorithm based on Bayesian decision theory. Zaidi (1998) proposed a heuristics based algorithm for color constancy (see also Funt and Lewis (2000)). Tappen et al. (2002) proposed several extensions to Horn's algorithm. They suggested training a classifier to distinguish between a change that is due to a change of reflectance from a change that is caused by shading. All algorithms that have been described in this and the previous chapter will be evaluated in detail in Chapter 13 and Chapter 14.

8

Learning Color Constancy

Color constancy algorithms can be derived by looking at the underlying physics of color image formation. An interesting question is whether color constancy algorithms are an acquired capability of the visual system, i.e. if they are learnt, or whether these algorithms are hardwired, i.e. evolved.

8.1 Learning a Linear Filter

Hurlbert and Poggio (1987, 1988) have shown that color constancy may be learnt from examples. Owing to computational constraints, they worked on individual lines of simulated color Mondrians. The color Mondrians were created by artificially illuminating a reflectance image of a color Mondrian with illumination gradients. In their model, the measurement made by the sensor I is proportional to the product of the illumination intensity L and the surface reflectance R.

$$I = RL \tag{8.1}$$

Next, the logarithm is applied which turns the product of the reflectance R and the illumination intensity L into a sum.

$$\log I = \log R + \log L \tag{8.2}$$

A set of training vectors was assembled from this data. Let n_s be the number of training samples. Each input vector $\mathbf{v}_{\mathbf{c}i}$ with $i \in \{1, \ldots, n_s\}$ corresponded to a horizontal line from the illuminated Mondrian. Since they worked with simulated data, the corresponding reflectances $\mathbf{v}_{\mathbf{R}i}$ were known. Let n_x be the width of the Mondrian image. Therefore, each vector $\mathbf{v}_{\mathbf{c}i}$ and $\mathbf{v}_{\mathbf{R}i}$ contained n_x data samples. The training samples were collected in a matrix of size $n_s \times n_x$. This resulted in two matrices $\mathbf{R}$ and $\mathbf{C}$. They assumed that the transform from input to output, i.e. from measured intensities to reflectances, is linear.

$$\mathbf{R} = \mathbf{TC}. \tag{8.3}$$

The solution of this equation, which is optimal in the least-squares sense is given by

$$\mathbf{T} = \mathbf{RC}^+ \tag{8.4}$$

Color Constancy M. Ebner
© 2007 John Wiley & Sons, Ltd

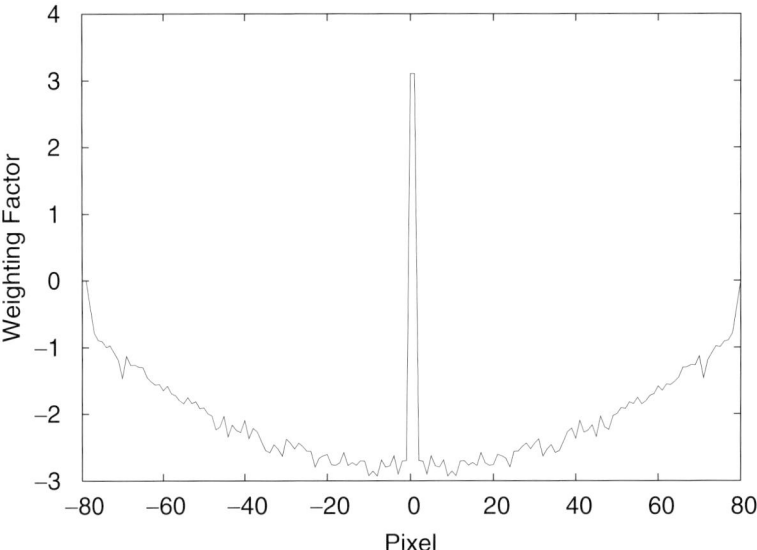

Figure 8.1 Filter obtained using the method described by Hurlbert and Poggio (1987, 1988). A linear operator and a sinusoidal illumination are assumed. Data kindly supplied by Anya Hurlbert, University of Newcastle, UK.

where $\mathbf{C}^+ = \mathbf{C}^T(\mathbf{C}\mathbf{C}^T)^{-1}$ is the Moore-Penrose pseudoinverse. Hurlbert and Poggio extracted a space variant filter from this matrix.

The first row of the matrix $\mathbf{T}$ is used to compute the reflectance of the first pixel, the second row is used to compute the reflectance of the second pixel, and so on. They assumed that the operator should approximate a convolution far from the image boundaries. That is, the operator should be space invariant. Inspection of the matrix $\mathbf{T}$ confirmed this. Therefore, if one takes a row at the center of the matrix, it should be similar to the next row except that it is shifted by one pixel to the right. They averaged the rows at the center to obtain the filter kernel. Figure 8.1 shows the resulting filter when either linear illumination gradients or slowly varying sinusoidal illumination with random wavelength, phase, and amplitude were used.

The resulting filter basically takes the value of the pixel at the center and subtracts an average of the surrounding pixel value from the center pixel. This filter, therefore, resembles the algorithms of Land (1986a) (see Section 7.1) and my own algorithms (see Chapter 10 and Chapter 11).

8.2 Learning Color Constancy Using Neural Networks

A number of researchers have tried training neural networks to achieve color constancy. A neural network basically consists of a set of nodes connected by weights (McClelland and Rumelhart 1986; Rumelhart and McClelland 1986; Zell 1994). Artificial neural networks are an abstraction from biological neural networks. Figure 8.2 shows a motor neuron in (a) and a network of eight artificial neurons on the right. A neuron may be in one of

LEARNING COLOR CONSTANCY

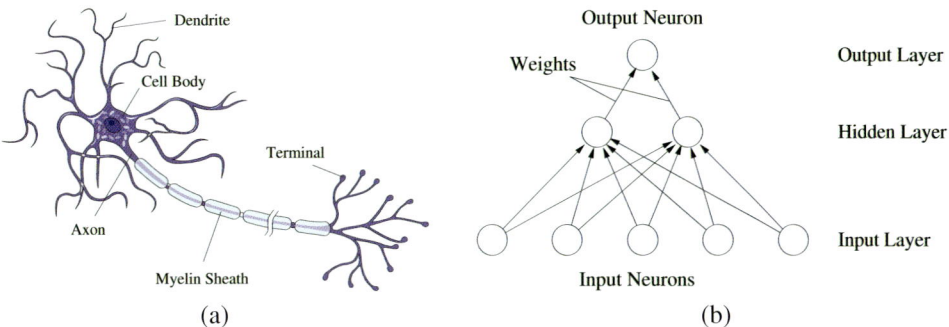

Figure 8.2 A motor neuron (a) and small artificial neural network (b). A neuron collects signals from other neurons via its dendrites. If the neuron is sufficiently activated, it sends a signal to other neurons via its axon. Artificial neural network are often grouped into layers. Data is entered through the input layer. It is processed by the neurons of the hidden layer and then fed to the neurons of the output layer. (Illustration of motor neuron from Life ART Collection Images © 1989–2001 by Lippincott Williams & Wilkins used by permission from SmartDraw.com.)

two states: activated or silent. If a neuron is activated, a signal is sent over the axon via synapses to other neurons. Each neuron collects the signals from other neurons to which it is connected. If the combined signal is sufficiently strong, then the neuron is said to fire. It sends its signal to other neurons.

A simple artificial neuron simply adds up the signals it receives from other neurons to which it is connected. Each input signal is multiplied by the weight of the connection. Let o_j be the output of neuron j and let $w_{i,j}$ be the weight between neurons i and j, then the activation a_i of neuron i is computed as

$$a_i = \sum_j w_{i,j} o_j. \tag{8.5}$$

The activation a_i can be used to compute an output value o_i for the neuron i. A threshold is used to define whether the neuron is active or not. Let θ_i be the threshold of neuron i, then the output is given by

$$o_i = \begin{cases} 1 & \text{if } a_i > \theta_i \\ 0 & \text{if } a_i \leq \theta_i \end{cases}. \tag{8.6}$$

Instead of computing a binary output, a sigmoidal function is frequently used to compute the output value.

$$o_i = \frac{1}{1 + e^{a_i - \theta_i}} \tag{8.7}$$

Artificial neural networks often have a layered structure as shown in Figure 8.2 (b). The first layer is the input layer. The second layer is the hidden layer. The third layer is the output layer. Learning algorithms such as back-propagation that are described in many textbooks on neural networks (Kosko 1992; Rumelhart and McClelland 1986; Zell 1994) may be used to train such networks to compute a desired output for a given input. The networks are trained by adjusting the weights as well as the thresholds.

Pomierski and Groß (1995) describe a method for color constancy which is similar to the algorithm color cluster rotation of Paulus et al. (1998). This algorithm has been described in detail in Section 6.6. Color cluster rotation views the image pixels as a cloud of points in a three-dimensional space. Assuming that the image was taken under a white illuminant and that the scene is sufficiently diverse, then the main or principal axis of the point cloud is aligned with the gray vector. A nonwhite illuminant causes the main axis to be tilted. It can be brought back into registration by rotating the cloud of points such that the main axis is again aligned with the gray vector.

The method of Oja (1982) is used to determine the principal axis. Let $\mathbf{c}$ be the original data points of the image. Let $\mathbf{w}' = [w_r, w_g, w_b]$ be the 3 element weight vector of the neural system. The following update equations are iterated until convergence:

$$\mathbf{w}' = \mathbf{w}' + f\mathbf{c}_w(\mathbf{c} - \mathbf{c}_w \mathbf{w}') \tag{8.8}$$

where f is a learning factor and $\mathbf{c}_w = \mathbf{c}^T \mathbf{w}'$ is simply the projection of $\mathbf{c}$ onto the vector $\mathbf{w}'$. After learning, i.e. iterating over the cloud of points until convergence, the main axis is stored inside the weight vector $\mathbf{w}'$. It can then be used to rotate the measured input data such that the main axis of the data $\mathbf{w}'$ is aligned with the gray vector $\mathbf{w}$. After aligning the cloud of points with the gray vector, the data is rescaled such that the color space is completely filled. Figure 8.3 illustrates the process. Figure 8.3(a) shows the cloud of points inside the RGB cube. The eigenvector that corresponds to the largest eigenvalue is also shown. Figure 8.3(b) shows the cloud of points after rotating the cloud of points onto the gray vector. The last step of the algorithm rescales all point such that the color space is completely filled, as shown in Figure 8.3(c).

Funt et al. (1996) have trained a neural network to estimate the chromaticity of the illuminant from an image. The pixels of the input image were first transformed to rg-chromaticity space. When the rg-chromaticities are plotted in a two-dimensional graph, one obtains a triangular shaped region. This region is sampled by the input layer of the neural network. Funt *et al.* used 1250 binary neurons as the input layer. Each neuron samples a specific part of the triangular shaped rg-chromaticity space as shown in Figure 8.4. The neuron is set to 1 if the corresponding rg-chromaticity occurs in the image. Otherwise, the

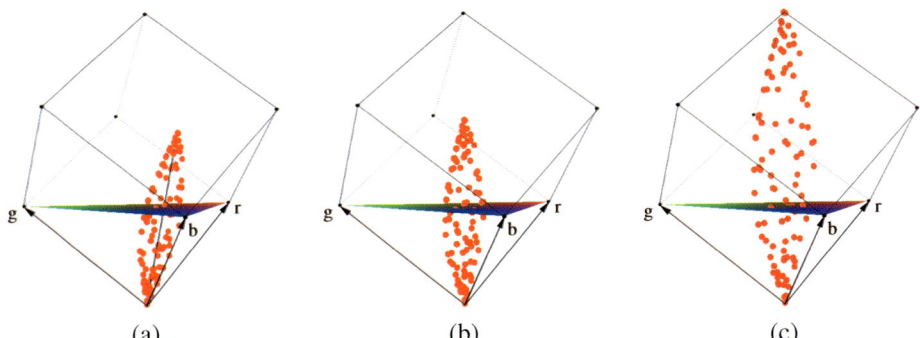

Figure 8.3 Algorithm of Pomierski and Groß (1995). (a) A cloud of points inside the RGB color cube. The main axis is also shown. (b) The cloud of points is rotated onto the gray vector. (c) The point cloud is rescaled such that the color space is completely filled.

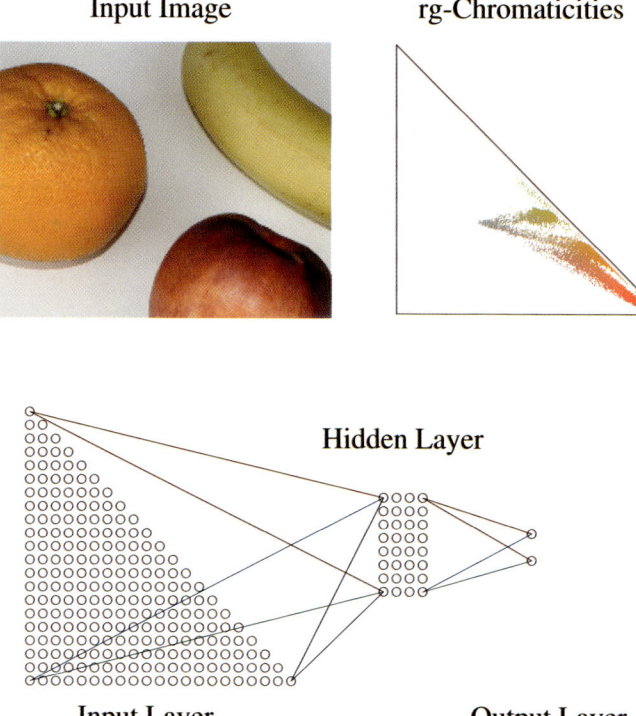

Figure 8.4 Funt et al. (1996) transform the colors of the input image to rg-chromaticity space. The input layer of the neural network samples the triangular shaped region of the chromaticity space. The network consists of an input layer, a hidden layer, and an output layer. The two output neurons estimate the chromaticity of the illuminant.

corresponding input neuron is set to 0. Funt *et al.* used 32 neurons for the hidden layer. The output layer consists of only two neurons for the estimated rg-chromaticity of the illuminant. Training was done using back-propagation. Funt et al. (1996) trained and tested their network on synthetic data. The illuminants were measured with a spectrophotometer around a university campus. Algorithm performance was judged by computing the root mean squared error between the estimated and the actual illuminant. The network performed better than the gray world assumption, the white patch retinex algorithm, and also the two-dimensional gamut constraint algorithm when no constraints are imposed on the illuminant. The two-dimensional gamut constraint algorithm used with illumination constraints resulted in the lowest root mean squared error.

Cardei and Funt (1999) suggested to combine the output from multiple color constancy algorithms that estimate the chromaticity of the illuminant. Their approach is called committee-based color constancy. By combining multiple estimates into one, the root mean squared error between the estimated chromaticity and the actual chromaticity is reduced. Cardei and Funt experimented with committees formed using the gray world assumption, the white patch retinex, and an estimate from a neural network. Each algorithm is used to estimate the chromaticity of the illuminant. A simple method to combine the three estimates is to average the chromaticities. Instead of averaging the chromaticities with equal weights, one can also form a weighted average of the data which is optimal in the least mean square sense. Cardei and Funt used a set of test images to obtain an optimal weighted average of the data for the test set. The third method to combine the estimates, which were tested by Cardei and Funt was to train a neural network. The network had six input neurons (two for the rg-chromaticities of each algorithm). Each committee was then tested on a different set of images. The weighted estimate that was optimal in the least mean squared sense performed best on the test set.

Instead of choosing a neural network and training the weights, one may also make an explicit proposal for a neuro-computational model to achieve color constancy.

8.3 Evolving Color Constancy

The human organism and its visual system in particular is a product of natural evolution. Since we do not know which color constancy algorithm is actually used by the human visual system, we can turn to artificial evolution for an answer to this question. We have used genetic programming to search for a color constancy algorithm (Ebner 2001, 2006). Genetic programming is a variant of an evolutionary algorithm, which is used to evolve computer programs (Banzhaf et al. 1998; Koza 1992, 1994; Koza et al. 1999).

Evolutionary algorithms are frequently used to find optimal solutions in many different problem areas. They are based on Darwin's principle of "survival of the fittest" (Darwin 1996; Maynard Smith 1993). A population of individuals has to compete with other individuals for access to food and mates. Only the successful ones are allowed to reproduce. This leads to the reproduction of certain inheritable traits into the next generation.

The theory of evolution provides answers to many questions pertaining to every day life. For instance, why are we afraid of certain animals but not of others? The simple answer is that some animals are more dangerous than others, i.e. certain types of spiders or snakes are very dangerous. Therefore, it makes sense that many people are afraid of these types of animals. People who like to play with spiders or snakes may have had a slightly lower reproduction rate because some of them died when playing with these animals.

Evolution basically tells us what life is all about. It can be summed up as: survive, eat, reproduce (Dennett 1995). The first and major goal is survival, i.e. not to get eaten by others. The second is finding food. Without food an individual cannot survive. The hunt for food induces a selection pressure on the prey population. If the other two goals are satisfied, i.e. there is no danger of getting eaten by others and the individual is not hungry, then the goal is to find a mate and to reproduce.

Evolutionary algorithms simulate evolution on a computer. The overall structure of an evolutionary algorithm is shown in Figure 8.5. The key ingredients are reproduction, variation, and selection. One also works with a population of individuals. Each individual represents a possible solution to the problem one is trying to solve. The individuals are evaluated on the basis of how good they are at solving the given problem. The individuals who are better at solving the given problem are allowed to reproduce more often than the individuals who are not as good at solving the problem. When an individual is reproduced, small variations are introduced. Genetic operators are applied, which slightly

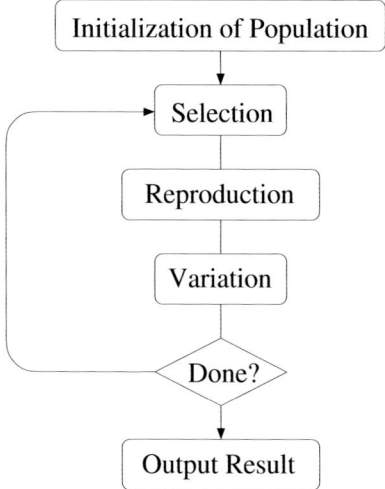

Figure 8.5 Main loop of an evolutionary algorithm. New individuals are created by selecting highly fit individuals and reproducing these individuals. The offspring, however, are usually not exact copies. Some individuals may be altered by genetic operators. In the course of time, the individuals adapt to their environment, i.e. the selection criterion.

change the genetic material of the individuals. The genetic operators are again similar to the mechanisms used by natural evolution. From time to time, an individual is mutated. Two individuals can also exchange genetic material. This is called a *crossover operation*. Owing to selection pressure, individuals adapt to the problem they have to solve.

Often, solutions are sought for simple parameter optimization problems. In this case, the individuals are simply bit strings where the parameters are coded into the bits of the individual (Goldberg 1989; Holland 1992; Mitchell 1996). Some evolutionary algorithms, called *evolution strategies*, work directly in parameter space (Rechenberg 1994; Schöneburg et al. 1994; Schwefel 1995). However, the real advantage of evolutionary algorithms becomes apparent when we recognize that arbitrary structures can be evolved. Evolutionary algorithms can be used to evolve three-dimensional shapes (Bentley 1996, 1999; Ebner 2003b), or even analog or digital circuits (Ebner 2004a; Higuchi et al. 1993; Koza et al. 1999; Miller et al. 2000; Thompson 1996).

In tree-based genetic programming the programs are represented as trees. Figure 8.6 shows a sample individual. An individual is evaluated by traversing its tree structure. The individual shown in Figure 8.6 represents a symbolic expression that evaluates to $2x - 1$. The tree can be viewed as the parse tree of an executable program. The inner nodes are called *elementary functions*. The outer nodes are called *terminal symbols*. In order to apply genetic programming to a given problem domain, one needs to define the set of elementary functions and terminal symbols. This defines the representation of the individuals. The individuals of the initial population are created randomly from the set of elementary functions and terminal symbols.

We have used genetic programming to evolve a program, which is executed by a processing element of a parallel architecture as shown in Figure 8.7. The processing elements

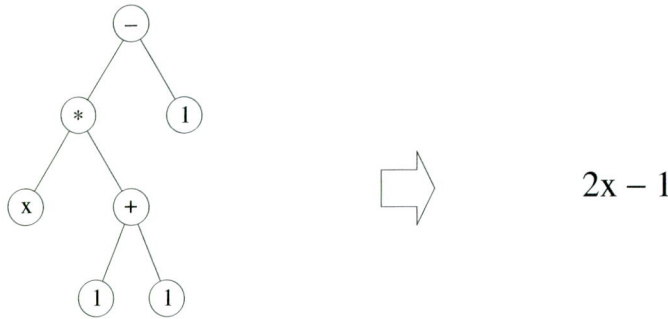

Figure 8.6 In tree-based genetic programming, individuals are represented as trees. A sample individual is shown on the left. This individual represents a symbolic expression that evaluates to $2x - 1$.

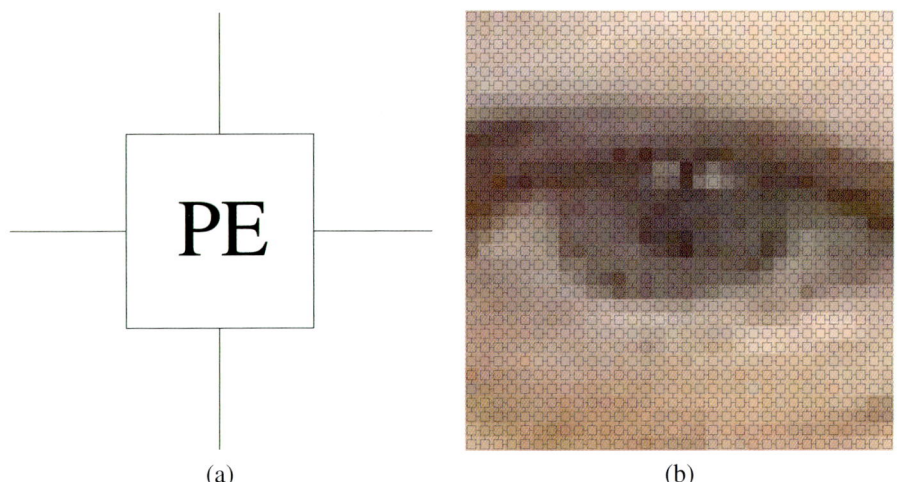

Figure 8.7 A single processing element (a) is connected to four other processing elements. The elements are able to perform simple calculations and can obtain data from the neighboring elements. By combining such processing elements, we construct an $n \times n$ matrix with one processing element per image pixel (b).

are arranged as an $n \times n$ matrix with one processing element per pixel. Each processing element is able to perform simple calculations and to obtain data from neighboring elements. A single instruction multiple data architecture is used. This type of architecture is called *single instruction multiple data* because each processing element carries out the same instruction as the other elements on the data stored locally (Almasi and Gottlieb 1994). The main incentive to use this type of architecture is because of its simplicity. It only allows local connections to neighboring elements. No information is shared globally. Thus, it is biologically plausible. Such massively parallel architectures are ideal for image processing

Table 8.1: Elementary functions used for the experiments.

Elementary Function	Arity	Symbol
Addition	2	+
Subtraction	2	−
Multiplication	2	*
Protected division	2	/
Multiply by 2	1	mul2
Divide by 2	1	div2

Table 8.2: Set of terminal symbols.

Terminal Symbol	Symbol
Constant one	1
Red input band $c_r(x, y)$	red
Green input band $c_g(x, y)$	green
Blue input band $c_b(x, y)$	blue
Current band $c_i(x, y)$	band
Estimate from current element $L_i(x, y)$	center
Estimate from left element $L_i(x - 1, y)$	left
Estimate from right element $L_i(x + 1, y)$	right
Estimate from element above $L_i(x, y - 1)$	up
Estimate from element below $L_i(x, y + 1)$	down

(Bräunl et al. 1995). The task of each processing element is to estimate the illuminant **L** for its image pixel. This estimate is then used to compute the output according to

$$o_i(x, y) = \begin{cases} \frac{c_i(x,y)}{L_i(x,y)} & \text{if } L_i(x, y) > 0.001 \\ 1 & \text{otherwise} \end{cases} \quad (8.9)$$

for color band $i \in \{r, g, b\}$. Table 8.1 shows the set of elementary functions. Table 8.2 shows the set of terminal symbols that were used for the experiments.

Each processing element has access to the color of its input pixel **c** and to the illuminant estimated by neighboring processing elements. Therefore, the terminal symbols are red, green, blue, and band which supply the information stored in the red, green, and blue channel as well as the information of the current color band. The estimated illuminant is available through the symbols center (from the current element), left, right, up, and down (from neighboring elements). The constant 1 was also used as a terminal symbol. Simple arithmetic functions like addition (+), subtraction (−), multiplication (*), and protected division (/) are used as elementary functions. In case of a division by zero or a value smaller than 10^{-10}, the result is replaced by 1.0. Division by two (div2) and multiplication by two (mul2) are also included in the set of elementary functions.

The illuminant is estimated by iteratively executing the following update equation for each color channel i.

$$L_i(x, y) = \text{program}(L_i(x, y), L_i(x-1, y), L_i(x+1, y), L_i(x, y-1), L_i(x-1, y+1), \mathbf{c}) \quad (8.10)$$

where program is defined by the tree-structure of the individual. Initially, the estimate $L_i(x, y)$ is set to the color of the input image, i.e. $L_i(x, y) = c_i(x, y)$. Reproduction, mutation, and crossover were used as genetic operators. Figure 8.8 visualizes the three operators used. Reproduction simply copies an individual into the next generation. Mutation replaces a randomly selected sub-tree by a newly created sub-tree. The crossover operation exchanges two randomly selected sub-trees of two individuals. These operators were applied to create a new population of individuals. Experiments were run for 50 generations.

Individuals were selected based on their performance on three fitness cases. The fitness cases were randomly created for each generation. Therefore, the individuals could not tune their response to a particular set of input data. Each fitness case consisted of a color Mondrian created by randomly drawing colored boxes on top of each other. This is our reflectance image or ground truth. Each Mondrian was artificially illuminated by randomly choosing an illuminant **L** and multiplying the reflectance image with the chosen illuminant component-wise. For each fitness case, we computed the deviation of the output of the individual from the correct reflectance image. Fitness was based on the worst performance of the three tests. An individual had to perform well on all three test cases to achieve high fitness. We performed 10 experiments with different seeds of the random number generator. The performance of the best evolved individual is shown in Figure 8.9. Its program code

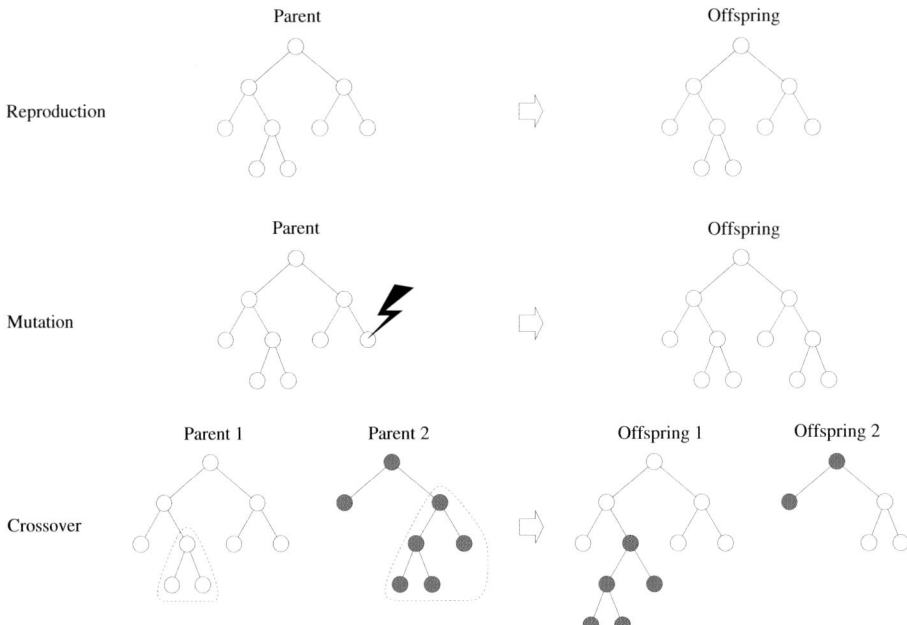

Figure 8.8 Genetic operators used in the genetic programming experiments.

LEARNING COLOR CONSTANCY

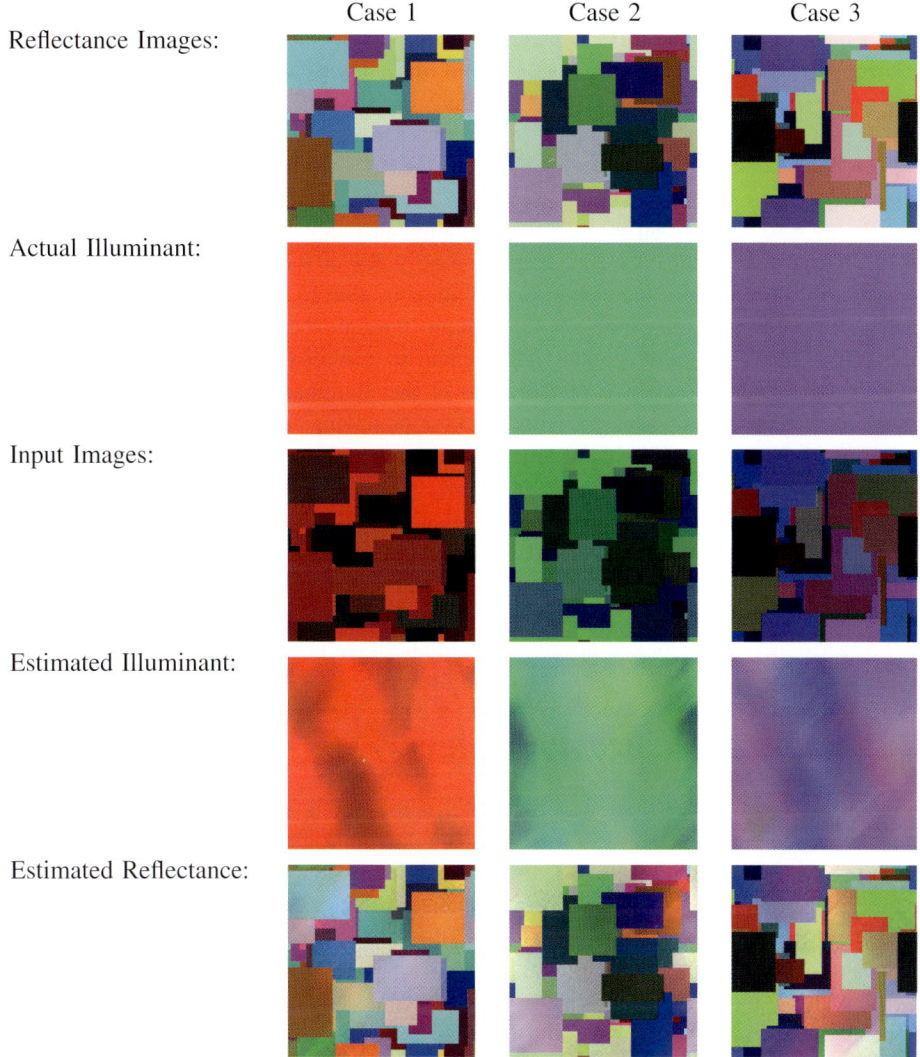

Figure 8.9 Performance of best evolved individual on three different fitness cases. The images in the first row show the reflectance images. The images in the second row show the virtual illuminant. The images in the third row show the input images presented to the individual. The images in the fourth row show the illuminant that was estimated by the evolved individual. In the last row, the estimated reflectance is shown. (Reproduced from Ebner M. 2006 Evolving color constancy. Special issue on evolutionary computer vision and image understanding of pattern recognition letters, Elsevier, 27(11), 1220-1229, by permission from Elsevier.)

is (div2 (+ (+ (div2 (+ (div2 (+ (div2 down) (div2 (+ (div2 (+ (div2 + band (div2 down))) (div2 (+ band (div2 (+ band (div2 center))))))) (div2 down))))) (* (div2 down) (/ right (div2 (+ down (+ (div2 down) (div2 (+ (div2 (+ band (div2 down))) (div2 (+ band (div2 (+ band (div2 center)))))))))))))) (div2 (+ (div2 (+ (div2 right) (div2 (+ band (div2 down))))) (* (div2 down) (/ right (div2 (+ down (+ (div2 down) (div2 (+ (div2 (+ band (div2 center))) (div2 (+ band band)))))))))) down)). The evolved individual uses mainly addition and division by 2 to solve the task. It essentially averages data from neighboring elements. From Figure 8.9, we see that the estimated illuminant is not quite constant. However, the evolved individual is able to estimate the overall color of the light sources for the three fitness cases. Thus, given an incentive, evolution is able to find an algorithm for color constancy. In this case, the incentive was given through the fitness function. In nature, an animal is likely to survive better if it is able to correctly judge the color of fruit or the color of a berry irrespective of the color of the illuminant. Color is a powerful cue. Imagine having to live in a world without color or having to live in a world where the colors constantly change. Classification and object recognition become a lot easier if color stays either constant or approximately constant.

So far, we have seen that given a neural architecture, the neural network may be trained to arrive at colors that are approximately constant. Evolution is also able to find a solution to the problem of color constancy. One of the remaining questions is about how a computational color constancy algorithm is actually mapped to the human visual system. Several researchers have looked at this problem and have proposed possible solutions.

8.4 Analysis of Chromatic Signals

D'Zmura and Lennie (1986) suggest that color constancy is achieved through eye movements. Owing to the movements of the eye, a single point on the retina samples light from many different points of the scene. They assume that an adaptation mechanism is used to discount the illuminant. The adaptation mechanism uses the space average color obtained through eye movements. In this case, at least in theory, a color constant descriptor would be available as early as the retina. However, as is known today, the first visual area where color constant cells were found is V4. The adaptation mechanism would have to be located inside the visual cortex. Given that color constancy also occurs using exposure times less than 0.01s (Land and McCann 1971), it seems unlikely that an adaptation mechanism that uses eye movements is actually used by the visual system. D'Zmura and Lennie suggest that the first color stage is a simple scaling of the color channels using space average color. The scaling of the color channels transforms the coordinate system such that the vector that describes the illuminant is aligned with the gray vector. Since the effect of a different illuminant can be modeled as a 3×3 matrix transformation as described in Section 3.7, a simple diagonal scaling will not align all axes of the coordinate system as was shown in Figure 3.23.

D'Zmura and Lennie assume that red–green as well as blue–yellow color opponent cells are used to arrive at stable hue descriptors. They represent the chromatic sensitivity of a cell using a three-dimensional chromaticity space as shown in Figure 8.10. The white point is denoted by W. The white point describes the steady state of adaptation of the cell. The vector **C** represents the most effective choice of chromatic contrast, which will activate the cell. The preferred hue of the cell can be found by projecting the vector **C** onto

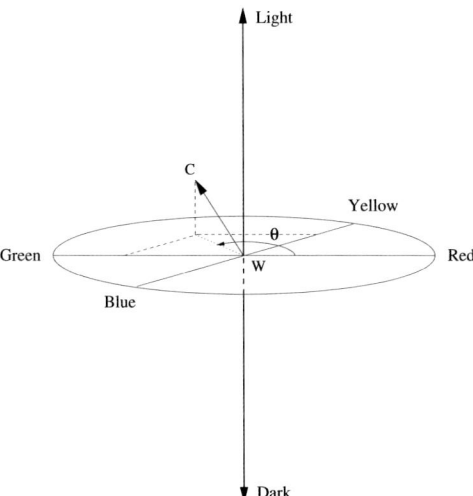

Figure 8.10 Three-dimensional chromaticity space of a color cell. (Reproduced from D'Zmura M and Lennie P 1986. Mechanisms of color constancy. Journal of the Optical Society of America A, 3(10), 1662-1672, by permission from The Optical Society of America.)

the chromaticity plane, which passes through the white point and is spanned by the red–green and blue–yellow color axes. D'Zmura and Lennie argue that because of the center surround characteristic of the color opponent cells, the response of these neurons will depend on the spatial frequency of the stimulus. As spatial frequency is increased, the best stimulus will become achromatic. The color vectors of different cells will rotate away from the chromaticity plane as the spatial frequency is raised. However, a stable hue descriptor would respond to color signals lying inside the chromaticity plane. Thus, the variation in lightness must be separated from the variation in chromaticity. D'Zmura and Lennie suggest that a stable hue descriptor may be arrived at by combining the output of several neurons linearly as shown in Figure 8.11. The responses from pairs of color opponent cells are added such that the opposing tendencies to respond to achromatic stimuli cancel. The result is a set of cells that respond best to purely chromatic modulations along either the red–green or the yellow–blue axis. The output of these cells can then be used to arrive at cells that respond to any desired hue using a linear combination of the cells from the previous layer.

D'Zmura and Lennie assume that several neurons, each of which is tuned to a particular direction inside the chromaticity plane, exist. The perceived hue would be determined by locating the neuron with the largest response.

8.5 Neural Architecture based on Double Opponent Cells

Dufort and Lumsden (1991) proposed a model for color categorization and color constancy. In their model, color constancy is achieved by using the output from double opponent cells

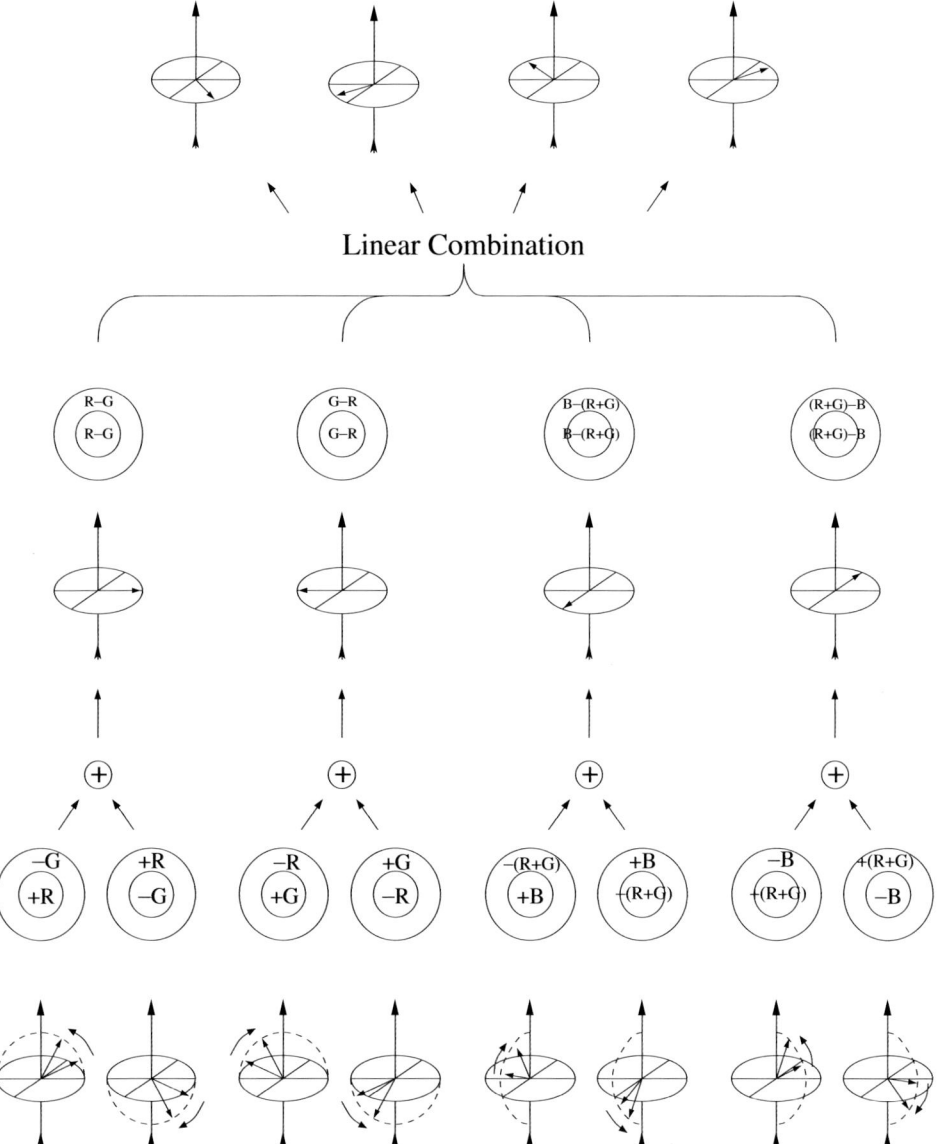

Figure 8.11 Linear combination of the output of several neurons in order to arrive at receptors that respond to any desired hue. (Reproduced from D'Zmura M and Lennie P. 1986 Mechanisms of color constancy. Journal of the Optical Society of America A, 3(10), 1662-1672, by permission from The Optical Society of America.)

that are found inside the blobs of the visual cortex. Dufort and Lumsden work with a neural network model that is described by a differential equation. The input n_i to a neuron i is the weighted sum of the firing rates f of other neurons to which neuron i is connected.

$$n_i = \sum_j w_{ij} f_j \qquad (8.11)$$

where f_j is the firing rate of neuron j and w_{ij} is the weight connecting neuron i with neuron j. Each neuron i has an internal cell potential E_i that varies proportional to the input n_i it receives. It also has an associated resting potential E_r. The cell membrane actively drives the current potential E_i toward the resulting potential E_r. The differential equation that describes both the activation of the neuron and the pull toward the resting potential is given by

$$\frac{dE_i}{dt} = n_i - s(E_i - E_r) \qquad (8.12)$$

where s is a scaling factor that can be conveniently set to 1. The cell potential is limited to the range $[-1, 1]$. The behavior of the cell is simulated using Euler's method

$$\Delta E_i = (n_i - E_i - E_r)\Delta t \qquad (8.13)$$

with $\Delta t = 1.0$. Dufort and Lumsden report convergence to within 0.01 in under five iterations. The firing threshold of cells afferent to the retinal ganglion cells was set to -1 in order to simulate slow wave signal transmission. It was set to 0.0 for the retinal ganglion cells and all other cells following the retinal ganglion cells.

Figure 8.12 shows how Dufort and Lumsden model red–green and blue–yellow color opponent cells as well as double opponent cells. Retinal ganglion cells receive their input either from bipolar or amacrine cells. The bipolar cells in turn receive their input from the cone receptors. Three types of receptors respond to light in the long-wave, middle-wave, and short-wave parts of the spectrum. The red–green double opponent cell is constructed using the output from the red–green color opponent cell in the center and subtracting the output from the red–green color opponent cell in the surround. The blue–yellow double opponent cell is constructed similarly.

The output from the red–green and the blue–yellow color opponent cells defines a two-dimensional coordinate system. A cell that responds maximally to red light can be constructed using the outputs from the red–green as well as the blue–yellow channel. Let x be the output of the red–green channel and let y be the output of the blue–yellow channel. A cell, which computes its output z according to

$$z = w_1 x - w_2 |y| \qquad (8.14)$$

for some weights w_1 and w_2, responds maximally if the red–green channel is at its maximum, i.e. $x = 1$ and no contribution is made from the blue–yellow channel, i.e. $y = 0$. This cell can be made to respond to arbitrary colors using a simple two-dimensional coordinate transform

$$\begin{pmatrix} x' \\ y' \end{pmatrix} \begin{pmatrix} \cos\theta & \sin\theta \\ -\sin\theta & \cos\theta \end{pmatrix} \begin{pmatrix} x \\ y \end{pmatrix}. \qquad (8.15)$$

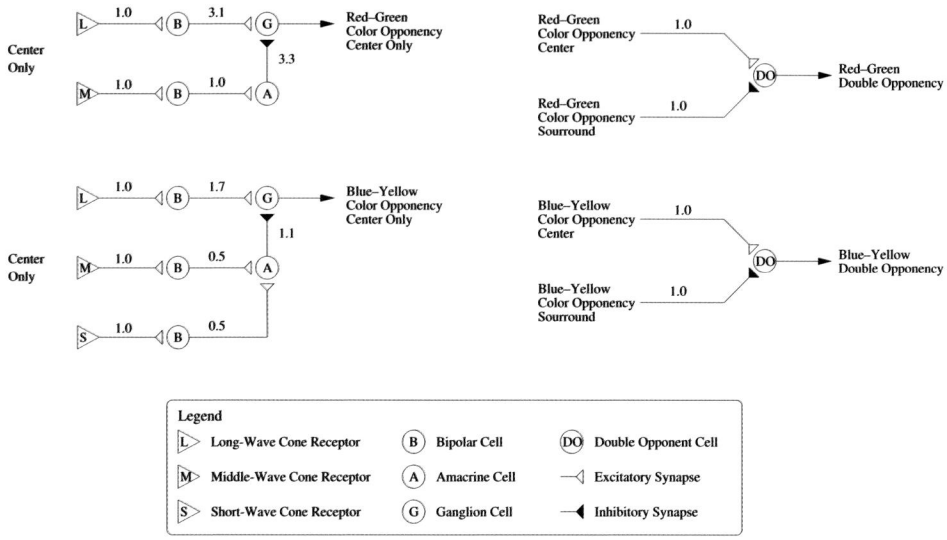

Figure 8.12 Modeling of color opponent and double opponent cells. (Redrawn from Figure 3 (page 298) from Dufort PA and Lumsden CJ 1991 Color categorization and color constancy in a neural network model of V4. Biological Cybernetics, Springer-Verlag, Vol. 65, pp. 293-303, Copyright Springer-Verlag 1991, with kind permission from Springer Science and Business Media.)

Thus, a cell that computes its output according to

$$z = w_1(x \cos \theta + y \sin \theta) - w_2 |y \cos \theta - x \sin \theta| \qquad (8.16)$$

will respond maximally to the color described by the angle θ. The first term is simply a weighted addition of x and y. This term can be realized using a single neuron that receives its input from neurons x and y. The absolute value can be computed using two neurons. One neuron computes the result if $y \cos \theta - x \sin \theta$ is positive, the other neuron computes the result if $y \cos \theta - x \sin \theta$ is negative.

Figure 8.13 shows how color classification may be achieved using a network of four neurons. The cell realizing the first term is basically a cell that describes the saturation of the input color. The two cells that are used to compute the absolute value are basically ratio detectors. If the color differs from the color to be detected, then one of the two cells will be activated which in turn will send an inhibitory signal to the color cell. The output from this network will not be color constant if the signals from the red–green and blue–yellow color opponent cells are used as x and y inputs. Dufort and Lumsden suggest that the output from the double opponent cells which are found in V1 are used as input for the network shown in Figure 8.13. Red–green double opponent cells may be constructed using excitatory input from the red–green color opponent cells in the center and inhibitory input from the red–green color opponent cells in the surround (Dufort and Lumsden 1991). The color cell of Figure 8.13 would be located in V4.

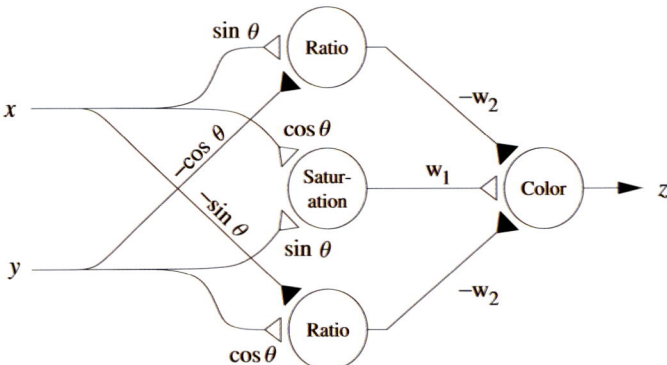

Figure 8.13 Neural network for color classification and color constancy. (Redrawn from Figure 6ab (page 299) from Dufort PA and Lumsden CJ 1991 Color categorization and color constancy in a neural network model of V4. Biological Cybernetics, Springer-Verlag, Vol. 65, pp. 293-303, Copyright Springer-Verlag 1991, with kind permission from Springer Science and Business Media.)

8.6 Neural Architecture Using Energy Minimization

Usui and Nakauchi (1997) proposed the neural architecture shown in Figure 8.14. They assume that the reflectances and illuminants can be described by a finite set of basis functions. In their work, they assume that three basis functions are sufficient for both reflectances and illuminants. The neural architecture tries to estimate the coefficients of the reflectance basis functions.

The architecture consists of four layers denoted by A, B, C, and D. The input layer is denoted by A. Two intermediate layers denoted by B and C receive input from the input layer A. The reflectance coefficients are estimated in the output layer D. The measured color signal, i.e. the input image is fed into the architecture at the bottom at layer A. Each layer consists of three sublayers that indicates that for each layer A, B, C, and D, data is processed for all three bands. Layer B is basically used to perform Horn's algorithm (Horn 1974) that was described in Section 7.2. This layer first applies the Laplacian to the input data and then thresholds the output. Before feeding the output to the output layer D, an inverse Laplacian is applied. Layer C applies a Gaussian blur to the input image. This layer is used to pull the reflectance toward the average input signal at edge positions. Layer B is used to signal layer C where edges are located.

The output reflectances are computed by minimizing an energy function that is computed using the data of layers B, C, and D. The energy function E consists of three components E_1, E_2, and E_3. The relative importance of the components is set by adjusting the weights w_1 and w_2.

$$E = E_1 + w_1 E_2 + w_2 E_3. \tag{8.17}$$

The first component is used to compute the reflectances according to the algorithm of Horn

$$E_1 = \sum_i \sum_{x,y} \left(\Theta(\nabla^2 c_i(x,y)) - \nabla^2 o_i(x,y) \right)^2 \tag{8.18}$$

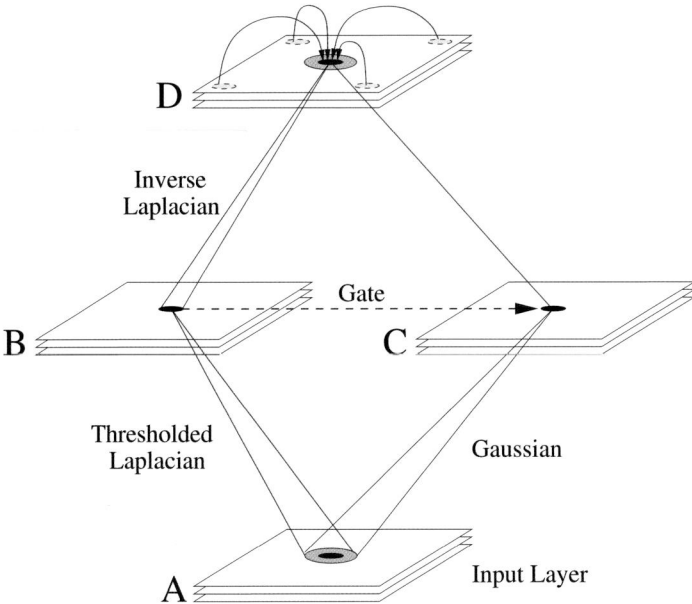

Figure 8.14 Neural architecture of Usui and Nakauchi (1997). The architecture consists of four layers denoted by A, B, C, and D. The input image is fed into the architecture at the input layer. Layer D is the output layer. (Redrawn from Figure 8.4.1 (page 477) Usui S and Nakauchi S 1997 A neurocomputational model for colour constancy. In (eds. Dickinson C, Murray I and Carded D), John Dalton's Colour Vision Legacy. Selected Proceedings of the International Conference, Taylor & Francis, London, pp. 475–482, by permission from Taylor & Francis Books, UK.)

where Θ is a threshold operation. In other words, the differentiated input signal of adjacent pixels and the differentiated estimated reflectance of adjacent pixels should be similar. The second component is used to pull the estimated reflectances toward the average of the input data at discontinuities of the image.

$$E_2 = \sum_i \sum_{x,y} \text{Gate}(\nabla^2 c_i(x,y)) \, (G \otimes c_i(x,y) - o_i(x,y))^2 \qquad (8.19)$$

where G is the Gaussian operator and $\text{Gate}(x)$ is a gate function which is equal to one if $|x| > \epsilon$ is for a threshold ϵ and zero otherwise. The third component is used to enforce the gray world assumption

$$E_3 = \sum_i \left(g_i - \frac{1}{n} \sum_{x,y} o_i(x,y) \right)^2 \qquad (8.20)$$

where g_i are the surface coefficients of a gray sample and n is the number of pixels in the input image. Color constancy is achieved by minimizing the energy function E with respect to the output coefficients o_i. Each coefficient is adjusted by performing a gradient descent

$$o_i(x, y) = o_i(x, y) - s \frac{\partial E}{\partial o_i(x, y)} \tag{8.21}$$

for some scaling parameter s. Usui and Nakauchi demonstrate the performance of the architecture on simulated two-dimensional Mondrian images.

Additional neural models for color constancy, which were not discussed in depth, were proposed by Courtney et al. (1995).

9

Shadow Removal and Brightening

So far, we have mostly assumed that objects are directly illuminated by a light source. But what about shadows? In real world scenes, one object may cast a shadow on another object. In this case, the area outside the shadow is directly illuminated, while the shadow area is diffusely illuminated by light that is reflected from other objects in the scene. Thus, for a natural scene containing shadows, we have at least two different illuminants, the light source and the illuminant of the shadow area. In Chapter 7, we have already seen a number of algorithms that can cope with an illuminant that varies smoothly over the image. A shadow, however, may have a sharp boundary. We discuss an algorithm for shadow removal in this chapter.

9.1 Shadow Removal Using Intrinsic Images

Finlayson et al. (2004, 2002) have shown how intrinsic images may be used to remove shadows from images. The method is a variant of Horn's color constancy algorithm (Horn 1974, 1986). His algorithm was already discussed in depth in Section 7.2. Horn suggested first taking the logarithm of the input colors and then applying the Laplace operator. A threshold operation is applied such that only values that are larger than a certain threshold remain. The threshold operation is used to distinguish between a change due to a change of the illuminant and a change due to a change of reflectance. Only changes due to a change of reflectance should be retained. A color image is obtained by reintegrating. Blake (1985) improved on the algorithm of Horn. He suggested applying the threshold operation to the first derivative. From the thresholded first derivative, the second derivative is formed and finally the image is reintegrated. The difficult part is deciding if the output of the Laplacian, or first derivative, was caused by a change of reflectance or a change of the illuminant. In practice, this is not easy, and it is quite likely that some small changes of reflectance are accidentally removed or that large changes of the illuminant are retained.

Finlayson et al. (2004, 2002) suggest using intrinsic images to determine if an edge of the input image is caused by a shadow, i.e. a change of the illuminant. Intrinsic images are free from shadows. They first segment the input image and then compute edges of the segmented input image and the RGB chromaticity intrinsic image. Pixels that have

---

Color Constancy M. Ebner
© 2007 John Wiley & Sons, Ltd

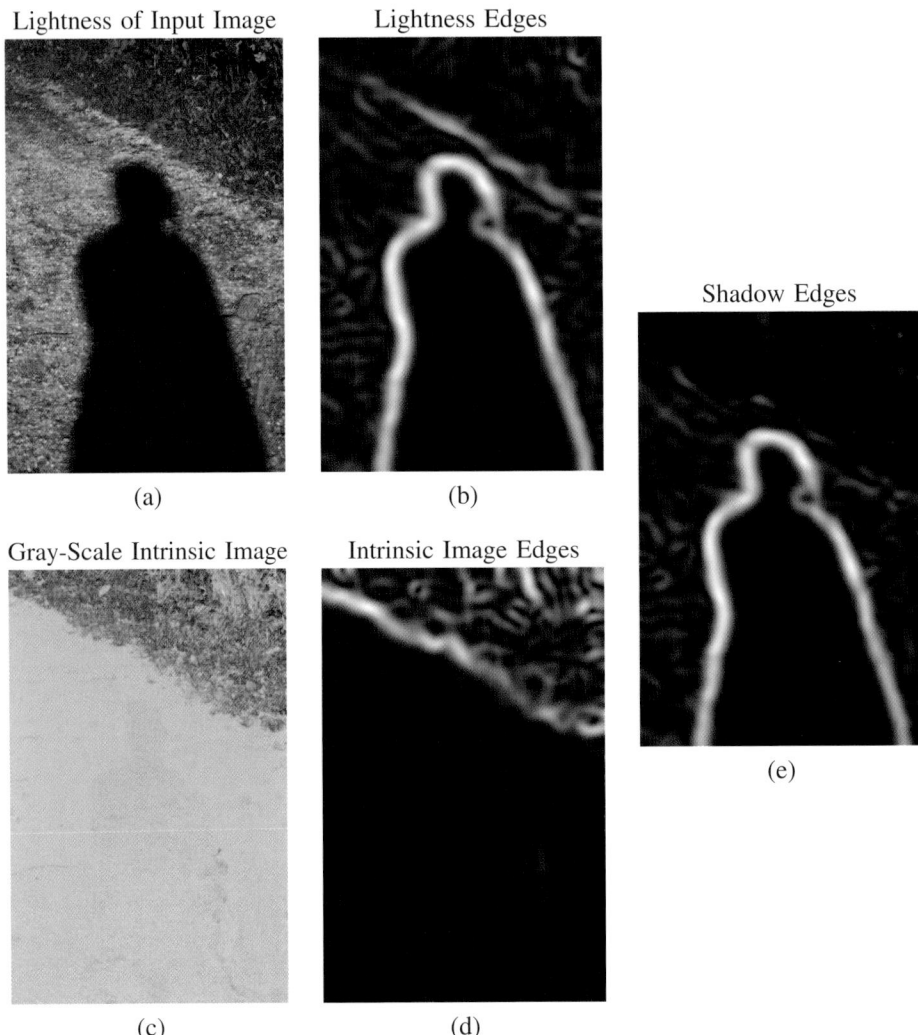

Figure 9.1 Extraction of shadow edges. The image in (a) shows the lightness of the input image. The image in (c) shows the gray-scale intrinsic image. The images (b) and (d) show the extracted edges. The images were blurred in order to extract only large-scale edges. The image in (e) was computed by subtracting the intrinsic edge image from the lightness edge image.

edge values higher than a threshold for any channel in the original segmented image and lower than another threshold in the RGB chromaticity intrinsic image are shadow edges. Thus, they first apply the first derivative to the logarithm of the input image. All edges that are not shadow edges are retained. Instead of simply zeroing the shadow edges, they suggest binarizing the shadow edges, thickening them using a morphological operator, and then iteratively replacing unknown derivative values on the boundary of the shadow edges by the median of known ones in the vicinity. Figure 9.1 illustrates the detection of shadow edges. Instead of segmenting the input image, we first compute the lightness of the input image. Then the gradient magnitude is computed from a blurred lightness image. We then compare the edges from the lightness image to the edges of the gray-scale intrinsic image. Figure 9.1 (a) shows the lightness of the input image. Figure 9.1 (c) is the gray-scale intrinsic image. Figure 9.1 (b) and (d) show the edge images. Before computing the edges, we also blurred the images to order to extract the large-scale edges. Figure 9.1 (e) was computed by subtracting the edges of the gray-scale intrinsic image from the edges computed for the input image. Only shadow edges remain. The binarized shadow edges are shown in Figure 9.2. A closing operation was also applied.

In order to obtain an illuminated image with shadows removed, we can now apply a method similar to the one used by Horn (1974, 1986) or Blake (1985) in their color constancy algorithms. Let $\mathbf{c}(x, y)$ be the pixel color at position (x, y). Let $S(x, y)$ be the binary image describing the location of shadow edges, i.e. the image in Figure 9.2(a). All derivatives at the position of shadow edges are set to zero. Then the Laplacian is computed from the derivatives. Thus, we have to solve the following equation:

$$\nabla^2 \log(\mathbf{c}(x, y)) = \nabla((1 - S(x, y))\nabla \log(\mathbf{c}(x, y))) \tag{9.1}$$

This is the familiar Poisson's equation (Bronstein et al. 2001; Kuypers 1990; Nolting 1992). We can use any of the methods described in Section 7.2 to compute a reintegrated image. The image in Figure 9.2(b) shows the resulting image. The pronounced shadow has been removed. Note that it also contains seemingly blurred areas along the original shadow edge. Thus, by zeroing the derivatives at the position of shadow edges, we also lose the reflectance information at these points. This is not a problem for sharply delineated shadows. However, a shadow edge may not be sharply delineated. Such shadow edges might not be removed. Finlayson et al. (2004) suggest replacing the unknown derivatives by the median of known ones. This is done iteratively. They take the binarized shadow image and replace all unknown derivatives along the border of the binarized shadow image by the median of known ones. Then the binarized shadow image is thinned by removing the border pixels. Again, all unknown derivatives along the border of the binarized shadow image are replaced by the median of known ones. This continues until the binarized shadow image has disappeared. The image in Figure 9.2(c) shows the reintegrated image obtained using the method of iteratively filling in derivatives. The area where the shadow edge is located looks much better now.

9.2 Shadow Brightening

Warnock (1999) describes a technique known among photographers for brightening shadows. If a subject is photographed against the sunlight, i.e. the background is very bright, then the subject may be underexposed. Photographers solved this problem in the darkroom by placing an out-of-focus lighter version of the negative in the path of the enlarger light. The same technique can also be applied to digital images. Warnock (1999) explains how this can be done using Photoshop. First, the input image is converted to a gray-scale image. Next, the image is blurred with a Gaussian such that all detail is lost. The blurred image can then be used to select shadow areas of the image. Areas that are not shadow areas

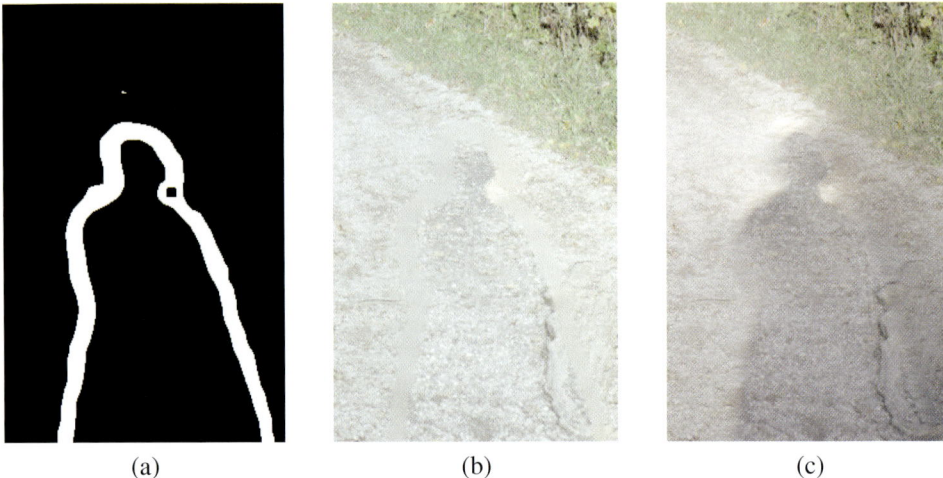

(a) (b) (c)

Figure 9.2 The image in (a) shows the binarized shadow edges. A closing operator was applied to close any gaps that may be present in the edge image. The image in (b) shows the resulting reintegrated image. Here, all gradients that are located on a shadow edge have been zeroed. The image in (c) also shows a reintegrated image. This time, unknown gradient values were filled in. The zeroed gradient values were iteratively replaced by the median gradient value of surrounding pixels.

are deselected manually. The shadow areas are brought out of the shadows by filling the image with 50% gray using the color dodge mode of Photoshop. Popular program such as Adobe's Photoshop Elements 2.0 include automatic routines for shadow brightening. A Photoshop plug-in for shadow adjustment is offered by the Eastman Kodak Company, Austin Development Center (2004). Moroney and Sobel (2001) describe a very general method to adjust the pixel values through a nonlinear operation. A mask image $m(x, y)$ is created by blurring the input image $c(x, y)$. This mask image is then used to adjust the color channels according to

$$o = c^{2^{\frac{k_1 m - k_2}{k_2}}} \qquad (9.2)$$

where o is the resulting pixel value and k_1 and k_2 are two constants.

Figure 9.3 shows results for a similar algorithm except that the blurred image is used to brighten the original image using a gamma correction. Output colors $o_i(x, y)$ for each image position (x, y) were computed from the original colors $c_i(x, y)$ of the image by applying

$$o_i(x, y) = (c_i(x, y))^{\frac{1}{1+H(x,y)}} \qquad (9.3)$$

where $H(x, y)$ is the blurred image. After the gamma correction was applied, the saturation of the colors was increased by 15%. The results are shown for a sample image in Figure 9.3. Instead of directly adjusting the RGB values, we could also transform the colors to a

SHADOW REMOVAL AND BRIGHTENING

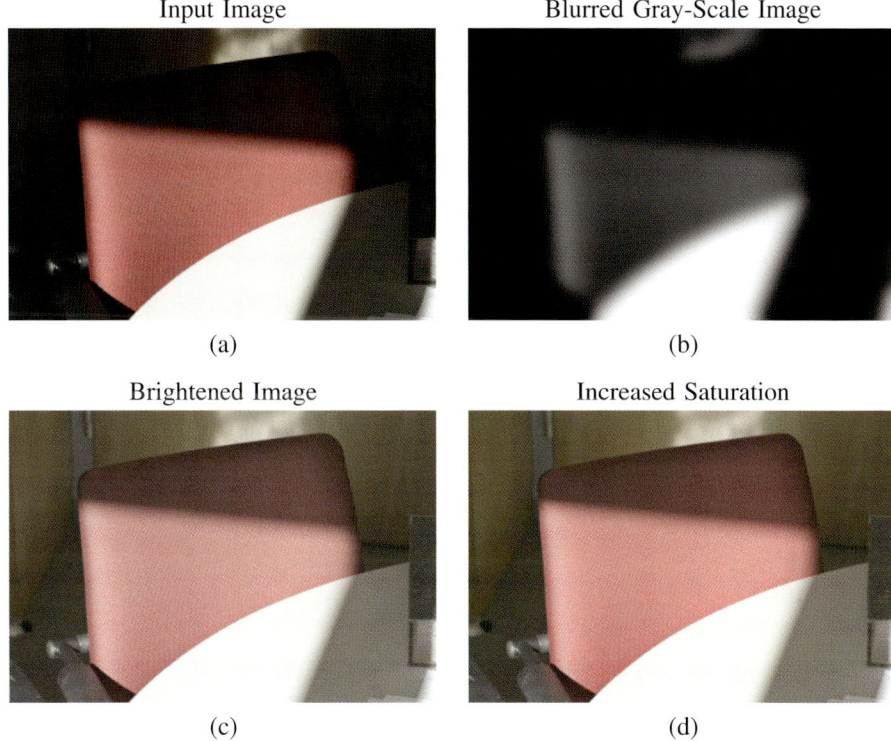

Figure 9.3 The image in (a) is the original input image that contains a strong shadow. The image in (b) shows the blurred gray-scale image. The blurred gray-scale image can be used to apply a gamma correction such that pixels inside dark areas are made brighter (c). The image in (d) shows the same brightened image except that the saturation was increased by 15%.

color space such as $L^*a^*b^*$ or HSI and then only adjust the intensity component while maintaining the original color of the pixels. In Chapter 10, we will see how local space average color can be computed using a resistive grid. The same technique can also be used to compute the blurred gray-scale image. This makes it possible to perform shadow brightening directly inside an imaging chip.

10

Estimating the Illuminant Locally

In Section 4.6, we have seen that a simple scaling of the color channels suffices if the response curves are very narrow banded and a single illuminant that is uniform over the entire image is assumed. If the response curves are not very narrow banded, they can be sharpened and again a diagonal transform suffices. But what do we do about the assumption that the illuminant is uniform? In practice, this assumption is frequently violated because of either one or multiple local light sources. A possible solution is to relax the assumption. If we want to develop color constancy algorithms that also work in the presence of spatially varying illuminants, we have to necessarily estimate the illuminant for each image pixel.

For a nonuniform illuminant, the intensity $I_i(x, y)$ measured by an image sensor for color band $i \in \{r, g, b\}$ at position (x, y) in the image is given by

$$I_i(x, y) = G(x, y) R_i(x, y) L_i(x, y) \tag{10.1}$$

where $G(x, y)$ is a factor due to the scene geometry at position (x, y), $R_i(x, y)$ is the reflectance, and $L_i(x, y)$ is the irradiance (see Equation 3.59 in Section 3.6 for a derivation). Assuming linearized image colors, we have $c_i(x, y) = I_i(x, y)$. If we know the illuminant $L_i(x, y)$ for each image pixel, we can compute a color-corrected image by dividing each image pixel value by the estimate of the illuminant for that pixel.

10.1 Local Space Average Color

In order to estimate the illuminant locally, we have to again make some assumptions. A reasonable assumption is that even though the illuminant is nonuniform, it varies smoothly over the image. In Section 6.2, we have seen that the illuminant can be approximated by global space average color. Global space average color is given by

$$a_i = \frac{1}{n} \sum_{x,y} c_i(x, y) \tag{10.2}$$

Color Constancy M. Ebner
© 2007 John Wiley & Sons, Ltd

where n is the number of image pixels. The globally uniform illuminant can then be estimated as

$$L_i = f a_i \tag{10.3}$$

where f is a factor, e.g. $f = 2$. If we apply the gray world assumption locally, then we need to do the averaging locally for each image pixel.

A simple and fast method to compute local space average color is to perform a convolution (Gonzalez and Woods 1992; Horn 1986; Jähne 2002; Jain et al. 1995; Parker 1997; Press et al. 1992; Radig 1993; Shapiro and Stockman 2001) with an extended kernel defining a neighborhood around the given pixel. Local space average color is then given by

$$a_i = k \int\int c_i g(x - x', y - y') \, dx' \, dy' \tag{10.4}$$

where $g(x, y)$ is a kernel function. The constant k is chosen such that

$$k \int\int g(x, y) \, dx \, dy = 1. \tag{10.5}$$

This constant k takes care of normalizing the result. For a large kernel that extends outside the image, this factor has to be computed for each pixel. A possible choice for a kernel is a Gaussian.

$$g(x, y) = e^{-\frac{x^2 + y^2}{2\sigma^2}} \tag{10.6}$$

A two-dimensional Gaussian is shown in Figure 10.1(a).

In order to estimate the illuminant locally, we have to use quite a large surround. The surround has to be made large enough such that the averaging is done over many different objects with different reflectances. This is required for the assumption that on average the world is gray to hold. The size of the surround is defined through the constant σ. A small value for σ will result in a small neighborhood, whereas a large value will result in a large neighborhood. We usually set this constant to $\sigma = 0.093 n_{\max}$ where $n_{\max} = \max\{n_x, n_y\}$ and n_x and n_y are the width and height of the input image. With such a choice for σ, the neighborhood extends almost half way across the entire image. Figure 10.2 shows the

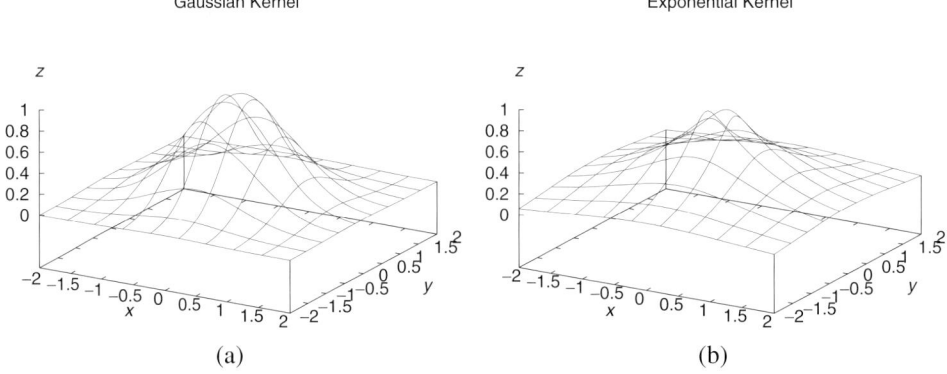

Figure 10.1 A Gaussian kernel is shown in (a), and an exponential kernel is shown in (b).

ESTIMATING THE ILLUMINANT LOCALLY

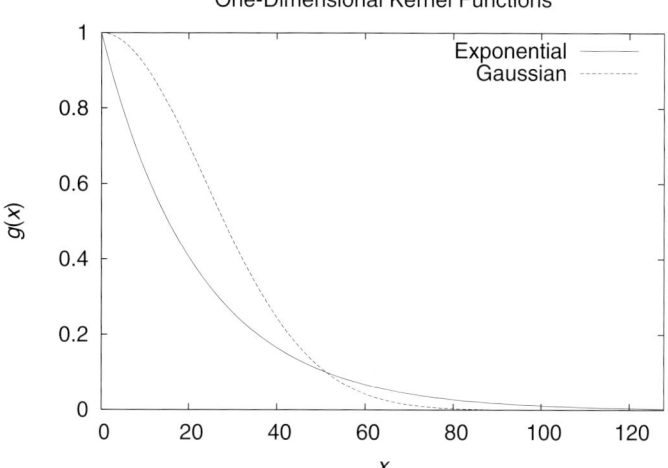

Figure 10.2 A Gaussian and an exponential kernel for an image of 256 pixels.

one-dimensional kernel for an image of 256 pixels. Obviously, care must be taken at the boundary of the image. A safe strategy is not to make any assumptions about pixels that are located outside of the image. Only image pixels are averaged.

Instead of using a Gaussian as a kernel we can also use an exponential kernel with

$$g(x, y) = e^{-\frac{r}{\sigma}}. \tag{10.7}$$

and $r = \sqrt{x^2 + y^2}$. This kernel is shown in Figure 10.1(b). A one-dimensional version is shown in Figure 10.2. An exponential kernel was used by Moore et al. (1991) for their color constancy algorithm, which is described in Section 7.3. When using an exponential kernel, we usually set this constant to $\sigma = 0.087 n_{\max}$ where $n_{\max} = \max\{n_x, n_y\}$ and n_x and n_y is the width and height of the input image. Compared to a Gaussian kernel, the pixels in the immediate neighborhood contribute less to the average whereas the pixels further away contribute more to the average. Figure 10.3 shows local space average color computed once using a Gaussian and once using an exponential kernel.

10.2 Computing Local Space Average Color on a Grid of Processing Elements

The illuminant has to be estimated for each image pixel. An algorithm that has to be executed for all pixels lends itself to parallelization. In the following discussion, we will be working with a grid of processing elements, i.e. the same type of architecture that we have used to evolve a color constancy algorithm in Section 8.3. Each element will locally estimate space average color. Together with the gray world assumption, we can then estimate the reflectances of the objects in view. Several methods for computing a color-corrected image are described in the next chapter.

Input Images

Gaussian Kernel

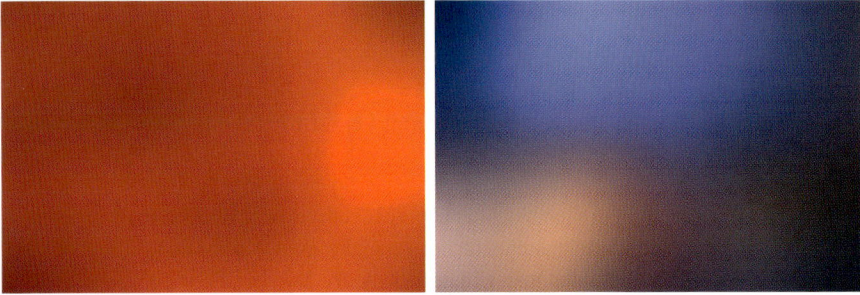

Exponential Kernel

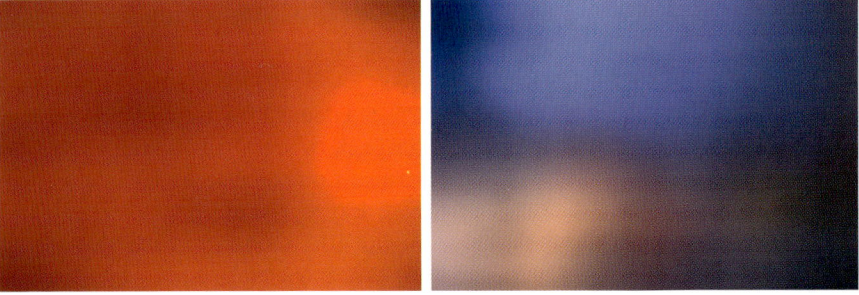

Figure 10.3 The input images are shown in the first row, the result of a convolution with a Gaussian kernel is shown in the second row, and the result of a convolution with an exponential kernel is shown in the last row.

Local space average color can be computed with a surprisingly simple algorithm (Ebner 2002, 2004c). We assume a two-dimensional mesh of processing elements. Each processing element is able to perform simple computations and is able to exchange data with neighboring processing elements. We have one processing element per pixel as shown in Figure 8.7. Each processing element is connected to four other processing elements above and below as well as to the processing elements on the left and right. Note that the elements can

ESTIMATING THE ILLUMINANT LOCALLY

only exchange data locally and not globally. Thus, the algorithm is scalable to arbitrary image sizes. We compute local space average color by assuming some value for space average color at each pixel position. Initially, this value can be arbitrary. Then we look at the space average color computed by neighboring elements. We average the estimates that were computed by the neighboring elements. The current pixel color is then slowly faded into the result. We now formalize the necessary calculations.

Each element has access to the color stored at position (x, y). We assume that we have three different color bands and that the color at position (x, y) is

$$\mathbf{c}(x, y) = [c_r(x, y), c_g(x, y), c_b(x, y)]^T. \qquad (10.8)$$

The computations are performed for all three color bands independently. Let the current estimate of local space average color be

$$\mathbf{a}(x, y) = [a_r(x, y), a_g(x, y), a_b(x, y)]^T. \qquad (10.9)$$

We can initialize a_i with $i \in \{r, g, b\}$ with an arbitrary value. The following updates are then iterated indefinitely for all three color bands:

$$a'_i(x, y) = \frac{1}{4}(a_i(x-1, y) + a_i(x+1, y) + a_i(x, y-1) + a_i(x, y+1)) \qquad (10.10)$$

$$a_i(x, y) = (1-p)a'_i(x, y) + pc_i(x, y) \qquad (10.11)$$

where p is a small percentage larger than zero.

This percentage depends on the size of the image. For our experiments, we usually set this value to $p = 0.0005$ for images of size 256×256. In the subsequent text, we discuss how this value may be chosen. The first step averages the data obtained from neighboring elements. The second step fades the color $\mathbf{c}$ at position (x, y) into the current average $\mathbf{a}$. This can be viewed as a biased average, e.g.

$$a_i(x, y) = \frac{(\frac{1}{p} - 1)a'_i(x, y) + c_i(x, y)}{\frac{1}{p}}. \qquad (10.12)$$

For $p = 0.0005$, this is equivalent to

$$a_i(x, y) = \frac{1999 a'_i(x, y) + c_i(x, y)}{2000}. \qquad (10.13)$$

We can view this as local space average color having a carrying capacity of 2000, where we have 1999 components of a'_i and one component of c_i.

The current average is always multiplied with a value smaller than one. Therefore, this component will slowly decay toward zero. This is the reason the initialization can be arbitrary. Sooner or later, nothing will be left of the original value. If the estimates that were computed by the neighboring elements are lower than the current estimate, then the current estimate will be pulled down. If the estimates of the neighboring elements are higher, the current estimate will be pulled up.

We have to give special attention to the elements located at the boundary of the images. These elements have less than four neighbors. The extension to a fewer number of neighbors

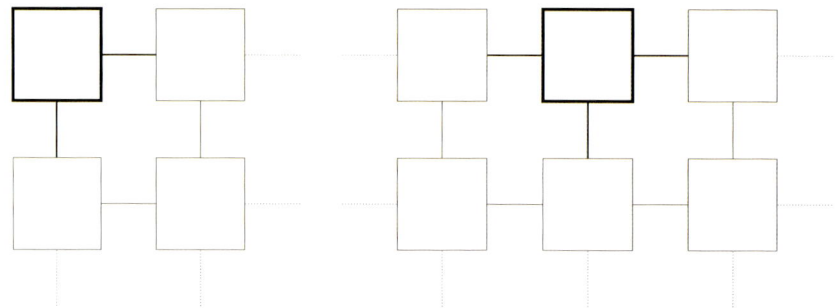

Figure 10.4 Corner elements have only two neighbors (left). Border elements have only three neighbors. Corner elements average data from two neighboring elements, while border elements average data from three neighboring elements.

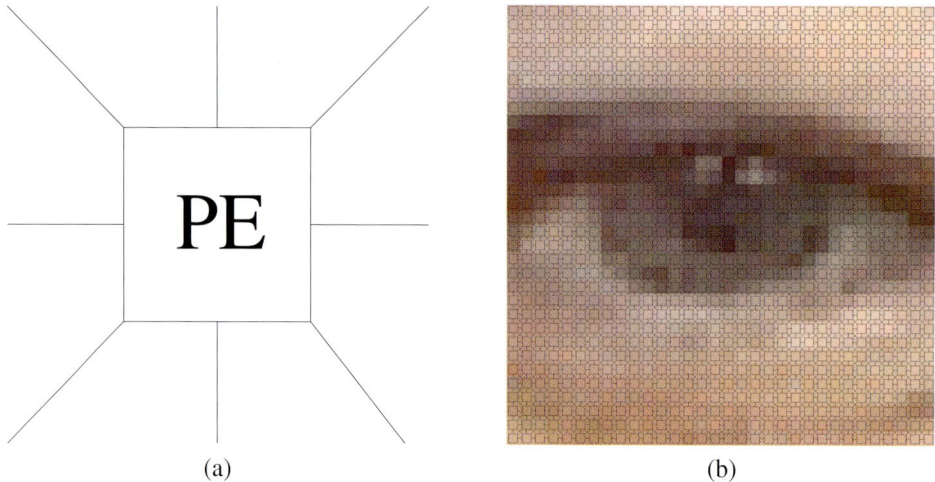

Figure 10.5 A processing element with diagonal connections (a) and a matrix of 32×32 processing elements (b).

is straightforward. If the current element is located at a border, then we have only three neighboring elements (Figure 10.4). The four corner elements only have two neighbors. If we have only three neighboring elements, then we average the data from these three neighboring elements. If we have only two neighboring elements, then we average the data from two elements. Similarly, we could also increase the neighborhood. We could also connect elements diagonally as shown in Figure 10.5. In this case, each element would have eight neighbors and we would average the data from these eight neighbors. In practice, it is sufficient to use only the four nearest neighbors. The use of eight neighbors produces a similar behavior.

ESTIMATING THE ILLUMINANT LOCALLY

Let $N(x, y)$ be the neighbors of the processing element located at (x, y).

$$N(x, y) = \{(x', y') | (x', y') \text{ is neighbor of element } (x, y)\} \tag{10.14}$$

Local space average color is then computed iteratively by averaging the data from all available neighboring elements.

$$a'_i(x, y) = \frac{1}{|N(x, y)|} \sum_{(x', y') \in N(x, y)} a_i(x', y') \tag{10.15}$$

$$a_i(x, y) = (1 - p)a'_i(x, y) + pc_i(x, y) \tag{10.16}$$

The first step averages the data from the neighboring element and the second step slowly adds the color measured at the current processing element.

Figure 10.6 shows how the averaging operation works. Suppose we initialize a small 3×3 mesh of processing elements with the numbers from 1 to 9. If we keep averaging the data from neighboring elements, we obtain the sequence shown in Figure 10.6. After a few iterations, we arrive at the correct average value of 5.0. Note that we have not included the center element in the averaging operation. In principle, we could carry out the following sequence of steps equally well.

$$a'_i(x, y) = \frac{1}{|N(x, y)| + 1} \sum_{(x', y') \in N(x, y) \cup \{(x, y)\}} a_i(x', y') \tag{10.17}$$

$$a_i(x, y) = (1 - p)a'_i(x, y) + pc_i(x, y) \tag{10.18}$$

In other words, we could average the data from the four neighboring elements as well as the current element. Omitting the current element from the averaging operation may lead to

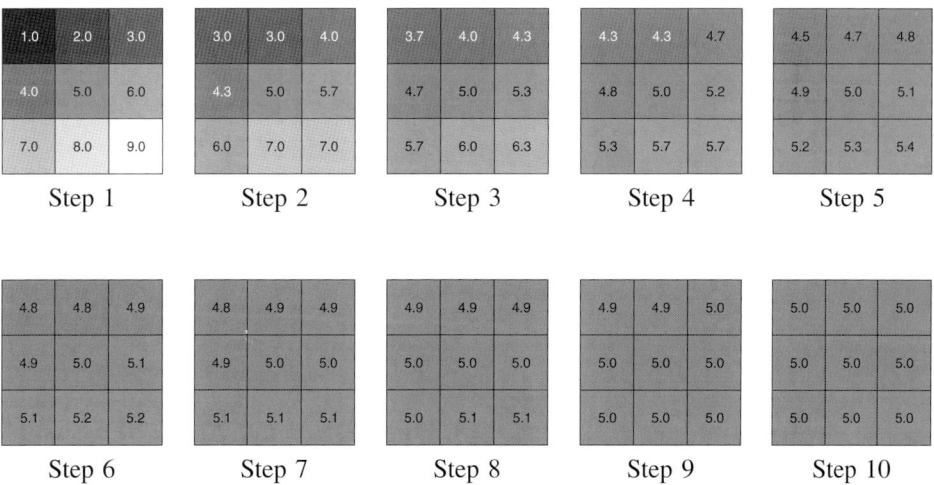

Figure 10.6 A 3×3 mesh of processing elements is initialized with numbers from 1 to 9. The sequence shows what happens if we continue to average the data from neighboring elements. The current element was not included in the averaging operation.

oscillatory behavior as shown in Figure 10.7. The oscillatory behavior disappears if we also average the current element. However, not including the current element in the averaging operation makes it easier to implement the algorithm in hardware.

The second step fades the color at the current element into the current average. This can also be viewed as a force that pulls the average into the direction of the color at the current element. Figure 10.8 shows the computed local average color for two different input images.

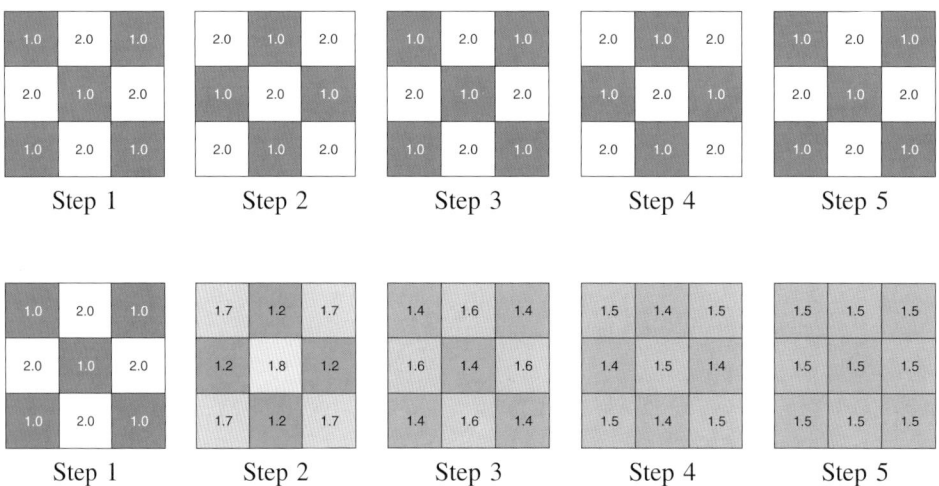

Figure 10.7 Oscillatory behavior might result if we do not include the current element into the averaging operation (top). If the current element is also averaged, the oscillatory behavior disappears (bottom).

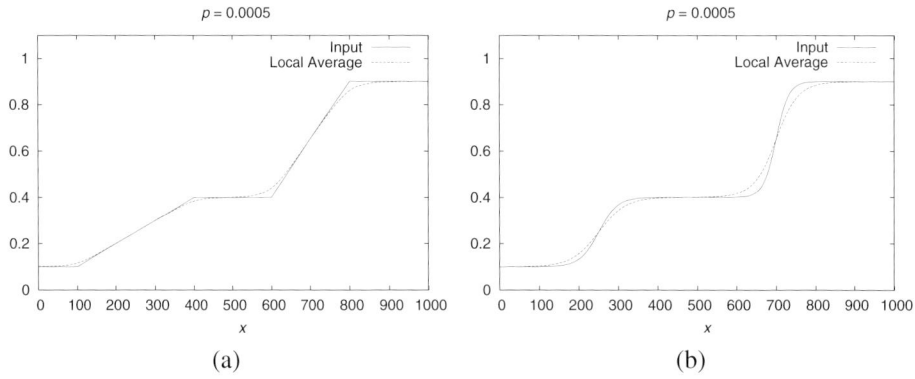

Figure 10.8 Result of the averaging operation where the current color is faded into the result. A linear sequence of processing elements is assumed. Only a single band is processed. The image in (a) shows the result for an input image that changes linearly over some parts of the image. The image in (b) shows the results for an input image with smooth sigmoidal changes.

ESTIMATING THE ILLUMINANT LOCALLY

To better visualize what happens, we only assume a linear sequence of processing elements and only one band is processed. The image in (a) shows the result for an input image where the input stays constant for some regions of the image and changes linearly in other regions. We can see that the input is closely approximated by the computed average. The match is almost perfect in case of a constant or a linear change of the input. Figure 10.8(b) shows an input image where the input stays constant in some regions of the image and changes smoothly in other regions. The smooth change was modeled using a sigmoidal function. Again, the computed average follows the input value. The match is very good for the constant regions and also follows the input inside the sigmoidal regions. However, the match is not as good when compared to the linear changes of the input. We see in Chapter 12, why linear changes are handled better by the algorithm.

Let us now focus on the parameter p. What happens if p is increased or decreased? If we set $p = 0.005$, the match improves considerably. The computed average is shown in Figure 10.9, where we have used the same linearly changing input function as in Figure 10.8. If we set $p = 0.00005$, then the computed average flattens and approaches the average value of the input values. The parameter p determines the extent over which the average will be computed. If p is large, then the average will be computed over a small area. If p is small, it will be computed over a large area.

Figure 10.10 shows how space average color is computed for an input image. Here, all three color bands are independently processed. The images show snapshots of local space average color computed after 1, 50, 200, 1000, 5000, and 20 000 iterations of the algorithm. We can clearly see how the algorithm refines its estimate of local space average color. In this case, space average color was initialized to be zero. Therefore, at the first time step, space average color is equivalent to the color at the current pixel. Note that initialization of space average color is arbitrary, as any initialization will decay toward zero after some time. At the second time step, this original estimate will be averaged and again distributed to the neighboring elements. Each processing element will take the data obtained from the neighboring elements and then slowly fade the current color into the result. With each

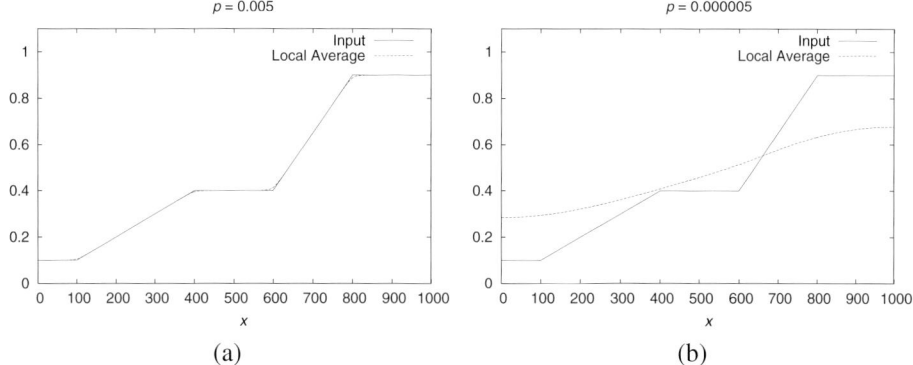

Figure 10.9 If p is set to a comparatively large value, then the computed average matches the input closely (a). If p is set to a comparatively small value, then the computed average approaches the global average of the input values (b).

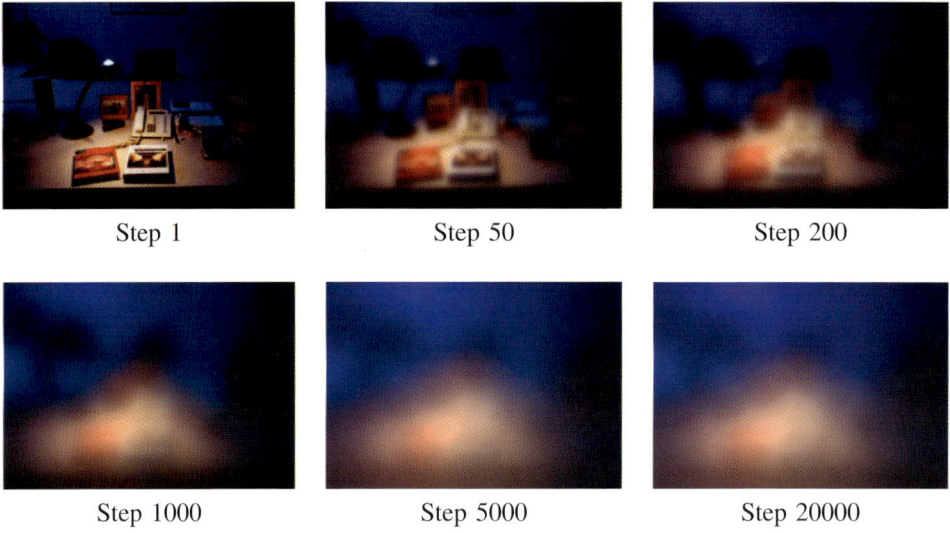

Figure 10.10 Snapshots of the computation of local space average color are shown after 1, 50, 200, 1000, 5000, and 20000 time steps.

Figure 10.11 If p is large, then local average color will only be computed for a relatively small area. If p is small, local average color will be computed for a large area. Eventually this area is bounded by the size of the image.

step, the elements receive data from further away and the estimate improves. This process continues until convergence.

Figure 10.11 shows what happens if we change the parameter p. Space average color is shown after convergence. If p is large, local average color will be computed for a small region around the current point. On the other hand, if p is small, local space average color will be computed for a large area.

The average value computed at a position (x, y) will be multiplied by $(1 - p)$ before it is passed to the neighboring elements. After n steps, the average will be reduced by a factor $(1 - p)^n$. If we assume a 256×256 image, space average color computed by an element on one side of the image will contribute by an amount reduced by $(1 - p)^{256} = 88\%$

ESTIMATING THE ILLUMINANT LOCALLY

(if we assume $p = 0.0005$) to the average color computed by an element on the other side of the image. Thus, the average computed by an element on the right side of the image will only contribute 88% to the average computed by an element on the left side of the image. The parameter p determines the area of support over which the average will be computed. For large values of p, the area of support will be very small. It will be contained completely inside the image. In this case, the assumption that on average the world is gray does not hold. If we choose very small values of p, then the area would, in principle, extend beyond the image boundary. Owing to the finite size of the image, the area of support always ends at the image boundary. We have to choose small values for p such that the gray world assumption holds even if this leads to a longer running time. Of course, the optimal setting of the parameter p depends on the image content.

The algorithm is run until space average color converges. For large values of p, convergence occurs earlier, and for small values of p, convergence occurs later. The color of the current pixel is added to the result by multiplying it with p. Therefore, we will need at least $\frac{1}{p}$ iterations to bring space average color into the range of the current pixel color. Time until convergence for different values of p is shown in Figure 10.12. For $p = 0.0005$, we need approximately 24 000 iterations until convergence. This may seem to be quite a large number of iterations. However, if the algorithm is realized in hardware, the number of iterations can be neglected. A significant speedup can be achieved using successive over-relaxation (SOR) (Bronstein et al. 2001; Demmel 1996). First the data from neighboring elements is averaged, and then the color of the current pixel is faded into the result. This gives us the direction in which to move, i.e. the corrective step. Instead of moving in this direction, we move even farther. Let $w \in (1, 2)$ be a parameter that describes the amount

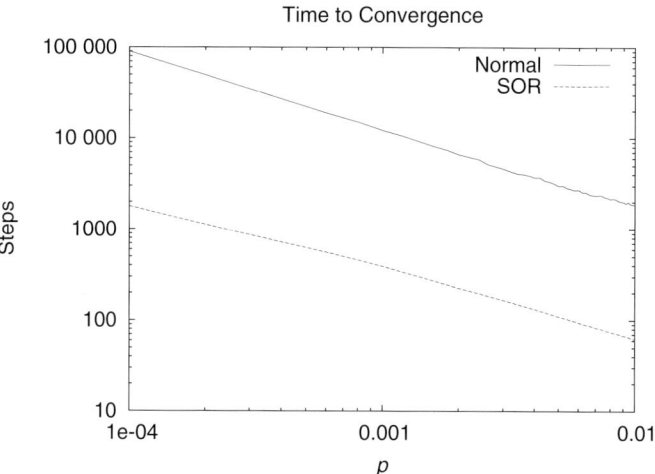

Figure 10.12 Time to convergence computed for different values of p. Convergence is considerably faster if successive over-relaxation is used.

of over-relaxation. Then, local space average color a_i^{SOR} is computed as

$$a_i'(x, y) = \frac{1}{|N(x, y)| + 1} \sum_{(x', y') \in N(x, y) \cup \{(x, y)\}} a_i^{SOR}(x', y') \qquad (10.19)$$

$$a_i(x, y) = (1 - p)a_i'(x, y) + pc_i(x, y) \qquad (10.20)$$

$$a_i^{SOR}(x, y) = (1 - w)a_i^{SOR}(x, y) + wa_i(x, y). \qquad (10.21)$$

Convergence time for $w = 1.999$ are shown in Figure 10.12. Note that the speedup is considerable compared to the straightforward implementation.

Since the algorithm contains only very simple operations, it can be implemented easily in hardware. The number and complexity of the required computations are the important factors. In fact, the algorithm can be implemented using a resistive grid (Horn 1986; Moore et al. 1991). Because of the exponential weighting of neighboring points in a resistive grid, the computed operation is almost equivalent to an exponential kernel $g(r) = e^{-\frac{r}{\sigma}}$, where σ denotes the area of support and $r = \sqrt{x^2 + y^2}$. If a resistive grid is used for the computation, then convergence will be almost instantaneous.

10.3 Implementation Using a Resistive Grid

Figure 10.13 shows a two-dimensional resistive grid. The resistors feeding the network have the resistance R_o. All other resistors of the network have the resistance R. According to Kirchhoff's law (Orear 1982), the current flowing into node at position (x, y) must equal the current flowing out of the node. Let $I_o(x, y)$ be the current flowing into the network at position (x, y) of the grid. Let I_{left}, I_{right}, I_{up}, and I_{down} be the current flowing to the left, right, up, and down. We have,

$$I_o(x, y) = I_{left}(x, y) + I_{right}(x, y) + I_{up}(x, y) + I_{down}(x, y). \qquad (10.22)$$

According to Ohm's law (Orear 1982), the current I flowing through a resistor R is given by $I = \frac{V}{R}$ where U is the voltage across the resistor. Let $V_o(x, y)$ be the voltage applied

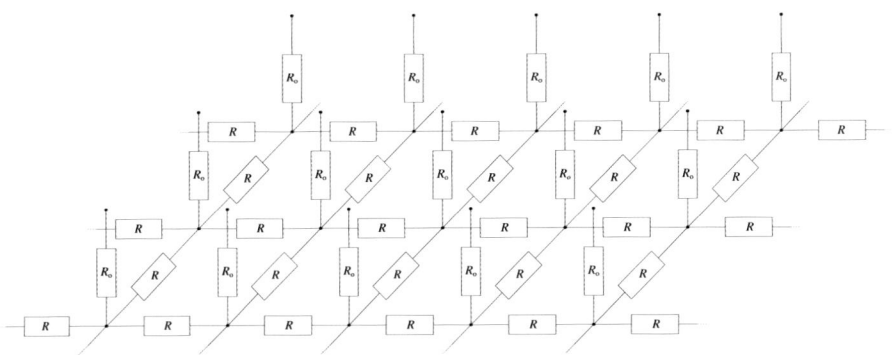

Figure 10.13 A two-dimensional resistive grid.

ESTIMATING THE ILLUMINANT LOCALLY

to the grid at node (x, y). Let $V(x, y)$ be the voltage at node (x, y). We have,

$$\frac{V(x, y) - V_o(x, y)}{R_o} = \frac{V(x-1, y) - V(x, y)}{R} + \frac{V(x+1, y) - V(x, y)}{R}$$

$$+ \frac{V(x, y-1) - V(x, y)}{R} + \frac{V(x, y+1) - V(x, y)}{R} \quad (10.23)$$

$$\frac{1}{R_o}(V(x, y) - V_o(x, y)) = \frac{1}{R}(V(x-1, y) + V(x+1, y) + V(x, y-1) + V(x, y+1))$$

$$- \frac{4}{R} V(x, y). \quad (10.24)$$

Solving for $V(x, y)$, we obtain

$$\left(\frac{1}{R_o} + \frac{4}{R}\right) V(x, y) = \frac{1}{R}(V(x-1, y) + V(x+1, y) + V(x, y-1) + V(x, y+1))$$

$$+ \frac{1}{R_o} V_o(x, y) \quad (10.25)$$

$$V(x, y) = \frac{R_o}{4R_o + R}(V(x-1, y) + V(x+1, y) + V(x, y-1) + V(x, y+1))$$

$$+ \frac{R}{4R_o + R} V_o(x, y) \quad (10.26)$$

which is equivalent to the update equations described in the preceding text. For convenience, let us repeat the update equation again here.

$$a_i(x, y) = \frac{1-p}{4}(a_i(x-1, y) + a_i(x+1, y) + a_i(x, y-1) + a_i(x, y+1)) + pc_i(x, y) \quad (10.27)$$

where p is a small percentage. We see that local space average color $a_i(x, y)$ corresponds to the voltage $V(x, y)$ at points (x, y) of the grid and the color of the input pixel $c_i(x, y)$ corresponds to the voltage $V_o(x, y)$ applied at the grid point (x, y). Given a percentage p, we can compute the relation between the resistances R_o and R of the resistive grid.

$$p = \frac{R}{4R_o + R} \quad (10.28)$$

Solving for $\frac{R_o}{R}$, we obtain

$$\frac{R_o}{R} = \frac{1-p}{4p}. \quad (10.29)$$

Thus, R_o has to be larger than R by a factor of $\frac{1-p}{4p}$.

Let us now take a look at the function computed by the resistive grid.

$$\frac{1}{R_o}(V(x, y) - V_o(x, y)) = \frac{1}{R}(V(x-1, y) + V(x+1, y) + V(x, y-1) + V(x, y+1))$$

$$- \frac{4}{R} V(x, y)$$

$$(10.30)$$

On the right-hand side, we have the discrete version of the Laplace operator. Moving to the continuous domain, we have

$$\frac{1}{R_o}(V(x,y) - V_o(x,y)) = \frac{1}{R}\nabla^2 V(x,y). \tag{10.31}$$

We now represent both $V(x,y)$ and $V_o(x,y)$ in terms of its Fourier transform. Let $V(w_x, w_y)$ and $V_o(w_x, w_y)$ be the Fourier transforms of $V(x,y)$ and $V_o(x,y)$, respectively. Thus, we have

$$V(x,y) = \iint V(w_x, w_y) e^{i(w_x x + w_y y)} dw_x dw_y. \tag{10.32}$$

and a similar expression for $V_o(x,y)$. Let $\sigma = \sqrt{\frac{R_o}{R}}$. Then, we obtain

$$V(x,y) - V_o(x,y) = \sigma^2 \nabla^2 V(x,y). \tag{10.33}$$

Moving to the Fourier domain, we have

$$\iint \left(V(w_x, w_y) - V_o(w_x, w_y) \right) e^{i(w_x x + w_y y)} dw_x dw_y$$

$$= -(w_x^2 + w_y^2)\sigma^2 \iint V(w_x, w_y) e^{i(w_x x + w_y y)} dw_x dw_y \tag{10.34}$$

or

$$\iint \left(V(w_x, w_y) - V_o(w_x, w_y) + \sigma^2(w_x^2 + w_y^2) V(w_x, w_y) \right) e^{i(w_x x + w_y y)} dw_x dw_y = 0. \tag{10.35}$$

Thus, we must have

$$V(w_x, w_y) = \frac{V_o(w_x, w_y)}{1 + \sigma^2(w_x^2 + w_y^2)}. \tag{10.36}$$

The output computed by the resistive grid is therefore

$$V(x,y) = \iint \frac{V_o(w_x, w_y)}{1 + \sigma^2(w_x^2 + w_y^2)} e^{i(w_x x + w_y y)} dw_x dw_y. \tag{10.37}$$

If we compute the output in simulation, we can apply a Fourier transform, multiply the result by

$$\frac{1}{1 + \sigma^2(w_x^2 + w_y^2)}, \tag{10.38}$$

and then transform the result back to the spatial domain by applying an inverse Fourier transform. The only drawback of this method is that toroidal boundary conditions are assumed. In other words, local space average color computed for a pixel at the lower border of the image will be a mixture of the actual local space average color at the lower border and the actual local space average color of the upper border of the image.

Let us consider the one-dimensional case again. Figure 10.14 shows a linear network of resistors. If we carry out the same derivation as mentioned earlier but now for a one-dimensional resistive grid, we arrive at the following equation (Jähne 2002):

$$\frac{1}{R_o}(V(x) - V_o(x)) = \frac{1}{R}\frac{\partial^2}{\partial x^2}V(x) \tag{10.39}$$

ESTIMATING THE ILLUMINANT LOCALLY

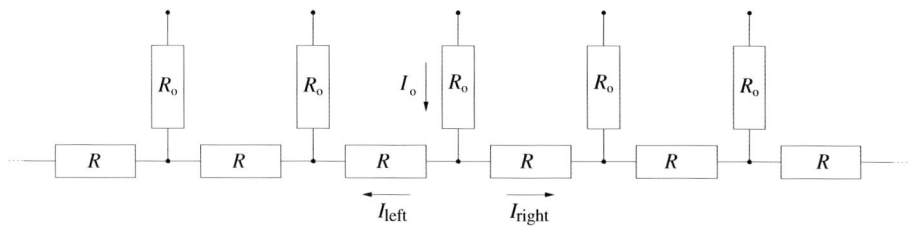

Figure 10.14 A one-dimensional resistive grid.

The solution of this equation is

$$V(x) = \int \frac{V_o(w)}{1+\sigma^2 w^2} e^{iwx} \, dw \tag{10.40}$$

where $V_o(w)$ is the one-dimensional Fourier transform of $V_o(x)$. To compute the output for the one-dimensional case, we multiply the Fourier transform by $\frac{1}{1+\sigma^2 w^2}$ in the frequency domain. Since a multiplication in the frequency domain is equivalent to a convolution in the spatial domain (Gonzalez and Woods 1992; Horn 1986; Jähne 2002; Jain et al. 1995; Parker 1997; Shapiro and Stockman 2001), the same result can be achieved by convolving $V_o(x)$ with the inverse Fourier transform of $\frac{1}{1+\sigma^2 w^2}$. The Fourier transform of $e^{-\frac{|x|}{\sigma}}$ is $\frac{2\sigma}{1+\sigma^2 w^2}$ (Bronstein et al. 2001). Therefore, the function computed by the resistive grid is simply a convolution with $\frac{1}{2\sigma} e^{-\frac{|x|}{\sigma}}$. For a one-dimensional grid, we can compute the parameter σ as

$$\sigma = \sqrt{\frac{R_o}{R}} = \sqrt{\frac{1-p}{4p}} \tag{10.41}$$

or if we compute p from the parameter σ, we obtain

$$p = \frac{1}{1+4\sigma^2}. \tag{10.42}$$

In order to compute the output of the two-dimensional resistive grid using a convolution, we have to find the inverse Fourier transform of

$$\frac{1}{1+\sigma^2(w_x^2 + w_y^2)}. \tag{10.43}$$

The Fourier transform of $e^{-\frac{r}{\sigma}}$ with $r = \sqrt{x^2 + y^2}$ is $\frac{2\pi\sigma^2}{(1+\sigma^2(w_x^2+w_y^2))^{\frac{3}{2}}}$ (Weisstein 1999b), which is almost the desired result. If we use $r = |x| + |y|$, then the Fourier transform of $e^{-\frac{r}{\sigma}}$ is $\frac{4\sigma^2}{(1+\sigma^2 w_x^2)(1+\sigma^2 w_y^2)}$. Again, this is not quite what we are looking for. In practice, the output of the resistive grid can be approximated by a convolution with $\frac{1}{4\sigma^2} e^{-\frac{|x|+|y|}{\sigma}}$. The advantage of the kernel $e^{-\frac{|x|+|y|}{\sigma}}$ is that this kernel is separable and can be applied in both the x and y directions independently. Figure 10.15 shows that local space average color computed with a resistive grid closely matches local space average color computed with an exponential kernel.

234 ESTIMATING THE ILLUMINANT LOCALLY

Resistive Grid

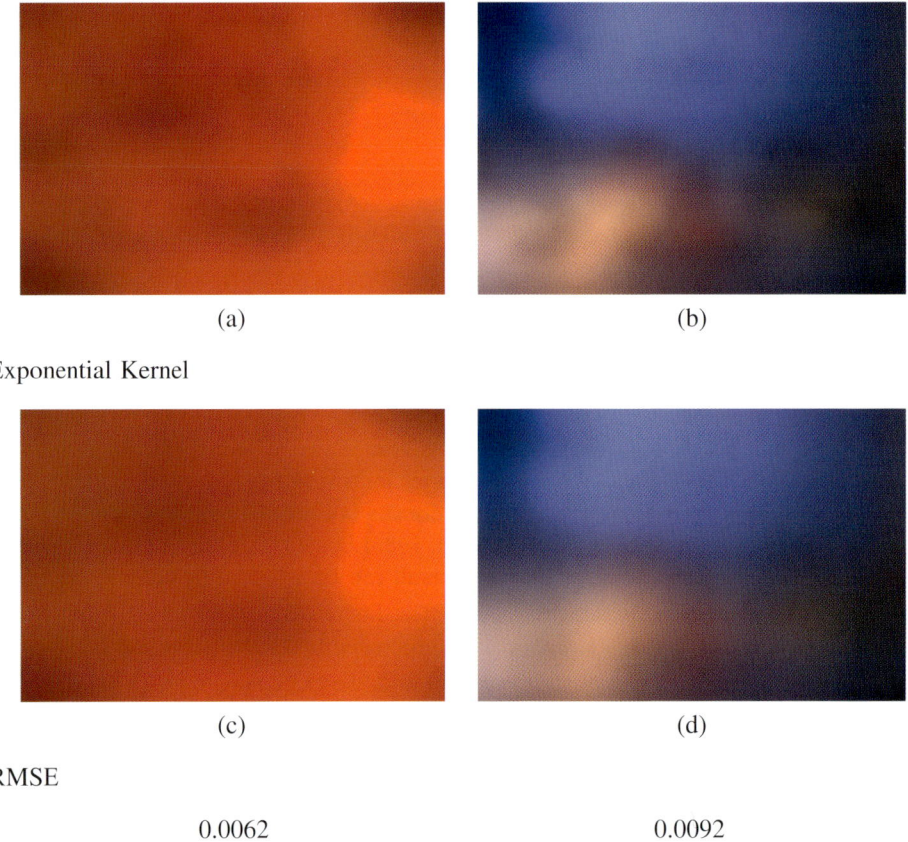

Exponential Kernel

(a)   (b)

(c)   (d)

RMSE

             0.0062   0.0092

Figure 10.15 Results for the resistive grid are shown in (a) and (b). Results for an exponential kernel $e^{-\frac{|x|+|y|}{\sigma}}$ are shown in (c) and (d). The percentage p was set to 0.0005. The root mean squared error between the two images is also shown.

Computation of local space average color on a grid of processing elements can be viewed as a diffusion process where color flows from one element to the next. Such a diffusion process can also be modeled with a resistive grid (Jähne 2002). If we remove the input resistors and connect each grid point to ground via a capacitor with capacity C, we obtain the circuit shown in Figure 10.16. The current $I_C(x, y)$ flowing through the capacitor is given by $\frac{1}{C}\frac{\partial}{\partial t}V(x, y)$. Thus, we arrive at the following equation for the voltage $V(x, y)$ at each grid point.

$$0 = \frac{V(x-1, y) - V(x, y)}{R} + \frac{V(x+1, y) - V(x, y)}{R}$$
$$+ \frac{V(x, y-1) - V(x, y)}{R} + \frac{V(x, y+1) - V(x, y)}{R} - C\frac{\partial}{\partial t}V(x, y) \quad (10.44)$$

ESTIMATING THE ILLUMINANT LOCALLY

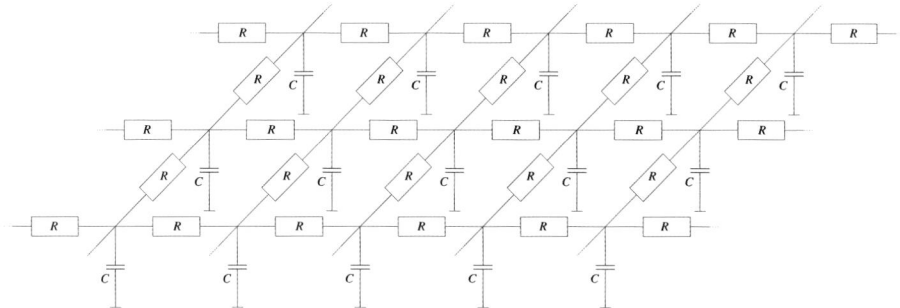

Figure 10.16 A resistive grid where each node point is connected to ground via a capacitor.

Moving to the continuous domain, we obtain

$$0 = \frac{1}{R}\left(\frac{\partial^2}{\partial x^2} + \frac{\partial^2}{\partial y^2}\right)V(x, y) - C\frac{\partial}{\partial t}V(x, y) \tag{10.45}$$

$$\frac{\partial}{\partial t}V(x, y) = D\left(\frac{\partial^2}{\partial x^2} + \frac{\partial^2}{\partial y^2}\right)V(x, y) \tag{10.46}$$

with $D = \frac{1}{RC}$. This equation is the description of a diffusion process (Bronstein et al. 2001; Jähne 2002; Weickert 1997). A flow **j** occurs in a direction opposite to the concentration gradient. Since the concentration is given by $V(x, y)$, the flow **j** is given by

$$\mathbf{j} = -D\nabla V(x, y) \tag{10.47}$$

where D is the diffusion coefficient. Together with the continuity equation (Jähne 2002; Nolting 1992)

$$\frac{\partial}{\partial t}V(x, y) + \nabla \mathbf{j} = 0, \tag{10.48}$$

which is just a reformulation of the fact that the net charge of the system is constant, we again arrive at the preceding equation

$$\frac{\partial}{\partial t}V(x, y) = \nabla(D\nabla V(x, y)) = D\nabla^2 V(x, y). \tag{10.49}$$

If we now assume somewhat arbitrarily that, at time $t = 0$, the voltage at the grid points is a sinusoidal pattern $V(x, y) = V \sin(2\pi x/\lambda)$, and substitute this into the preceding equation, we find that the voltage decays exponentially.

$$V(x, y) = Ve^{-\frac{t}{\tau}}\sin(2\pi x/\lambda) \tag{10.50}$$

The time constant τ can be found by substituting $V(x, y)$ into the preceding equation. It is related to the wavelength λ of the sinusoidal pattern as follows:

$$\tau = \frac{RC}{4\pi^2}\lambda^2. \tag{10.51}$$

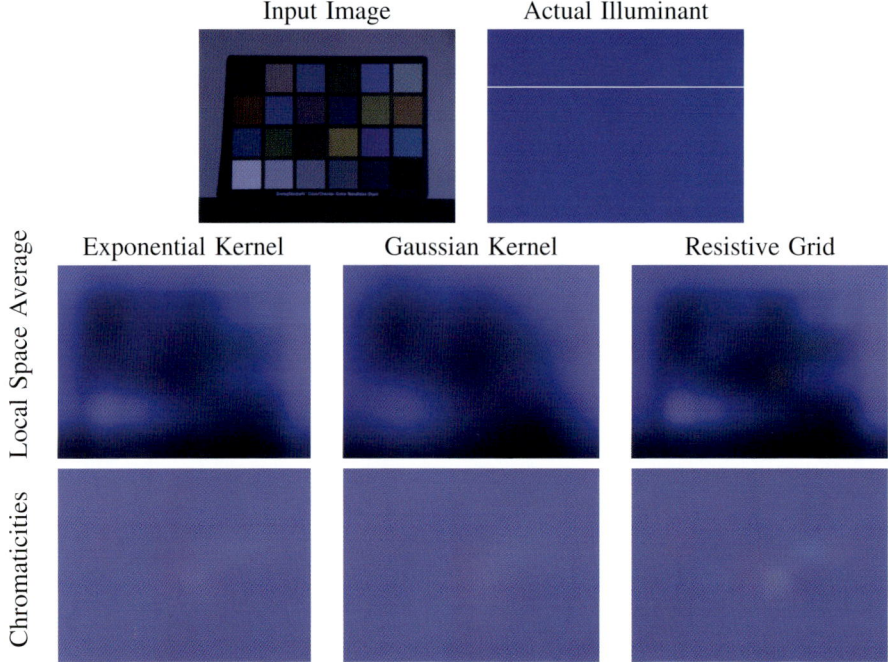

Figure 10.17 Local space average color computed using an exponential kernel, a Gaussian kernel, and a resistive grid. The input image is shown on top. The second image shows the actual chromaticities of the illuminant. The bottom row of images shows the normalized color (chromaticities) of the local space average color. The input image and the actual illuminant are from a database for color constancy research created by Barnard et al. (2002c). (Original image data from "Data for Computer Vision and Computational Colour Science" made available through http://www.cs.sfu.ca/~colour/data/index.html. See Barnard K Martin L Funt B and Coath A 2002c A data set for color research, Color Research and Application, Wiley Periodicals, 27(3), 147–151. Reproduced by permission of Kobus Barnard.)

Thus, the time to convergence increases quadratically with the wavelength λ of the pattern. If we double the wavelength, the time to convergence is four times longer.

The general solution to the homogeneous diffusion equation $\frac{\partial}{\partial t} V(x, y) = D \nabla^2 V(x, y)$ with $V(x, y)|_{t=0} = f(x, y)$ is given by (Bronstein et al. 2001; Jähne 2002; Weickert 1997)

$$V(x, y) = G_{D\sqrt{2t}}(r) \otimes f(x, y) \tag{10.52}$$

where $\otimes$ denotes convolution and $G_\sigma(r)$ with $r = \sqrt{x^2 + y^2}$ is a Gaussian kernel

$$G_\sigma(r) = \frac{1}{2\pi\sigma^2} e^{-\frac{r^2}{2\sigma^2}}. \tag{10.53}$$

ESTIMATING THE ILLUMINANT LOCALLY

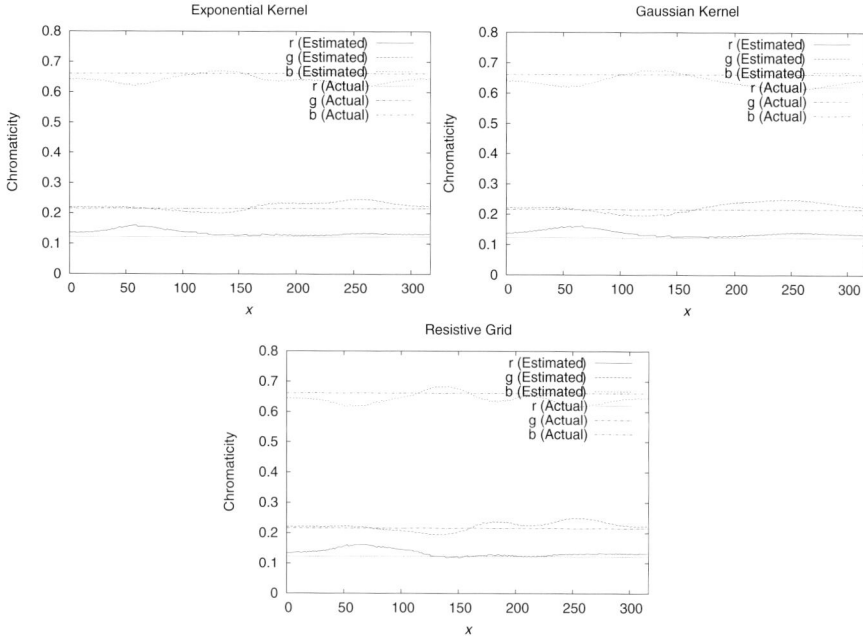

Figure 10.18 A comparison between the computed and the actual illuminant along a horizontal line of the image (marked by a white line in Figure 10.17). Local space average color is computed using an exponential kernel, a Gaussian kernel, or a resistive grid.

10.4 Experimental Results

Barnard et al. (2002c) have created a series of images to test color constancy algorithms.[1] The images appear very dark because they are linear (no gamma correction) and they were purposely under-exposed to avoid any clipped pixel values. Barnard et al. also supply detailed information on the spectral power distribution of all illuminants used and the response functions of the camera sensor. Before an image was taken, the color of the illuminant was measured through the camera sensor using a white reference standard. Given the actual color of the illuminant, we can compare local space average color to the color computed by the different methods described in this chapter. Figure 10.17 shows the results for an image from the database of Barnard et al. The input image (a Macbeth color checker) is shown at the top of the figure. The image on the right shows the actual color of the illuminant as measured by the camera. The second row of images shows local space average computed using an exponential kernel, a Gaussian kernel, and a grid of processing elements or a resistive grid. The third row shows the normalized colors (chromaticities) for each method.

A comparison between the computed chromaticities and the actual chromaticities of the illuminant along a horizontal line of the image is shown in Figure 10.18. The line where

---

[1] http://www.cs.sfu.ca/~color/data/index.html

the chromaticities were extracted is marked by a white line in Figure 10.17. In all three cases, the computed chromaticities closely match the chromaticities of the actual illuminant. Results for the Gaussian kernel are a little smoother compared to those for the exponential kernel and the resistive grid.

11

Using Local Space Average Color for Color Constancy

Now that we have calculated local space average color as described in Chapter 10, we can use it to adjust the colors of the input image. We can distinguish between several different methods to calculate the colors of the output image. In each case we will use the assumption that, on average, the world is gray.

11.1 Scaling Input Values

According to the basic gray world algorithm, discussed in Section 6.2, we can estimate the light illuminating the scene as twice the space average color. Let L_i be the intensity of the illuminant at wavelength λ_i with $i \in \{r, g, b\}$.

$$L_i \approx f a_i = \frac{f}{n} \sum_{x,y} c_i(x, y) \tag{11.1}$$

where f is a constant factor that depends on the object geometry, $c_i(x, y)$ is the intensity in channel i of the image pixel at position (x, y) and n is the number of image pixels. The gray world assumption uses all pixels of the image to estimate the illuminant. If the scene is sufficiently rich, we can apply the gray world assumption locally and use local space average color to estimate the illuminant locally (Ebner 2002, 2004c, 2006). Let $a_i(x, y)$ denote local space average color at position (x, y) computed by any of the methods described in the previous chapter. Let $L_i(x, y)$ be the intensity of the illuminant at wavelength λ_i, which illuminates the surface seen at pixel position (x, y) in the image. We can estimate the intensity of the illuminant as

$$L_i(x, y) \approx f a_i(x, y). \tag{11.2}$$

This assumes, of course, that there are no changes of the illuminant due to the surface geometry and that the amount of light reflected by the objects does not depend on the

Color Constancy M. Ebner
© 2007 John Wiley & Sons, Ltd

position of the viewer, i.e. we only have Lambertian surfaces. In addition, we also assume that the sensors are linear with no bias. No bias means that a surface that reflects none of the incident light produces a sensor output of zero. If the sensors are not linear, then we need to linearize the input values. Computer images obtained from a scanner or digital camera are usually stored using the sRGB color space that assumes a gamma correction factor of 1/2.2. The images look correct when viewed on a standard monitor. If we want to apply our color constancy method to such images, we have to transform the images to a linear intensity space by undoing the gamma correction. Then we perform the color correction step as described in the subsequent text and finally, we again apply a gamma correction of 1/2.2.

After we have estimated the illuminant, we can use this information to calculate the reflectances of the objects in view. In Chapter 3 we have seen that, for the given assumptions, pixel intensity c is given by the product of a geometry factor G, the object reflectance R, and the illuminant L. For a nonuniform illuminant, we obtain

$$c_i(x, y) = G(x, y) R_i(x, y) L_i(x, y). \tag{11.3}$$

Therefore, the product between the geometry factor G and the reflectances can be calculated as

$$G(x, y) R_i(x, y) = \frac{c_i(x, y)}{L_i(x, y)} \approx \frac{c_i(x, y)}{f a_i(x, y)}. \tag{11.4}$$

Figure 11.1 shows the complete algorithm using a notation similar to the one used by Horn (1974). If we compare this algorithm to Figure 7.11, we see that computationally, it is much simpler. Also, one of the major difficulties, the selection of a suitable threshold, has disappeared. Instead of selecting a threshold, we now need to determine the extent over which local space average color will be computed. Since the algorithm is highly parallel, it can be implemented easily in hardware. The computation of local space average color can be done using a resistive grid as was described in Chapter 10. Koosh (2001) gives a nice overview on performing analog computations in very large scale integration (VLSI).

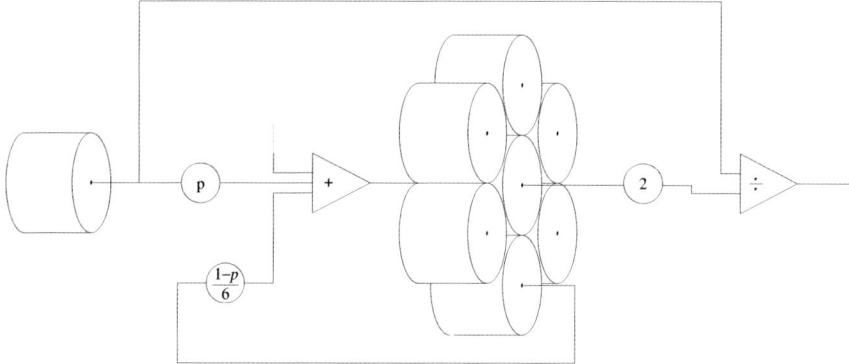

Figure 11.1 Implementation of color constancy algorithm that uses local space average color to scale the input values. Data from surrounding pixels is iteratively averaged. Input values are divided by twice the local space average color.

Figure 11.2 Output images that were generated by dividing the input values by twice the local space average color.

By integrating such a color constancy algorithm directly into the imaging chip, we would obtain a color corrected image before the data is stored in memory. No external processor would be required. Efforts to create artificial retina chips have already tried to include functions such as edge detection, elimination of noise, and pattern matching directly into the image acquisition chip (Dudek and Hicks 2001; Kyuma et al. 1997).

The output of this simple algorithm can be seen in Figure 11.2 for two input images. Note that quite a large number of assumptions were made:

- The factor due to the scene geometry does not depend on the position of the viewer relative to the object, i.e. we have a Lambertian surface.
- The sensors are delta functions.
- The sensor's response is linear with no bias.
- On average, the world is gray.
- The light illuminating the scene can be approximated by a constant factor times the local space average color.

Most of these assumptions are not correct. Scene geometry does have an influence on the intensity of the image pixels. The response functions of the sensors are not delta functions, more likely, they are Gaussian shaped. The sensors are not necessarily linear and may have a bias. If the scene only contains a few colors, then the gray world assumption does not hold true. The light illuminating the scene cannot necessarily be approximated by twice the local space average color. The area of support depends on the scene viewed. Nevertheless, the algorithm performs surprisingly well. Fleyeh (2005) have used this algorithm for detection and segmentation of traffic signs and report very good results.

11.2 Color Shifts

Let us now have a different look on the use of local space average color for color constancy using color shifts. Consider the standard RGB color space. A unit cube is spanned by the

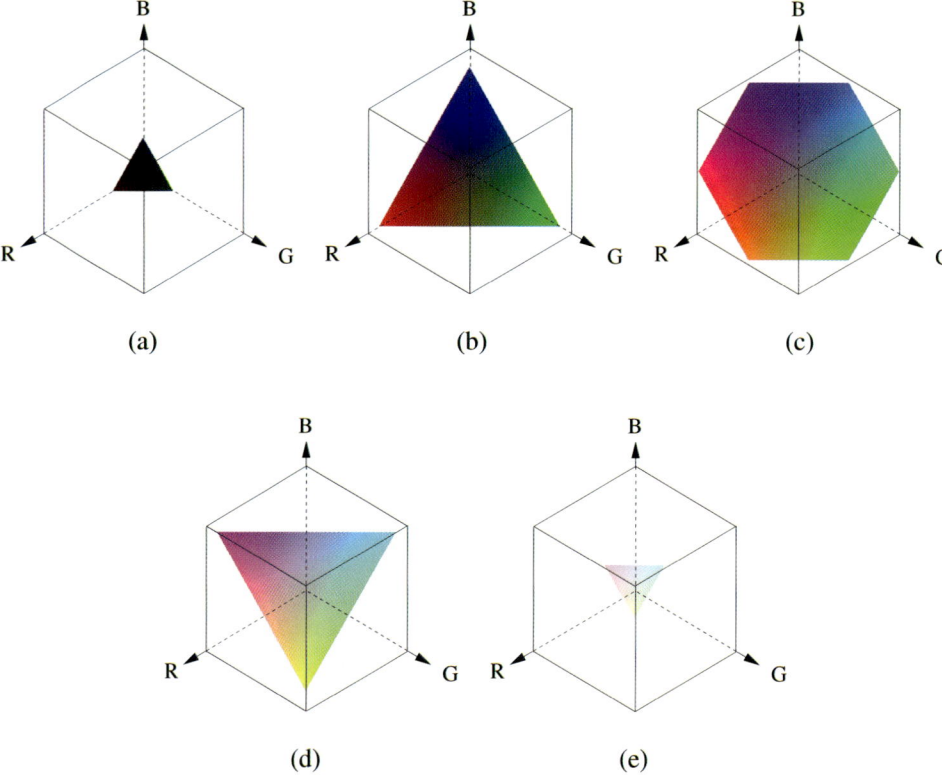

Figure 11.3 Cuts through the red, green, and blue (RGB) cube at a distance of (a) 10%, (b) 30%, (c) 50%, (d) 70%, and (e) 90% along the line from black to white.

three colors red, green, and blue. The corners of the cube can be labeled with the colors black, red, green, blue, yellow, magenta, cyan, and white. The gray vector runs through the middle of the cube from black [0, 0, 0] to white [1, 1, 1]. Figure 11.3 shows five cuts through this color cube at different intensities. The cuts are made perpendicular to the gray vector. Saturated colors are located away from the center of the cuts. For high and low intensities, we only have a few different colors. More colors are available for intermediate intensities. If we know that local space average color is not gray, then we can use a shift perpendicular to the gray vector to move the color back to the center (Ebner 2003c, 2004b). The distance between local space average color and the gray vector is a measure by how much local space average color deviates from the assumption that, on average, the world is gray. This is illustrated in Figure 11.4.

Let $\mathbf{r} = [1, 0, 0]^T$, $\mathbf{g} = [0, 1, 0]^T$, and $\mathbf{b} = [0, 0, 1]^T$ be the three color vectors red, green, and blue, which span the unit color cube. We can calculate the required shift by first projecting local space average color onto the gray vector. If we subtract the result from local space average color, we obtain the component perpendicular to the gray vector. This vector points from the gray vector to the local space average color and its length is equal to

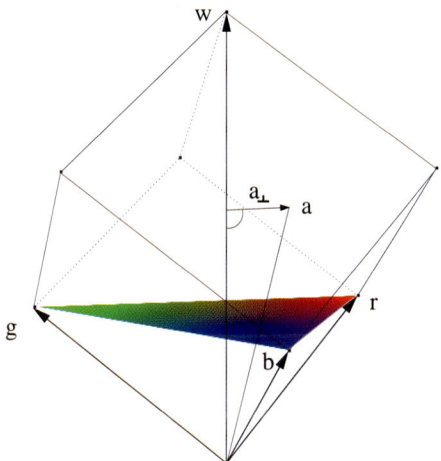

Figure 11.4 The gray vector passes from [0, 0, 0] to [1, 1, 1] directly through the middle of the color cube. If local space average color is located away from the gray vector, we can use a shift perpendicular to the gray vector to move the color back to the center.

the distance between the gray vector and local space average color. Let $\mathbf{w} = \frac{1}{\sqrt{3}}[1, 1, 1]^T$ be the normalized gray vector. Let $\mathbf{a} = [a_r, a_g, a_b]^T$ be local space average color that was calculated by the method described in Chapter 10. Then the component of local space average color, which is perpendicular to the gray vector, is calculated as

$$\mathbf{a}_\perp = \mathbf{a} - (\mathbf{a}^T \mathbf{w})\mathbf{w}. \tag{11.5}$$

If we subtract this vector from the current color, we move local space average color back to the gray vector. This is visualized in Figure 11.6. Let $\mathbf{c} = [c_r, c_g, c_b]^T$ be the color of the input pixel. Thus, output colors can be calculated by subtracting the component of local space average color, which is perpendicular to the gray vector.

$$\mathbf{o} = \mathbf{c} - \mathbf{a}_\perp. \tag{11.6}$$

Let us look at the individual components of this equation. The components o_i with $i \in \{r, g, b\}$ are given by

$$o_i = c_i - a_i + \frac{1}{3}(a_r + a_g + a_b). \tag{11.7}$$

If we define $\bar{a} = \frac{1}{3}(a_r + a_g + a_b)$ to be the intensity of the local space average color, we obtain

$$o_i = c_i - a_i + \bar{a}. \tag{11.8}$$

Output images that were generated with this operation are shown in Figure 11.5.

The data flow is shown in Figure 11.7. Computations are shown only for a single color channel. In the top layer, we have the input pixels. Input pixel values are stored in the

Figure 11.5 For these images, output pixels were calculated by subtracting the component of local space average color, which is perpendicular to the gray vector, from the current pixel color.

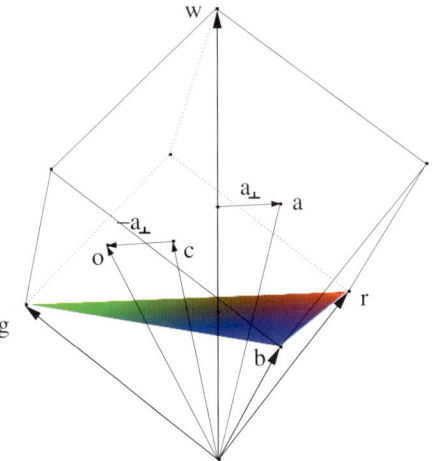

Figure 11.6 First we calculate the component $\mathbf{a}_\perp$ that is perpendicular to the gray vector $\mathbf{w}$ and points to local space average color. If we subtract this vector from the color of the input pixel $\mathbf{c}$, we essentially move the local space average color onto the gray vector.

cells denoted c_r, c_g, and c_b. The second layer calculates local space average color. The third layer calculates the colors of the output pixels. In order to perform the color shift, we also need the intensity of local space average color. This information is stored in cell $\bar{a}$, i.e. $\bar{a} = \frac{1}{3}(a_r + a_g + a_b)$. We only require addition, subtraction, and multiplication with constant values. A division operation is not needed. Thus, the algorithm can be implemented easily in hardware. Since only local interconnections are required, the algorithm is scalable to arbitrary image sizes. It can be integrated directly into charge-coupled devices (CCD) or complemented metal oxide semiconductor (CMOS) imaging chips. Again, the simplest way to compute local space average color, as shown in the second layer, is to use a resistive grid.

USING LOCAL SPACE AVERAGE COLOR FOR COLOR CONSTANCY

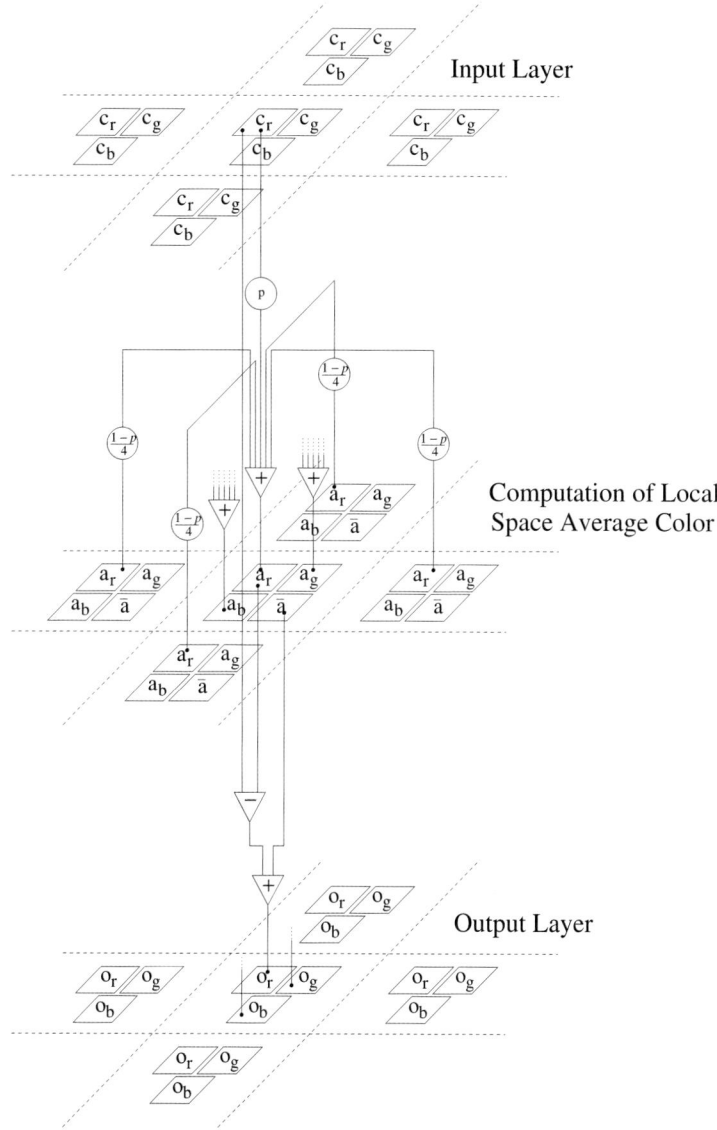

Figure 11.7 The first layer is the input layer. The second layer computes local space average color. The output is computed in the third layer. The measured red, green, and blue components are stored in cells c_i with $i \in \{r, g, b\}$. Space average color is stored in cells a_i, the average of these three components is stored in $\bar{a}$. Output cells are denoted by o_i (from Ebner 2003c).

11.3 Normalized Color Shifts

Notice that the color of the current pixel and local space average color may be located at different levels. It was already shown in Figure 11.3 how the color levels differ. If the intensity of the current pixel is very low and the intensity of local space average color is very high, then it may happen that the color vector is moved too much. Or it may be moved too little if the intensity of local space average color is very low and the intensity of the current pixel is very large. In order to perform the color shift on the same intensity level, we can normalize both colors first (Ebner 2003c, 2004b). Let $\hat{\mathbf{c}}$ be the normalized color of the current pixel and let $\hat{\mathbf{a}}$ be the normalized space average color.

$$\hat{\mathbf{c}} = \frac{1}{c_r + c_g + c_b}[c_r, c_g, c_b]^T. \tag{11.9}$$

$$\hat{\mathbf{a}} = \frac{1}{a_r + a_g + a_b}[a_r, a_g, a_b]^T. \tag{11.10}$$

That is, we project both colors onto the plane with $r + g + b = 1$. We can calculate the component perpendicular to the gray vector by projecting local space average color onto the gray vector and subtracting the resulting vector from local space average color.

$$\hat{\mathbf{a}}_\perp = \hat{\mathbf{a}} - (\hat{\mathbf{a}}^T \mathbf{w})\mathbf{w}. \tag{11.11}$$

Output colors can be calculated by subtracting this vector from the color of the current pixel. This operation is visualized in Figure 11.8. It essentially moves the local space average color back into the gray vector.

$$\hat{\mathbf{o}} = \hat{\mathbf{c}} - \hat{\mathbf{a}}_\perp. \tag{11.12}$$

If we look at how the individual components are transformed, we obtain

$$\hat{o}_i = \hat{c}_i - \hat{a}_i + \frac{1}{3} \tag{11.13}$$

which is a very simple equation. The normalization can be undone by using the intensity of the current pixel. We can scale the output pixels back to the original intensity.

$$o_i = (c_r + c_g + c_b)\hat{o}_i \tag{11.14}$$

$$= c_i - (c_r + c_g + c_b)(\hat{a}_i - \frac{1}{3}) \tag{11.15}$$

$$= c_i - \frac{c_r + c_g + c_b}{a_r + a_g + a_b}(a_i - \frac{1}{3}(a_r + a_g + a_b)) \tag{11.16}$$

$$= c_i - \frac{\bar{c}}{\bar{a}}(a_i - \bar{a}). \tag{11.17}$$

Output images that were generated with this operation are shown in Figure 11.9. Note that here, the output colors may not have the same lightness. Of course, we could also adjust the intensity of the output pixel such that the lightness is identical to the input pixel. Alternatively, the color shift could also be performed in a different color space such as the $L^*a^*b^*$ color space.

USING LOCAL SPACE AVERAGE COLOR FOR COLOR CONSTANCY

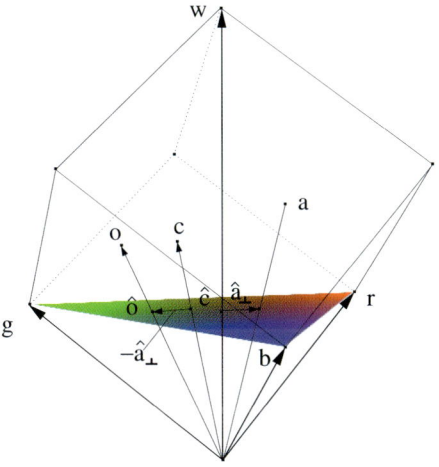

Figure 11.8 The color of the current pixel **c**, and local space average color **a** are projected onto the plane $r + g + b = 1$. Let $\hat{\mathbf{c}}$ and $\hat{\mathbf{a}}$ be the normalized points. Now, normalized local space average color is projected onto the gray vector **w**. The projection is subtracted from $\hat{\mathbf{a}}$, which gives us $\hat{\mathbf{a}}_\perp$. The component $\hat{\mathbf{a}}_\perp$ is orthogonal to the gray vector **w**. This component is subtracted from the color of the current pixel that gives us the normalized output color $\hat{\mathbf{o}}$. Finally, the output color is scaled back to the intensity of the input pixel.

Figure 11.9 For these images, output pixels were calculated by subtracting the component of local space average color, which is perpendicular to the gray vector, from the current pixel color.

The data flow is shown in Figure 11.10. Again, computations are only shown for a single color channel. In the top layer, we have the input pixels. The second layer calculates space average color. The third layer calculates the colors of the output pixels. The red, green, and blue components of each color are stored in the cells denoted c_r, c_g, and c_b. The average of the red, green, and blue components is stored in cell $\bar{c}$. Space average color is stored in cells a_i, the average of these three components in stored in $\bar{a}$. Output cells are denoted by o_i.

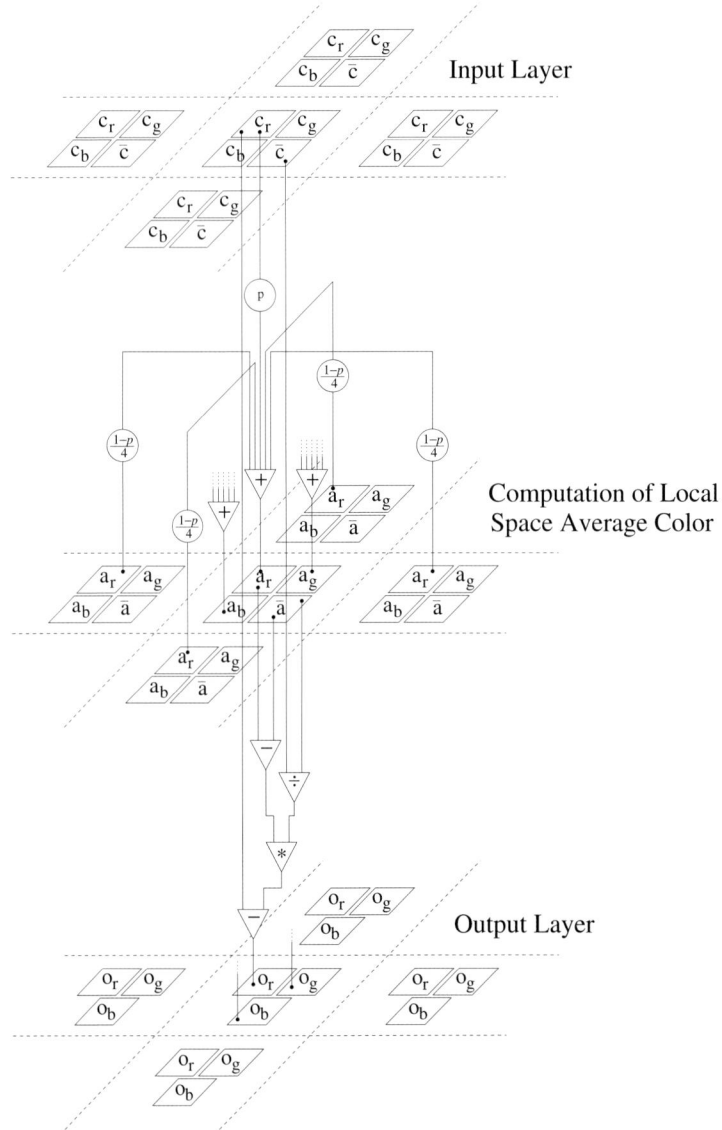

Figure 11.10 Computation of color constant descriptors using normalized color shifts. The first layer is the input layer. The second layer computes local space average color. The output is computed in the third layer. The measured red, green, and blue components are stored in cells c_i with $i \in \{r, g, b\}$. The average of these three components is stored in cell $\bar{c}$. Space average color is stored in cells a_i, the average of these three components in stored in $\bar{a}$. Output cells are denoted by o_i (from Ebner 2003c).

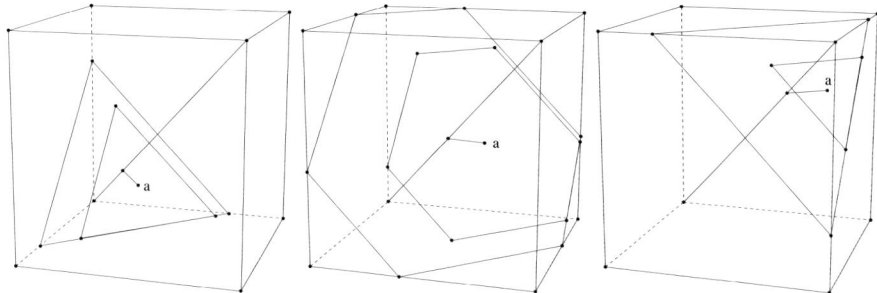

Figure 11.11 Color gamut for three different values of local space average color. If we assume that the space of possible colors is shifted toward local space average color **a**, this also implies a smaller color gamut.

11.4 Adjusting Saturation

If local space average color is not located on the gray vector, we assume that the color gamut is shifted toward local space average color and we have to undo this shift as explained in the preceding text. Notice that, with this model, the color gamut has to be necessarily smaller than the maximum possible color gamut for the given intensity. Figure 11.11 shows three different scenarios for a shifted white point. One with a space average color of low intensity, the second with a space average color of intermediate intensity, and the third with a space average color of high intensity. If we assume that no pixels have been clipped, then the gamut of colors will be smaller than it could be. In order to use the full color gamut available, we have to adjust the saturation of the colors after the shift.

A simple way to adjust the saturation of the output pixels would be to increase the saturation of all pixels by a certain amount. However, we could also increase the saturation based on the size of the color shift. Figure 11.12 illustrates this process. First local space average color is computed. All colors are shifted by the component of local space average color that lies perpendicular to the gray vector. Now the colors are centered around the gray vector. However, the gamut of colors is not as large as it could be. Now, we rescale all colors such that the entire color space is filled. Let $\mathbf{a}_\perp$ be the component of local space average color, which is perpendicular to the gray vector. Let **l** be the vector which points from the gray vector through local space average color to the border of the color space. Then the colors could be rescaled by a factor of $\frac{|\mathbf{l}|}{|\mathbf{l}-\mathbf{a}_\perp|}$ such that the whole color space is filled. For certain images, this increase in saturation may be too much. An alternative way would be to perform the increase in saturation based on the size of the shift.

Figure 11.13 shows the results obtained by increasing the saturation. The images (a) and (b) were computed using the algorithm described in Section 11.2. After subtracting the perpendicular component of local space average color from the current color, the saturation is slightly increased. The increase is based on the deviation of local space average color from the gray vector. The increase is performed such that the minimum deviation corresponds to no increase in saturation whereas the maximum deviation corresponds to an increase of 30%. The bottom two images were computed using the algorithm described in Section 11.3. This algorithm first normalizes both the color of the pixel and local space

250 USING LOCAL SPACE AVERAGE COLOR FOR COLOR CONSTANCY

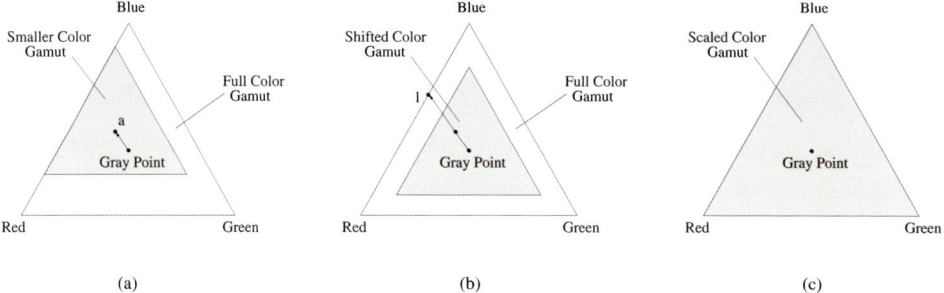

Figure 11.12 Shifted color gamut (a). The white point lies at the position **a**. A color correction can be performed by shifting the colors toward the gray vector (b). Now the color gamut is centered around the gray vector. In order to fully use the available color space, we can increase the color gamut as shown in (c).

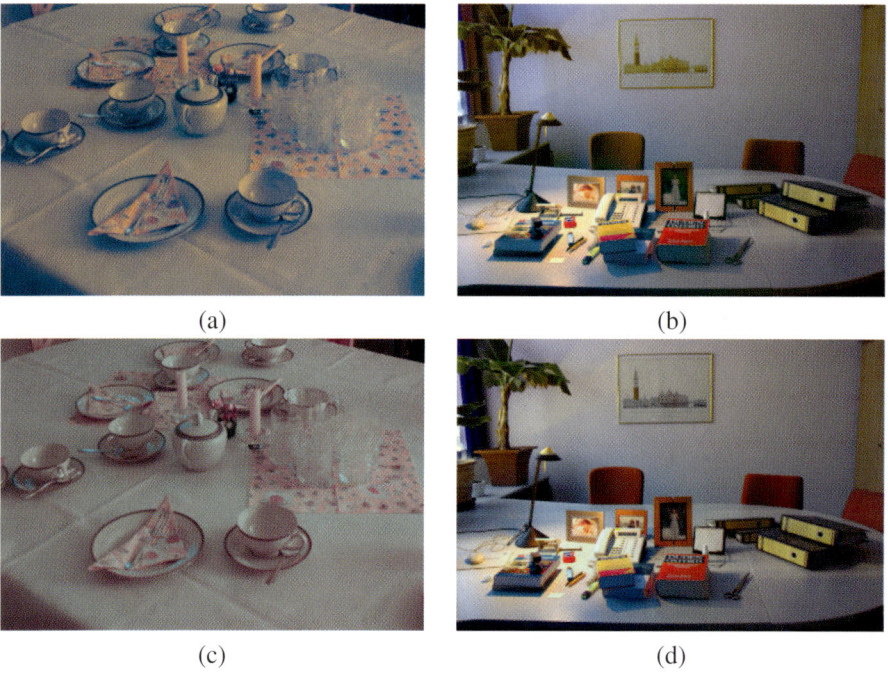

Figure 11.13 The images (a) and (b) were computed using the algorithm described in Section 11.2. The images (c) and (d) were computed using the algorithm described in Section 11.3. For all the images, the saturation was increased based on the size of the color shift.

average color before subtracting the perpendicular component. After subtracting the perpendicular component, the output color is rescaled such that the intensity is equivalent to the original intensity. The saturation is increased in the same way as described in the preceding text.

11.5 Combining White Patch Retinex and the Gray World Assumption

Another possibility is to subtract space average color in order to adjust the colors of the image (Ebner 2003a). This method of color constancy was also used by Moore et al. (1991).

$$o'_i(x, y) = c_i(x, y) - a_i(x, y). \tag{11.18}$$

If we subtract local space average color, we will obtain zero if the intensity of the image pixel is equivalent to the intensity of local space average color for the given band. We will obtain positive values if the pixel values are larger, and negative values if the pixel values are smaller. Thus, we also need to perform an additional transformation to move the output colors back into the range [0, 1]. Moore et al. use a global transformation. They first calculate the maximum and minimum output over all pixels and bands. They use the following transformation to scale the result back to the range [0, 1]:

$$o_i(x, y) = \frac{o'_i(x, y) - \min}{\max - \min} \tag{11.19}$$

where

$$\max = \max\{o'_i(x, y) | \text{ for all } (x, y) \text{ and } i \in \{r, g, b\}\} \tag{11.20}$$

$$\min = \min\{o'_i(x, y) | \text{ for all } (x, y) \text{ and } i \in \{r, g, b\}\}. \tag{11.21}$$

Note that this transformation does not necessarily move the average to 0.5. If the pixel values are nonlinear, then the average of the transformed values may be different from 0.5.

Ebner (2003a) applies a local transformation, which moves the average back to 0.5 and also rescales the data points. First, local space average color is subtracted from the input intensity. Next, we determine the absolute deviation between input intensity and the local space average color. The maximum deviation in a local neighborhood is used to scale either the maximum intensity to 1 or the minimum intensity to -1. Finally the result is brought back to the range [0,1]. The difference between the scaling by the inverse of twice the local space average color (discussed in Section 11.1) and the subtract and scale method (described here) is shown in Figure 11.14. A set of data points is shown in Figure 11.14(a). The same data set scaled by 0.3 is shown in (b). The algorithm described in Section 11.1 would transform this set by computing the average and then dividing all points by twice the average value. The result is shown in (c). The new positions of the data points in (c) match the positions shown in (a) closely. The set of data points shown in (b) can also be transformed by moving the average to 0.5 and then rescaling all data points around 0.5 by the maximum deviation in either direction such that a maximum range of [0,1] is obtained. The result of this operation is shown in (d). Again, the data points are brought closer to their original position. The rescaling works nicely if there is no bias in the input data. Figure 11.14(e) to (g) shows what happens if there is a systematic bias in the input data. If we subtract the average from the data, we also subtract the bias. Figure 11.14(f) shows the biased data points rescaled by twice the average value. Figure 11.14(g) shows the biased data points shifted by the average and then rescaled by the maximum deviation.

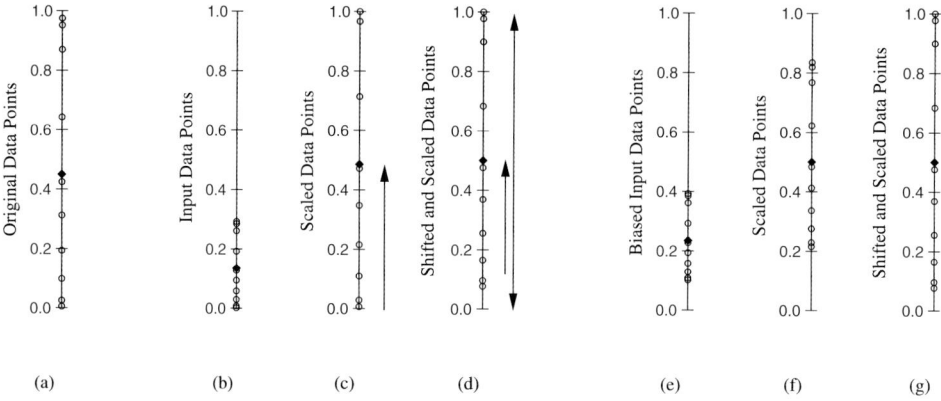

Figure 11.14 (a) Original data points. The filled diamond marks the average of the data points. (b) Input data points. All points were scaled by 0.3. (c) Result of scaling algorithm (d) Result of subtract & scale algorithm. (e)–(g) same as (b)–(d) except that a bias was applied to the input values.

The shift and rescale method was implemented in parallel as follows. Local space average color was calculated on a grid of processing elements as described in the preceding text. Let
$$\mathbf{c}(x, y) = [c_r(x, y), c_g(x, y), c_b(x, y)]^T \qquad (11.22)$$
be the color at image pixel (x, y). Let
$$\mathbf{a}(x, y) = [a_r(x, y), a_g(x, y), a_b(x, y)]^T \qquad (11.23)$$
be the local space average color computed by element (x, y). The absolute difference between local space average color and the intensity of the image pixel is also determined in parallel. Let $d_i(x, y)$ be the estimate of the maximum deviation at position (x, y) for band $i \in \{r, g, b\}$. We then take the maximum deviation across neighboring elements:
$$d_i'(x, y) = \max\{|a_i(x, y) - c_i(x, y)|, d_i(x-1, y), d_i(x, y-1), d_i(x+1, y), d_i(x, y+1)\}. \qquad (11.24)$$
This value is reduced by a small percentage p_d.
$$d_i(x, y) = (1 - p_d) d_i'(x, y) \qquad (11.25)$$
As the maximum deviation is passed from element to element, it is multiplied by $(1 - p_d)$. Therefore, over time, the value $d_i(x, y)$ will reduce to zero. The factor p_d determines how soon this reduction will occur.

Given the maximum deviation between local space average color and the pixel intensity, the output pixels are calculated as follows. First, local space average color is subtracted from the current intensity. Then, the deviation d_i is used to rescale the difference. Henceforth, we drop the coordinate of the current processing element.
$$o_i' = \frac{c_i - a_i}{d_i}. \qquad (11.26)$$

The values o'_i are now in the range $[-1, 1]$ and we need to transform them to the range $[0, 1]$.

$$o_i = \frac{1}{2}(1 + o'_i). \tag{11.27}$$

We also experimented with a sigmoidal activation function to transform the values o'_i to the range $[0, 1]$.

$$o_i = \frac{1}{1 + e^{-\frac{o'_i}{\sigma}}}. \tag{11.28}$$

The sigmoidal activation function produced better results. If $o'_i(x, y)$ is almost zero, i.e. local space average color and the color of the current pixel are similar, then the output will be 0.5, i.e. gray. For small deviations around the average color, the output will vary linearly. The output will be saturated at 1 for large positive deviations. It will approach 0

Linear Output:

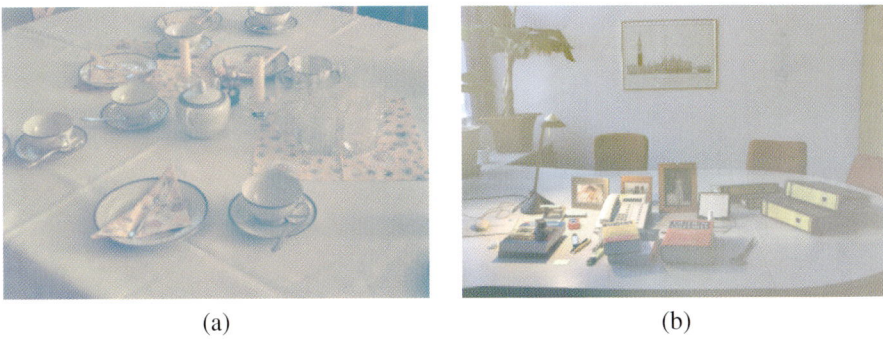

(a) (b)

Sigmoidal Output:

(c) (d)

Figure 11.15 The difference between local space average color and the color of the current pixel was used to compute the output color. The linear transformation was used for the images shown in (a) and (b). The sigmoidal transformation was used for the images shown in (c) and (d).

in the limit for large negative deviations. Two output images that were computed with this algorithm are shown in Figure 11.15. The Images in Figure 11.15(a) and (b) show the output images using the linear transformation and those in Figure 11.15(c) and (d) show the output images using the sigmoidal transformation.

12

Computing Anisotropic Local Space Average Color

Let us now have a look at what happens if the illuminant changes smoothly over the input image (Ebner 2004d). Figure 12.1 illustrates a smooth horizontal change of the illuminant. It is assumed that the illuminant changes linearly from one side of the image to the other side. If we look at the average computed by the processing element in the center of Figure 12.1, we see that the average computed by the element on the left will be a little lower and the average computed by the element on the right will be a little higher than the current average. On the other hand, space average color computed by the elements above and below will be the same as the one computed by the element in the center. If we average the data obtained from the four neighboring elements, then the value that is computed by the element on the left, which is a little lower, will cancel the value that is computed by the element on the right, which is a little higher. This is illustrated in Figure 12.1(b). The new average will be in equilibrium and we obtain a series of vertical bands. Thus, we will be able to estimate a linearly changing illuminant using the methods described in Chapter 10.

12.1 Nonlinear Change of the Illuminant

Suppose that we have a smooth but nonlinear change of the illuminant. This is illustrated in Figure 12.2. Again, we assume that the illuminant varies horizontally over the image. The average computed by the processing element above and below the current processing element will be the same average that is computed by the current processing element. The average computed by the processing element on the left will be much lower than the average computed by the current processing element. However, the average computed by the processing element on the right will only be a little higher. Therefore, the average computed by the processing element on the left and the processing element on the right will not cancel. The estimate of the illuminant will be underestimated. This is illustrated Figure 12.2(b). Therefore, the illuminant will not be estimated correctly if the change of the illuminant is nonlinear.

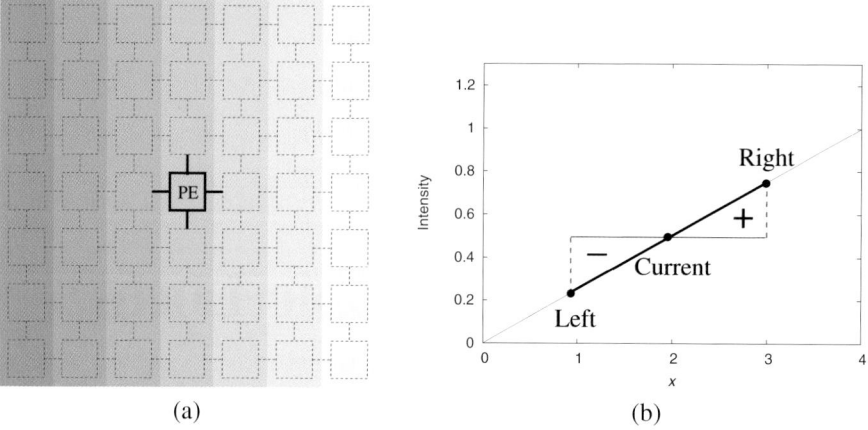

Figure 12.1 Consider a processing element that is located at a position where the illuminant changes linearly from left to right. The plot in (b) shows the smooth linear curve of the illuminant. Local space average color computed by the element on the left will be a little lower and local space average color computed by the element on the right will be a little higher. Both effects cancel each other and space average color will be in equilibrium, which will result in vertical bands. The positive and negative contributions cancel each other.

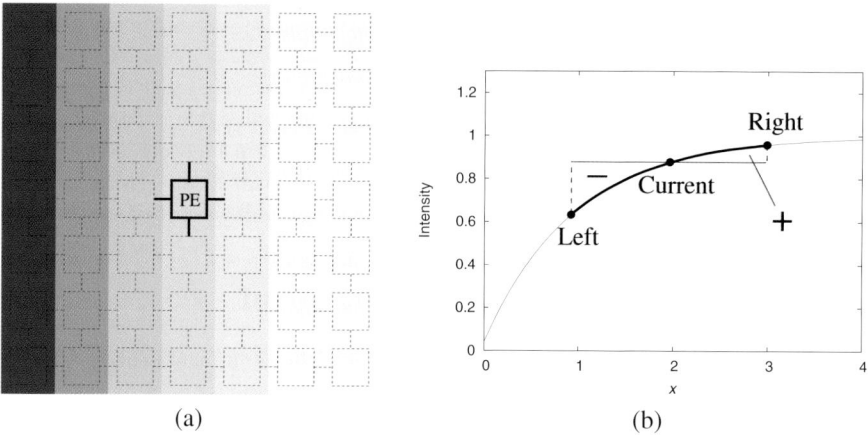

Figure 12.2 Consider a processing element that is located at a position where the illuminant changes smoothly but nonlinearly from left to right. The plot in (b) shows the smooth nonlinear curve of the illuminant. Local space average color computed by the element on the left will much lower than the local space average color computed by the element on the right. The positive contribution is much smaller than the negative contribution. Therefore, the estimate of the illuminant will be pulled to a lower value.

COMPUTING ANISOTROPIC LOCAL SPACE AVERAGE COLOR

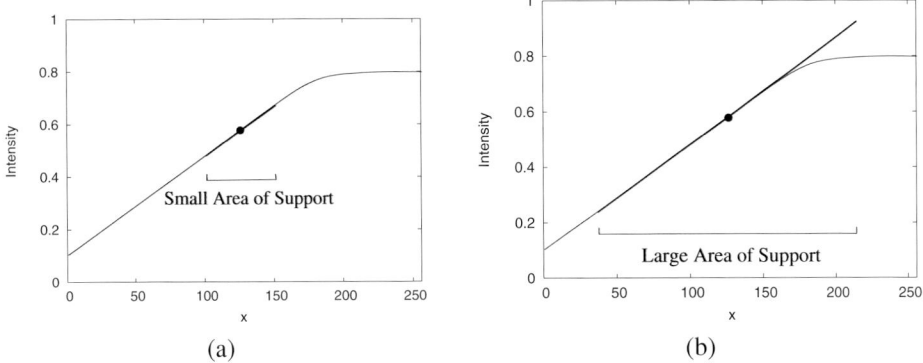

Figure 12.3 Even a nonlinear change of the illuminant can be considered to be linear provided the area of support is sufficiently small (a). However, in practice, we use quite large areas of support. Otherwise the gray world assumption would not be correct. In this case, the area of support will probably contain nonlinearities and the estimated local space average color will be incorrect (b).

The estimate of the illuminant is computed by averaging the data from neighboring elements and then adding the color of the current pixel to this average using a small percentage p. This percentage p determines the area of support over which the local space average color will be computed. If p is large, then the area of support will be small and vice versa. We have already discussed this in Chapter 10. For a sufficiently small area of support, the assumption that the illuminant varies linearly may be correct (Figure 12.3a). However, we usually use quite large areas of support. In this case, the area of support will probably straddle a nonlinear transition of the illuminant (Figure 12.3b). The large area of support is necessary for the gray world assumption to hold.

12.2 The Line of Constant Illumination

A more accurate estimate of the illuminant can be computed by averaging the data nonuniformly. A similar nonuniform smoothing operation is frequently used to segment noisy images and is known as *anisotropic diffusion* (Didas et al. 2005; Jähne 2002; Monteil and Beghdadi 1999; Weickert 1997; Weickert et al. 1998). Let us again assume that we have a nonlinear horizontal transition of the illuminant. The illuminant is constant along the vertical. Let us call this the line of constant illumination. Each processing element has its own line of constant illumination. For this simple case, each processing element has a vertical line of constant illumination. If we average only the data from the elements above and below, we will obtain vertical stripes. We can estimate the illuminant by nonuniform averaging provided the scene is sufficiently complex along the line of constant illumination, i.e. the number of object colors that correspond to the pixels along the line of constant illumination must be sufficiently large. If the line of constant illumination is aligned with the vertical or horizontal, we can simply average the data from the corresponding processing elements. However, in practice, the line of constant illumination can have an arbitrary

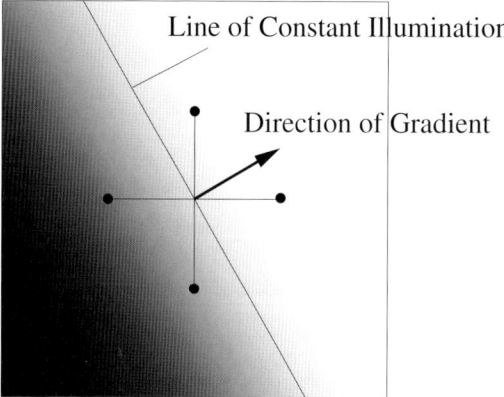

Figure 12.4 Nonlinear change of the illuminant. The arrow shows the direction of the gradient. The line of constant illumination is perpendicular to the direction of the gradient.

direction. It may not even be a line but may be a circle. A circle can result if a spotlight is used to illuminate the scene.

Curved lines of constant illumination will be treated in the following text. Let us first address the case of a straight line of constant illumination. Figure 12.4 shows a nonlinear transition of the illuminant. The arrow shows the direction of the gradient. The line of constant illumination runs perpendicular to the gradient. If the gray values are averaged along a direction that is perpendicular to the gradient, we obtain an estimate of the illuminant along the line of constant illumination. In this case, the result will be independent of the size of the area of support. This is a very important point. Note that if we compute isotropic local space average color, i.e. we compute local space average color without any directional preference, then the area of support (the scale of the averaging) depends on the image content. If we had some information regarding the lines of constant illumination, then we would not have to tune the area of support to the image content. The parameter p would only depend on image size.

If the number of different colors along the line of constant illumination is sufficiently large, then the estimate of the illuminant will be correct even for a nonlinearly changing illuminant. For a nonlinear changing illuminant, we should be averaging the data along the line of constant illumination. In order to determine the line of constant illumination, we have to calculate the direction of the gradient of the illuminant. The problem is that, in order to compute the gradient, the illuminant must be known. So we have come full circle. A possible solution is to first estimate the illuminant by computing local space average color as described in Section 10. This estimate can then be used to compute the illumination gradient, which in turn can be used to compute a more accurate estimate of the illuminant. Let $a_i(x, y)$ be the estimate of the illuminant (computed using one of the methods described in Section 10) at position (x, y) for channel i. Then the gradient of color channel i is given by

$$\nabla a_i = \begin{bmatrix} dx_i \\ dy_i \end{bmatrix} = \begin{bmatrix} \frac{\partial a_i}{\partial x} \\ \frac{\partial a_i}{\partial y} \end{bmatrix}. \tag{12.1}$$

COMPUTING ANISOTROPIC LOCAL SPACE AVERAGE COLOR

Since we usually work with the three color bands red, green and blue, we may have three different gradients. The three gradients do not necessarily point to the same direction. Computation of local space average color should be done along the same direction for all three color bands. Therefore, we have to find a way to combine the three gradients into one. Let (dx, dy) be the combined gradient. In order to compute the combined gradient, we could compute the average of the three gradients.

$$\begin{bmatrix} dx \\ dy \end{bmatrix} = \frac{1}{3} \sum_{i \in \{r,g,b\}} \nabla a_i \qquad (12.2)$$

Another possibility would be to choose the maximum gradient.

$$\begin{bmatrix} dx \\ dy \end{bmatrix} = \nabla a_j \quad \text{with} \quad j = \text{argmax}_i \{|\nabla a_i|\}. \qquad (12.3)$$

We could also compute a weighted average where the weights are based on the magnitude of the gradient.

$$\begin{bmatrix} dx \\ dy \end{bmatrix} = \frac{1}{|\nabla a_r| + |\nabla a_g| + |\nabla a_b|} \sum_{i \in \{r,g,b\}} |\nabla a_i| \nabla a_i. \qquad (12.4)$$

Once we have computed the combined gradient (dx, dy), we can compute the angle α of the gradient as

$$\alpha = \tan^{-1}\left(\frac{dy}{dx}\right). \qquad (12.5)$$

12.3 Interpolation Methods

Let us assign a local coordinate system based on the direction of the gradient. The directions front/back are pointing along the gradient and the directions left/right are pointing along the line of constant illumination as shown in Figure 12.5. In order to average the values

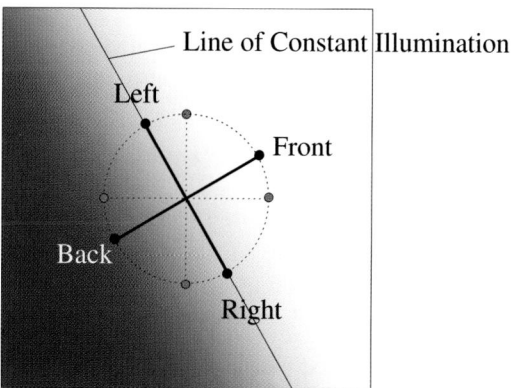

Figure 12.5 Rotated coordinate system. The directions front/back point along the direction of the gradient. The directions left/right point along the line of constant illumination.

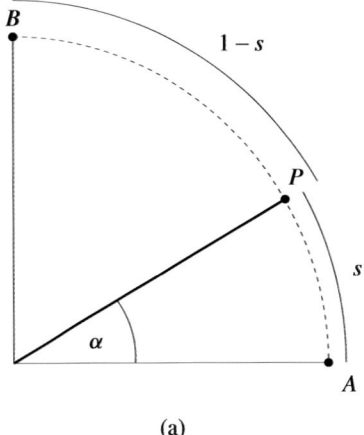 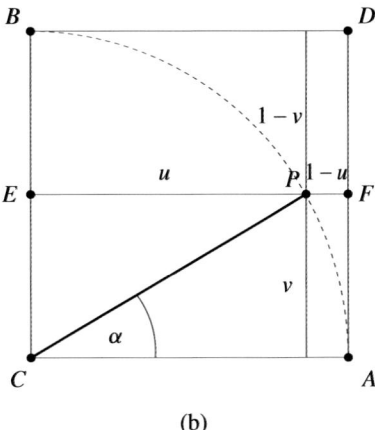

Figure 12.6 Two methods of interpolation: (a) The color at point P is obtained by interpolating the colors from points A and B. (b) If we have a grid of processing elements with diagonal connections, then we can use bilinear interpolation to calculate the color at point P using the data from points A, B, C, and D.

along the line of constant illumination we have to obtain the pixel values at these rotated coordinates. Interpolation can be used to calculate the values from the values obtained from the original coordinate system. We can use a variety of techniques to interpolate the original data. Some examples are shown in Figure 12.6.

The first interpolation method (a) calculates the color $\mathbf{c}_P$ at point P by interpolating the data from positions A and B. Let $\mathbf{c}_A$ and $\mathbf{c}_B$ be the pixel color at positions A and B. Let α be the direction of the gradient at the current element. Then the color at position P can be calculated as

$$\mathbf{c}_P = (1-s)\mathbf{c}_A + s\mathbf{c}_B \qquad (12.6)$$

with $s = \frac{2\alpha}{\pi}$. Instead of computing the angle α from the gradient, which requires a trigonometric operation, we could also use the following simplified interpolation, which has almost the same effect:

$$\mathbf{c}_P = \frac{1}{|dx|+|dy|}(|dx|\mathbf{c}_A + |dy|\mathbf{c}_B). \qquad (12.7)$$

The difference between the two methods is quite small:

$$\left| \frac{|dy|}{|dx|+|dy|} - \frac{2}{\pi}\tan^{-1}\left(\frac{dy}{dx}\right) \right| < 0.05. \qquad (12.8)$$

If a hardware implementation of the first calculation is too difficult or too expensive, we can use the second method. In the following text we will denote this interpolation method by $(\tilde{a})$.

If we also use the diagonal directions to connect our processing elements, then we can use bilinear interpolation (Akenine-Möller and Haines 2002), which is shown in

COMPUTING ANISOTROPIC LOCAL SPACE AVERAGE COLOR

Figure 12.6(b). Let $\mathbf{c}_A$, $\mathbf{c}_B$, $\mathbf{c}_C$, $\mathbf{c}_D$ be the color stored at the surrounding processing elements. Then, the color at position P can be calculated as

$$\mathbf{c}_P = u\mathbf{c}_E + (1-u)\mathbf{c}_F \tag{12.9}$$

with $\mathbf{c}_E = (1-v)\mathbf{c}_B + v\mathbf{c}_C$ and $\mathbf{c}_F = (1-v)\mathbf{c}_D + v\mathbf{c}_A$, $u = \cos(\alpha)$, and $v = \sin(\alpha)$. In other words, we first interpolate linearly along the vertical and then interpolate the values along the horizontal.

Let $\mathbf{a}(\text{front})$, $\mathbf{a}(\text{back})$, $\mathbf{a}(\text{left})$, and $\mathbf{a}(\text{right})$ be the interpolated colors of the rotated coordinate system. Let $\mathbf{c}(x, y)$ be the color at the current element. Now we can calculate local space average color a_i for color band i by averaging the colors obtained from the left and the right along the line of constant illumination:

$$a'_i(x, y) = \frac{1}{2}(a_i(\text{left}) + a_i(\text{right})) \tag{12.10}$$

$$a_i(x, y) = (1-p)a'_i(x, y) + pc_i(x, y) \tag{12.11}$$

where p is a small percentage. We can also include the data from the front/back direction by introducing an additional parameter q with $q \in [0, 0.25]$.

$$a'_i(x, y) = \left(\frac{1}{2} - q\right) a_i(\text{left}) + \left(\frac{1}{2} - q\right) a_i(\text{right}) + qa_i(\text{front}) + qa_i(\text{back}) \tag{12.12}$$

If q is equal to zero, we only average data along the line of constant illumination. For small values of q, the averaging operation is mainly performed along the line of constant illumination but some data also flows from the front and back virtual processing elements to the current processing element. In practice, we found that better results were obtained if the current element was included in the averaging operation. We can use either of the two methods to calculate a'_i and then include the current element using

$$a''_i(x, y) = \frac{1}{3}a_i(x, y) + \frac{2}{3}a'_i(x, y). \tag{12.13}$$

Finally, we slowly fade the color of the input image to the computed local space average color.

$$a_i(x, y) = (1-p)a''_i(x, y) + pc_i(x, y) \tag{12.14}$$

If we iterate long enough, we obtain local space average color, which is computed along the line of constant illumination.

Again, a significant speed-up can be achieved using successive over-relaxation (SOR) (Bronstein et al. 2001; Demmel 1996). Let $w \in (1, 2)$ be a parameter that describes the amount of over-relaxation. Then, local space average color a_i^{SOR} is computed by

$$a'_i(x, y) = \left(\frac{1}{2} - q\right) a_i^{\text{SOR}}(\text{left}) + \left(\frac{1}{2} - q\right) a_i^{\text{SOR}}(\text{right})$$
$$+ qa_i^{\text{SOR}}(\text{front}) + qa_i^{\text{SOR}}(\text{back}) \tag{12.15}$$

$$a''_i(x, y) = \frac{1}{3}a_i(x, y) + \frac{2}{3}a'_i(x, y) \tag{12.16}$$

$$a_i(x, y) = (1-p)a''_i(x, y) + pc_i(x, y) \tag{12.17}$$

$$a_i^{\text{SOR}}(x, y) = (1-w)a_i^{\text{SOR}}(x, y) + wa_i(x, y). \tag{12.18}$$

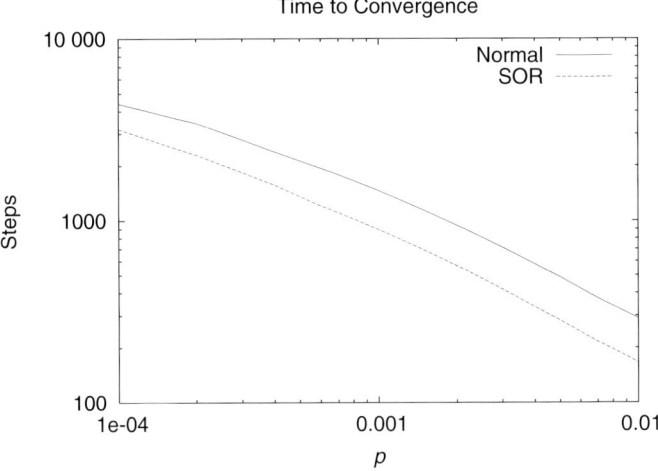

Figure 12.7 Time to convergence computed for different values of p. Convergence is considerably faster if successive over-relaxation is used.

Convergence time for $w = 1.999$ are shown in Figure 12.7. The speed-up is again considerable compared to the straightforward implementation.

So far we have only considered nonborder elements. Special attention has to be paid to the processing elements that are located at the border of the image. It is not sufficient to simply duplicate the pixels along the border of the image. Suppose that we have a diagonal transition of the illuminant. Duplicating the border pixels produces an S-shaped region. Since we are using very small values for p, the extent over which local space average color is computed will be quite large. The duplicated border pixels will affect all pixels of the image. Therefore, it is better to not make any assumptions about pixel values outside of the image. We only average the pixel values that are available. This is shown in Figure 12.8. Only the pixel at the center and the interpolated value along the line of constant illumination are averaged.

12.4 Evaluation of Interpolation Methods

In order to evaluate the different interpolation methods, three input images were created. All three images only contain a single color band. The images are shown in Figure 12.9. The first image (a) contains a gradient along the horizontal. The second image (b) shows the same gradient except that it is rotated by $30°$. The third image (c) models a spotlight at the center of the image. Each image is used both as an input image from which we compute local space average color as well as an image that defines the line of constant illumination. Ideally, we require that, when we compute local space average color along the line of constant illumination, the computed average should be equivalent to the input image. Therefore, we compute the root mean squared error between the input image and the image with the computed local space average color. The results are shown in Table 12.1. The output images for interpolation method (b) are shown in Figure 12.10. The table lists

COMPUTING ANISOTROPIC LOCAL SPACE AVERAGE COLOR

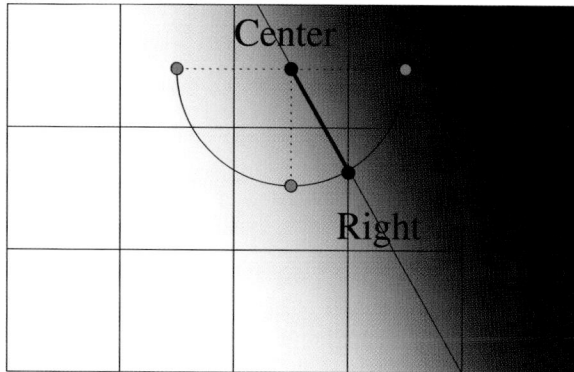

Figure 12.8 Care must be taken at the border elements. The pixel values outside of the grid of processing elements are unknown. No assumption should be made about these pixel values. Therefore, we only average the data of the current element and the interpolated value along the line of constant illumination.

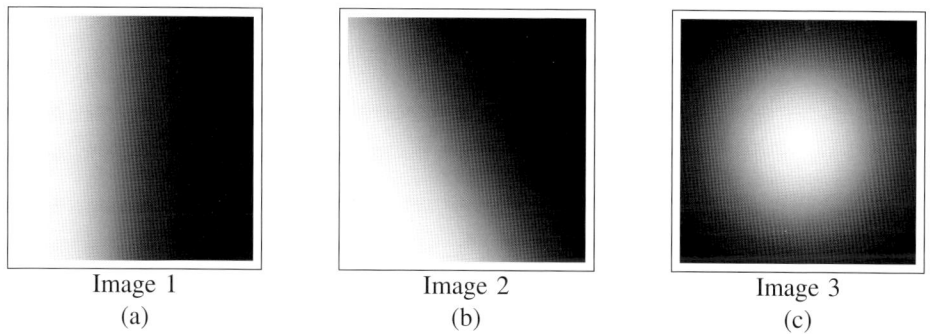

Figure 12.9 Images that were used to test the different interpolation methods.

Table 12.1: Root mean squared error between the input image and the image with the computed local space average color. Local space average color was computed by averaging the data obtained along the line of constant illumination.

	Image 1	Image 2	Image 3
Interpolation method (a)	0.011454	0.072972	0.136447
Interpolation method ($\tilde{a}$)	0.011454	0.070499	0.137794
Interpolation method (b)	0.011454	0.046182	0.189964

the root mean squared error between the input image and the computed local space average color.

All three methods of interpolation show the same performance on the horizontal gradient (Image 1). Results for the second image show that the performance of the approximative

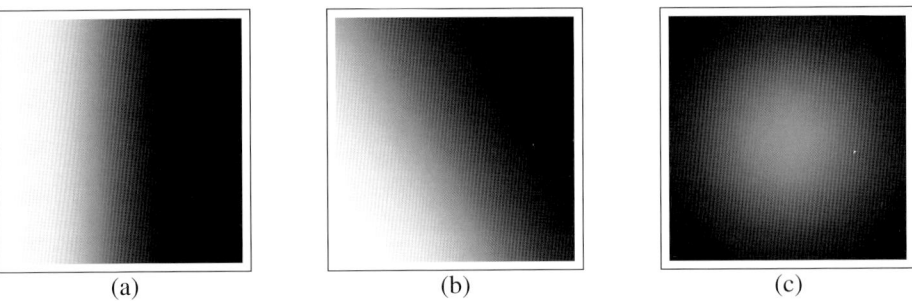

Figure 12.10 Output images for interpolation method (*b*). The first two images (a) and (b) are very similar to the input images shown in Figure 12.9. The third image (c) differs because here the line of constant illumination is curved. For this image, the values obtained along the line of constant illumination are too low. The spotlight is blurred and comes out too dark.

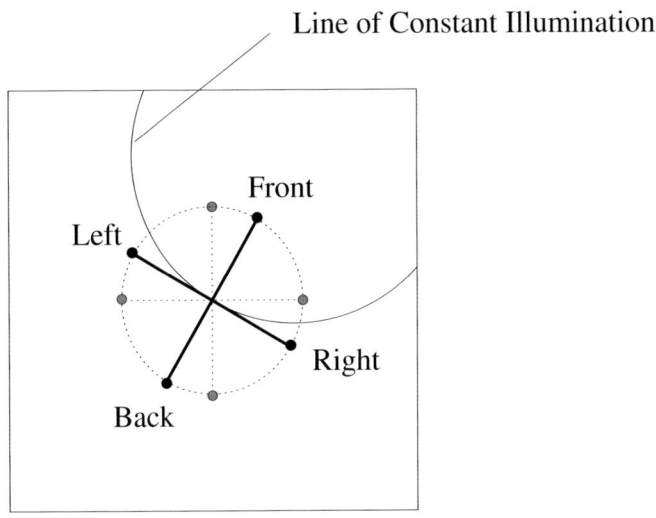

Figure 12.11 If the line of constant illumination is curved, then the values, which are obtained in a direction perpendicular to the gradient, are not correct.

calculation of method ($\tilde{a}$) is about the same as that of the exact method (*a*) in terms of root mean squared error. However, interpolation method ($\tilde{a}$) is slightly better while interpolation method (*b*) is the best. The results for image 3 show a different picture. Now method (*b*) performs the worst and methods (*a*) and ($\tilde{a}$) perform better. The reason for the bad performance on image 3 is clear. For this image, the line of constant illumination is curved. This is illustrated in Figure 12.11. The values obtained in a direction perpendicular to the gradient are too low for this particular image.

12.5 Curved Line of Constant Illumination

The solution to this problem is also clear. We need to calculate the local curvature of the illuminant for each processing element. Then, we can average the values along the line of constant illumination even if the shape of the change of the illuminant is curved. The curvature K of a point (x, y) on a surface $F(x, y)$ is defined as (Bronstein et al. 2001)

$$K = \frac{\begin{vmatrix} F_{xx} & F_{xy} & F_x \\ F_{yx} & F_{yy} & F_y \\ F_x & F_y & 0 \end{vmatrix}}{(F_x^2 + F_y^2)^{3/2}} \tag{12.19}$$

with $F_x = \frac{\partial F}{\partial x}$, $F_y = \frac{\partial F}{\partial y}$, $F_x = \frac{\partial F}{\partial x}$, $F_{xy} = \frac{\partial F}{\partial x \partial y}$, $F_{yx} = \frac{\partial F}{\partial y \partial x}$, $F_{xx} = \frac{\partial^2 F}{\partial x^2}$, and $F_{yy} = \frac{\partial^2 F}{\partial y^2}$. Thus, if we have an estimate of the illuminant, we can compute the local curvature for each processing element. We can then average the data along the curved lines of illumination. The curvature of the illuminant is computed by setting $F_x = dx$, $F_y = dy$, $F_{xy} = \frac{\partial}{\partial x} dy$, $F_{yx} = \frac{\partial}{\partial y} dx$, $F_{xx} = \frac{\partial}{\partial x} dx$, and $F_{yy} = \frac{\partial}{\partial y} dy$, where (dx, dy) is the combined gradient of the estimated illuminant. Given the curvature K, we can calculate the radius r of the curve: (Bronstein et al. 2001)

$$r = \left|\frac{1}{K}\right|. \tag{12.20}$$

Therefore, the radius r of the curvature at point (x, y) is given by

$$r = \left|\frac{(F_x^2 + F_y^2)^{3/2}}{F_x F_{xy} F_y + F_x F_{yx} F_y - F_x^2 F_{yy} - F_y^2 F_{xx}}\right|. \tag{12.21}$$

The sign of the curvature K tells us on which side of the curve the center of the curvature lies. If $K > 0$, then the center of the curvature lies on the positive side of the curve normal. It lies on the negative side of the curve normal if $K < 0$. If $K = 0$, then the line of constant illumination is really a straight line.

Now that we know the position of the center of the curvature, we can calculate the points where the unit circle around the current element and the circle that describes the curvature intersect. This is illustrated in Figure 12.12. Let us assume that the center of the curvature is located at point $(0, r)$. This simplifies the calculation of the two points of intersection. We have to calculate the two point of intersection P_1 and P_2. The following two equations describe both circles.

$$x^2 + y^2 = 1 \tag{12.22}$$

$$(x - r)^2 + y^2 = r^2 \tag{12.23}$$

By substituting the first equation into the second equation, we can solve for x.

$$(x - r)^2 + y^2 = r^2 \tag{12.24}$$

$$x^2 - 2xr + r^2 + y^2 = r^2 \tag{12.25}$$

$$2xr = 1 \tag{12.26}$$

$$x = \frac{1}{2r} \tag{12.27}$$

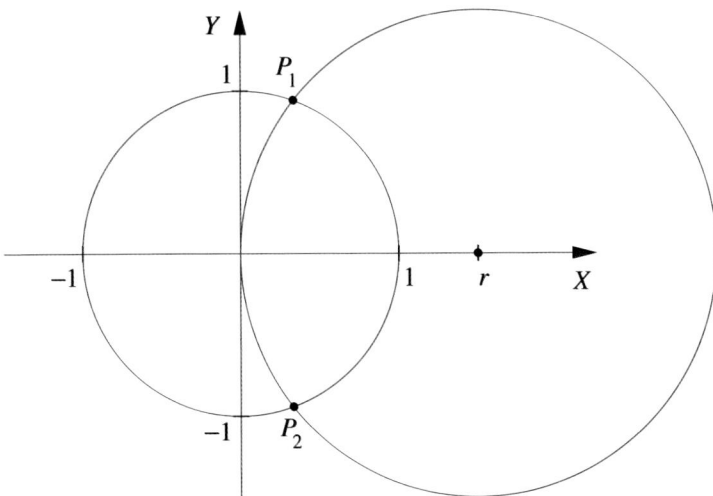

Figure 12.12 The values along the line of constant illumination can be obtained by calculating the intersection between the unit circle and the circle that describes the curvature at the current processing element. We have to obtain pixel values at positions P_1 and P_2 if the center of the curvature lies at position $(0, r)$.

The two y coordinates are then obtained using $x^2 + y^2 = 1$. This gives us

$$y_{1/2} = \pm\sqrt{1 - \frac{1}{4r^2}}. \tag{12.28}$$

If the center of the curvature does not lie on the X-axis, we can simply perform an appropriate rotation of the coordinate system. Now that we know the points of intersection, we can obtain pixel values at these positions using one of the methods of interpolation as described earlier. Let $\check{a}_i$ be the previous estimate of local space average color along the line of constant illumination. If only the two values along the line of constant illumination are averaged, then we compute

$$\check{a}'_i(x, y) = \frac{1}{2}(\check{a}_i(P_1) + \check{a}_i(P_2)). \tag{12.29}$$

We can also include data from the front/back directions. In this case, some data is allowed to flow in the direction of the gradient.

$$\check{a}'_i(x, y) = \left(\frac{1}{2} - q\right)\check{a}_i(P_1) + \left(\frac{1}{2} - q\right)\check{a}_i(P_2) + q\check{a}_i(\text{front}) + q\check{a}_i(\text{back}) \tag{12.30}$$

where $q \in [0, 0.25]$ determines the extent of the averaging operation that is performed in the direction of the gradient. Next, the current element is included by computing

$$\check{a}''_i(x, y) = \frac{1}{3}\check{a}_i(x, y) + \frac{2}{3}\check{a}'_i(x, y). \tag{12.31}$$

Table 12.2: Root mean squared error between the input image and the image with the computed local space average color. Local space average color was computed by averaging the data obtained along the line of constant illumination. The line of constant illumination was computed from the curvature at each pixel.

	Image 1	Image 2	Image 3
Interpolation method (a)	0.011454	0.073196	0.089363
Interpolation method ($\tilde{a}$)	0.011454	0.070766	0.084406
Interpolation method (b)	0.011454	0.046365	0.056084

Figure 12.13 Output images for interpolation method (b) when the local curvature is calculated in order to obtain the correct values along the line of constant illumination. The center of the spotlight is now much brighter.

Finally, the color from the current element is slowly faded into the result

$$\check{a}_i(x, y) = (1 - p)\check{a}_i''(x, y) + pc_i(x, y) \qquad (12.32)$$

where p is again a small value larger than zero. These update equations are iterated until convergence. By computing the curvature and then averaging the data along the line of constant illumination, we can also handle curved nonlinear transitions of the illuminant. Curved nonlinear transitions are a result of local illuminants that may be used to illuminate the scene.

We again tested all three methods of interpolation. The results are shown in Table 12.2. The output images for interpolation method (b) are shown in Figure 12.13. As expected, the results are similar for the linear change of the illuminant. Results for the spotlight stimulus have improved considerably.

12.6 Experimental Results

The two images at the bottom of Figure 12.14 show a strong intensity and color gradient. The images were created by taking two images of the same object from the databases created by Barnard et al. (2002c). The illuminant of the first image is almost white whereas the illuminant of the second image is bluish. The images were overlayed using the gradient

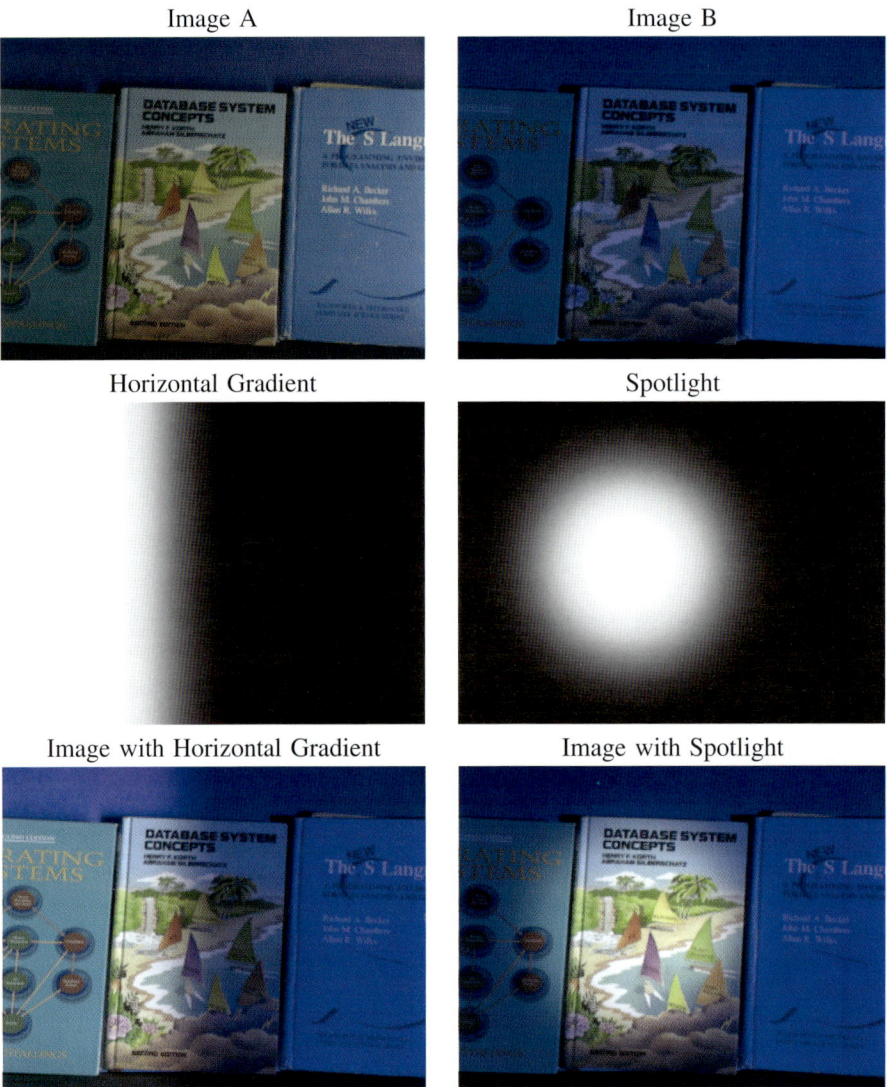

Figure 12.14 Two images from the database of Barnard et al. (2002c) were used to simulate intensity and color gradients. The gray-scale image shown on the left shows a horizontal gradient. This gradient was used to simulate a horizontal intensity and color gradient. The gray-scale image shown on the right shows a circular gradient. This gradient was used to simulate a spotlight. (Original image data from "Data for Computer Vision and Computational Colour Science" made available through http://www.cs.sfu.ca/~colour/data/index.html. See Barnard K, Martin L, Funt B and Coath A 2002 A data set for color research, Color Research and Application, Wiley Periodicals, 27(3), 147–151. Reproduced by permission of Kobus Barnard.)

COMPUTING ANISOTROPIC LOCAL SPACE AVERAGE COLOR

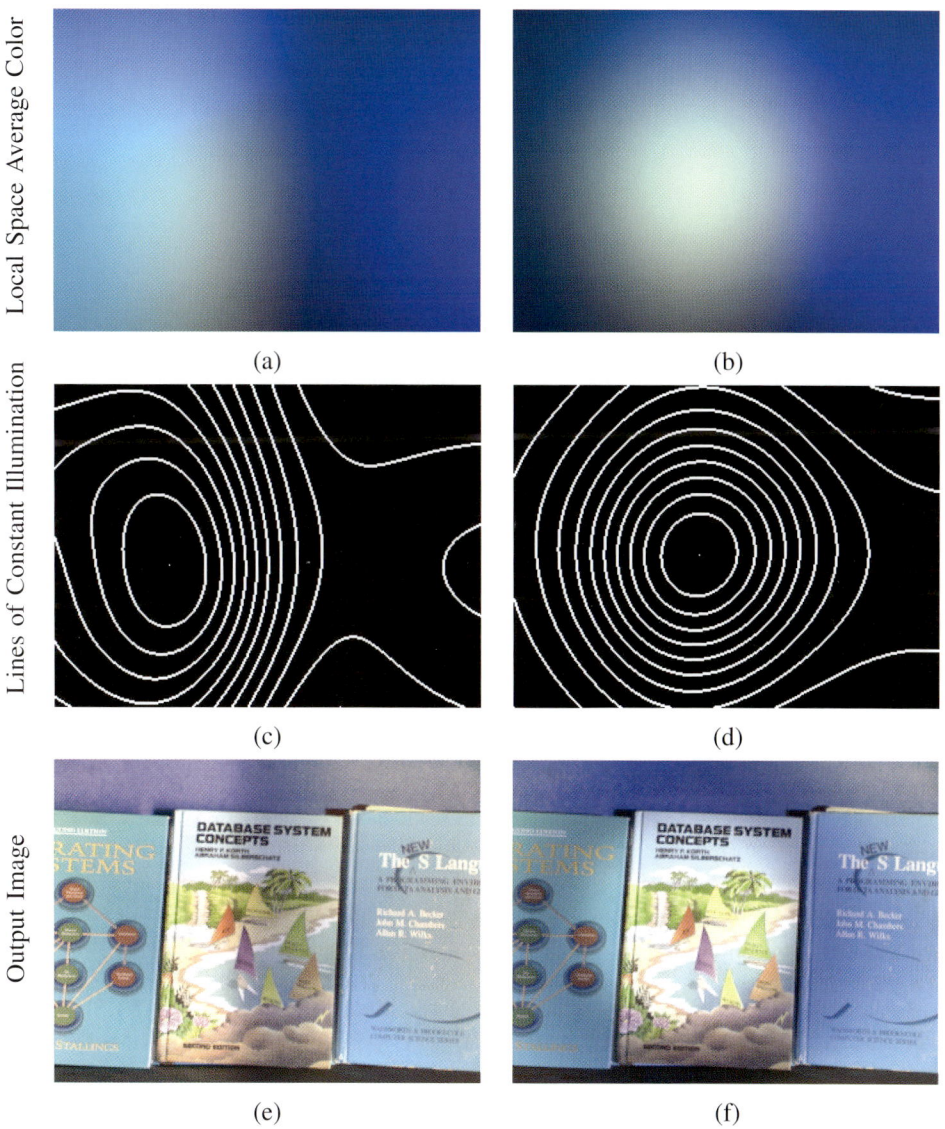

Figure 12.15 The two images (a) and (b) show local space average color computed using a Gaussian kernel with $\sigma = 0.14 n_{\max}$, where $n_{\max} = \max\{n_x, n_y\}$ and n_x and n_y are the width and height of the input image. The two images (c) and (d) show the estimated lines of constant illumination. The two images (e) and (f) are the output images using the algorithm of Section 11.1 except that anisotropic local space average color was used to estimate the illuminant. Anisotropic local space average color was computed with $p = 0.000013$.

Anisotropic L.S.A. Color

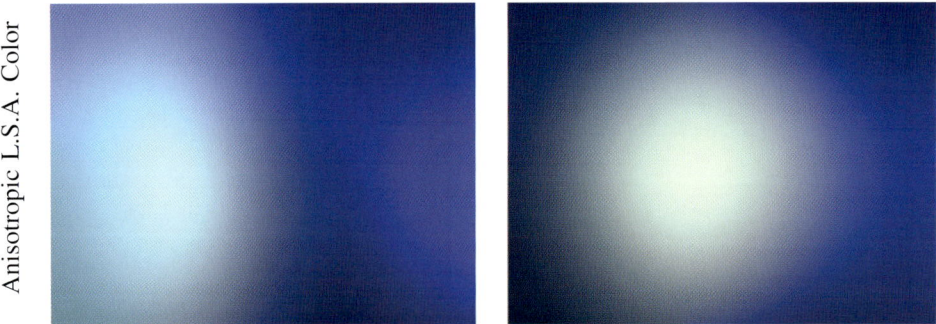

Figure 12.16 Anisotropic local space average color that was computed along the lines of constant illumination.

and spotlight masks shown in the middle of Figure 12.14. The intensity of the image that provides the illumination gradient was increased by a factor of 3. The bottom two images show the two overlayed images.

The strong illumination gradient can be removed by computing anisotropic local space average color along an estimate of the illuminant. Here, we will use the intensity of local space average color in order to estimate the illuminant. Figure 12.15 shows local space average color computed uniformly over the entire image. A Gaussian kernel with $\sigma = 0.14 n_{\max}$, where $n_{\max} = \max\{n_x, n_y\}$ and n_x and n_y are the width and height of the input image, was used. The lines of constant illumination are also shown. For the image in (a), a horizontal gradient was correctly estimated, whereas for the image in (b), a circular gradient was correctly estimated. The two images (e) and (f) in Figure 12.15 show the output images. Here, anisotropic local space average color is used to estimate the illuminant. The smoothing parameter was set to $p = 0.000013$. Anisotropic local space average color is shown in Figure 12.16. Both output images were computed using the method described in Section 11.1. The output color is computed by dividing the input color by the estimate of the illuminant.

Barnard et al. (2002c) measured the color of the illuminant for each image. Since we know the actual color of the illuminant and we also know how the two original images were overlayed, we can compare the estimated color to the ground truth data. Figure 12.17 shows the actual color of the illuminant. The two images in (c) and (d) show the chromaticities of the estimated illuminant. The two graphs in (e) and (f) show the data for a single horizontal line of the image.

Figure 12.18 shows the results obtained using isotropic local space average color. The two images in (a) and (b) show the output images using the algorithm described in Section 11.1. Each image pixel was divided by twice the local space average color. The two images in (c) and (d) show the chromaticities of the estimated illuminant. The smoothing parameter was set to $p = 0.000013$. The two graphs in (e) and (f) show the chromaticities of the actual and the estimated illuminant for a single horizontal line of the image. When looking at the two graphs of Figure 12.17 and Figure 12.18, we see that

COMPUTING ANISOTROPIC LOCAL SPACE AVERAGE COLOR

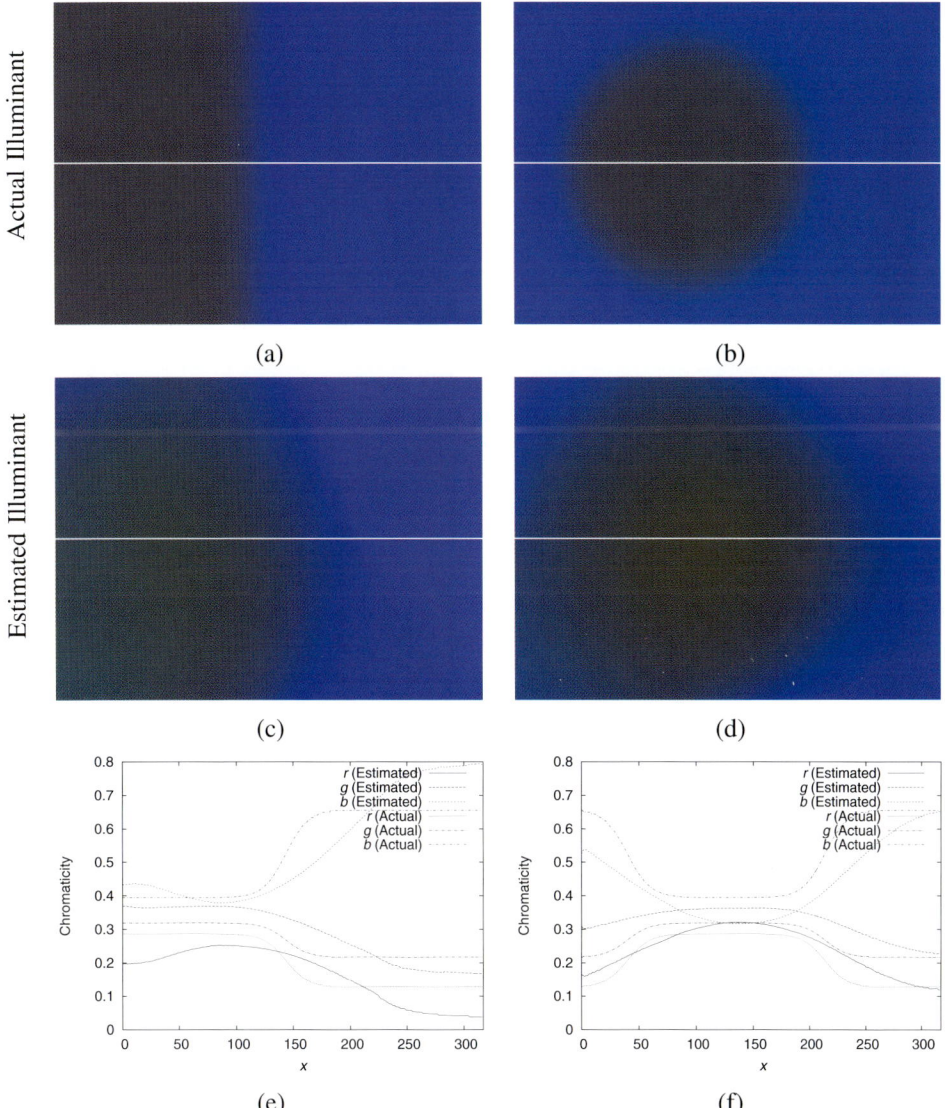

Figure 12.17 The images in (a) and (b) show the chromaticities of the actual color of the illuminant. The images in (c) and (d) show the estimated chromaticities. The two graphs in (e) and (f) show the estimated and the actual color of the illuminant for a horizontal line of the image.

the smoothing is much larger when isotropic local space average color is computed even though the same constant p was used in both cases.

Owing to the extent of the smoothing, the output computed using local space average color is not as good when compared with the output computed using anisotropic local space

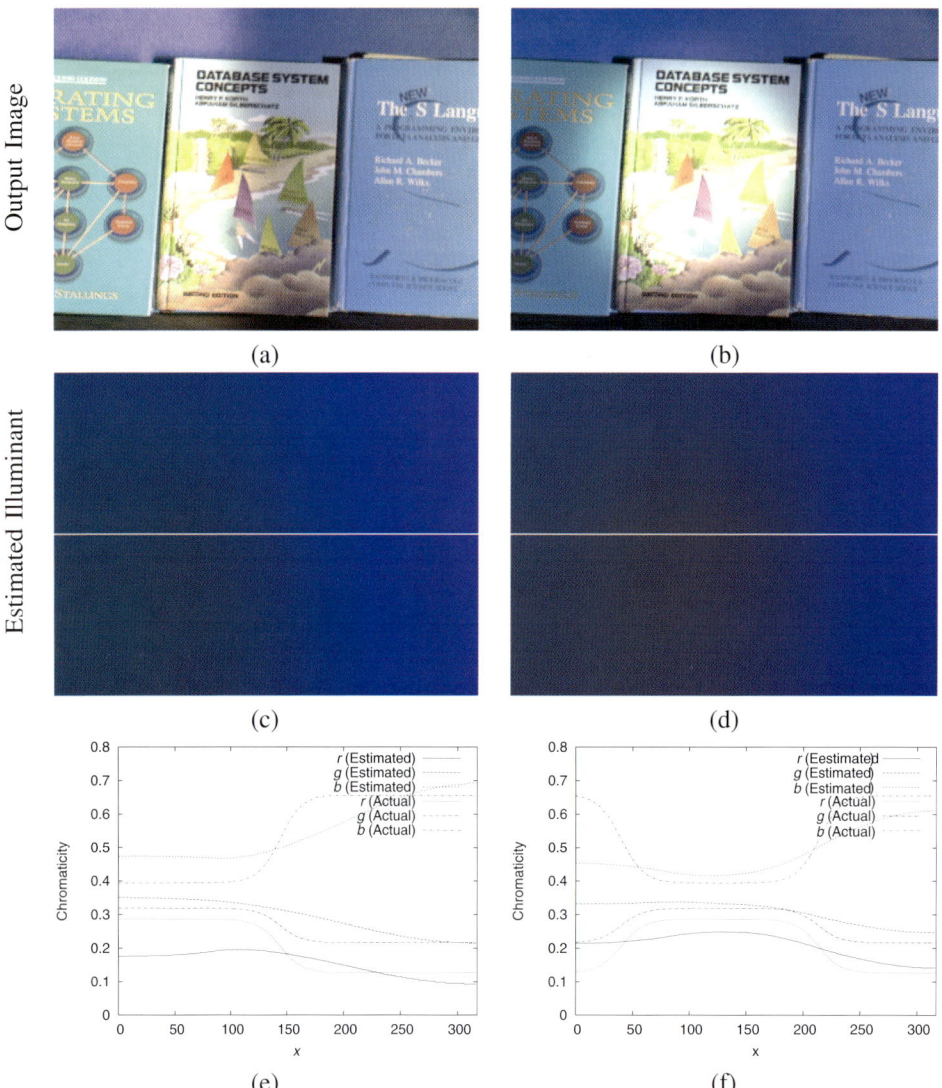

Figure 12.18 The two images in (a) and (b) show the computed output images using isotropic local space average color. The algorithm described in Section 11.1 was used. The two images in (c) and (d) show the estimated illuminant. The two graphs in (e) and (f) show the actual and the estimated illuminant for a single horizontal line of the image.

average color. Several image pixels are clipped. In most cases, simply computing isotropic local space average color may be the best thing to do. However, we cannot set the parameter p at some very small value if we are using isotropic smoothing. The parameter p has to be tuned to the given image. In contrast, computing local space average color along the line of constant illumination has the advantage that this algorithm is almost independent of the

COMPUTING ANISOTROPIC LOCAL SPACE AVERAGE COLOR 273

Anisotropic Local Space Average Color (1)

Anisotropic Local Space Average Color (2)

Anisotropic Local Space Average Color (3)

Figure 12.19 Output images for the different methods described in Chapter 11. The numbers 1 through 3 refer to the methods described in Sections 11.1 through 11.3.

smoothing parameter p. If the line of constant illumination can be accurately estimated, then the parameter p can simply be set to a tiny value.

Figure 12.19 shows the results for our standard two test images. For each image, output colors were computed by applying the algorithms described in Chapter 11 except that local space average color was computed along the estimated line of constant illumination. The numbers 1 through 3 refer to the methods described in Sections 11.1 through 11.3. A thorough evaluation of all algorithms follows in the next chapter.

13

Evaluation of Algorithms

In order to evaluate color constancy algorithms, a number of different methods can be applied. One goal is to compute reflectances for each pixel of the input image. Another goal may be, that an image processed by a color constancy algorithm is perceptually equivalent to what a photographer saw when he took the image. We have already seen in Section 2.5 that human color perception correlates with integrated reflectances of viewed objects. However, experiments by Helson (1938), described in Section 14.1, show that the human visual system does not in fact compute the reflectances of objects. If a color constancy algorithm is used for object recognition , then it makes sense to choose a color constancy algorithm that tries to estimate reflectances. If accurate color reproduction for consumer photography is the goal, then a color constancy algorithm that closely mimics the features of the human visual system is more desirable.

Some authors also take the approach to compute images that look as if they were taken under some canonical illuminant (e.g. Forsyth (1992)). In general, it is important that the output computed by a color constancy algorithm is not affected by changes of the illuminant. In other words, the computed descriptor should be constant irrespective of the illuminant. Otherwise, we would not have a color constancy algorithm. But at the same time, the output has to be nontrivial. If we would not make this assumption, a trivial color constancy algorithm could be found, which returns the same constant value for all input images. This was also pointed out by Forsyth (1992).

We will be using color-based object recognition to evaluate the quality of the color constancy algorithms. Where ground truth data is available, we can also compare the output of the different algorithms to the measured ground truth data.

13.1 Histogram-Based Object Recognition

Let us discuss color-based object recognition first. A highly successful method for object recognition is based on color histograms. The method was introduced by Swain and Ballard (1991). A color histogram is computed for all input images. The color histograms are compared using a method called *histogram intersection*. Let us now briefly review the

---
Color Constancy M. Ebner
© 2007 John Wiley & Sons, Ltd

method of Swain and Ballard. Let

$$\mathbf{c}(x, y) = [c_r(x, y), c_g(x, y), c_b(x, y)]^T \quad (13.1)$$

be the color of a pixel located at position (x, y). For each input image a color histogram is computed. Let H be the color histogram. The buckets of the histogram are indexed by the three-dimensional vector $\mathbf{c}$. The count of a bucket is denoted by $H(\mathbf{c})$. If each color band is quantized using n_q values per band, then the histogram contains n_q^3 buckets. In order to be independent of image size, one divides each bucket by the number of pixels that were used to compute the histogram. Let n be the number of image pixels. Then the normalized histogram $\tilde{H}$ is defined as

$$\tilde{H}(\mathbf{c}) = H(\mathbf{c})/n. \quad (13.2)$$

Each bucket $\tilde{H}(\mathbf{c})$ is simply the probability that the color vector $\mathbf{c}$ occurs in the original image.

Histogram-based object recognition compares only the color distributions of the image. In order to establish a match between a test and a model image, we only have to compare the color distributions. If the two distributions are similar, they may contain the same object. The main advantage is that the color distributions can be computed easily. Two color distributions can be compared quickly to see if a match can be found. No actual correspondences of image parts have to be computed. Color histograms are approximately invariant to a slight change of orientation. If the database does not contain too many objects, then the objects can be quickly distinguished based on their color distribution. A yellow folder with black writing on it can be distinguished easily from a red ball, as the color distributions are completely different. Things get more complicated as the database grows and more objects with similar color distributions are added. Additional features such as oriented lines can be included in the histogram to improve the recognition rate. However, since our focus is on the evaluation of color constancy algorithms, we will only use color histograms here.

Figure 13.1 shows some sample color histograms of different fruits. A banana, an orange, and a peach are shown. Below each image, the color histogram is shown. A quantization of 10 was used for each channel. Therefore, the color histogram consists of 1000 buckets. The first bucket contains colors $[r, g, b]^T$ with $r \in [0, 0.1]$, $g \in [0, 0.1]$ and $b \in [0, 0.1]$. In order to remove the influence of the background, the diagonal buckets located on the gray vector were set to zero. Since the color distributions are hard to grasp from such a linear histogram, the same data is shown in the last row in three dimensions. Each bucket of the three-dimensional histogram has a unique position inside the color cube. Each nonzero bucket is visualized by drawing a cube at the position of the bucket. The cube is drawn in the color of the bucket it represents. The size of the cube is based on the probability stored in the bucket. We see that in this case, the three different fruits can be recognized easily based on their different color distributions. Figure 13.2 shows the peach at two different scales and also slightly rotated. When we compare the histograms of the three images, we see that the color histogram is not quite invariant to changes in scale and orientation.

Swain and Ballard have proposed a method called *histogram intersection* to compare two color distributions. Let H_I be the histogram of the input image. Let H_M be the histogram of the model object. The given histogram is compared to the histogram of the model

EVALUATION OF ALGORITHMS

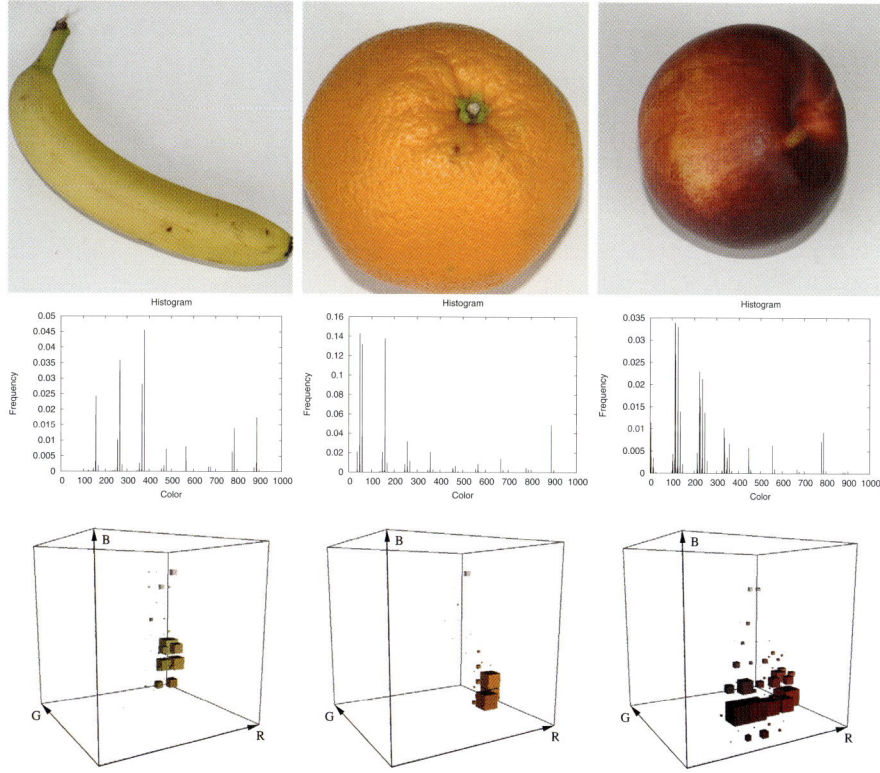

Figure 13.1 The first row shows images of different fruits. A banana, an orange, and a peach are shown. The color histogram is shown in the second row. The last row shows the color histogram in three dimensions.

object by computing the intersection $H_\cap$.

$$H_\cap = \sum_{\mathbf{c}} \min(H_I(\mathbf{c}), H_M(\mathbf{c})). \tag{13.3}$$

The intersection simply counts the number of pixels of a particular color in the model image, which are also found in the test image. In order to obtain a match value with the range [0, 1], the intersection is normalized by the total number of pixels in the model.

$$\tilde{H}_\cap = \frac{\sum_{\mathbf{c}} \min(H_I(\mathbf{c}), H_M(\mathbf{c}))}{\sum_{\mathbf{c}} H_M(\mathbf{c})}. \tag{13.4}$$

If we are working with normalized histograms where the buckets contain the probability that a particular color is found in the image, then the normalization can be omitted because we have $\sum_{\mathbf{c}} H_M(\mathbf{c}) = 1$. This particular method of comparing histograms was defined by Swain and Ballard because it does not require that the object has to be segmented

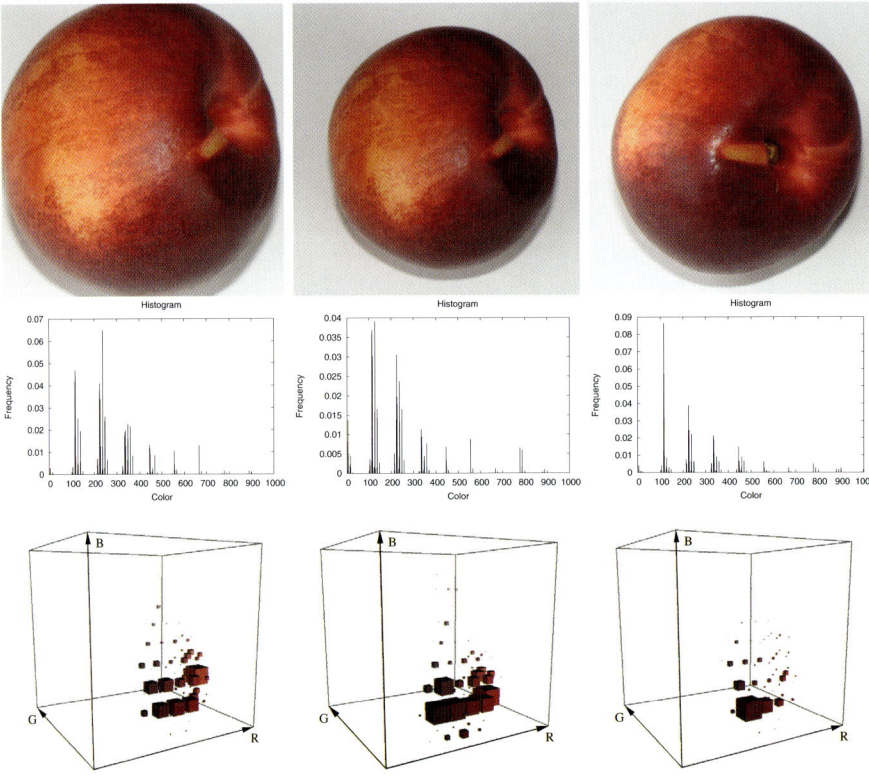

Figure 13.2 The first row shows the peach at different two different scales and in a rotated position. The color histograms are shown below each image. We see that the histogram is not quite invariant to changes in scale or orientation.

from the background. The measure is approximately invariant to changes in background composition. If the model object is indeed present in the input image, then the expression $\min(H_I(i), H_M(i))$ will be approximately equal to $H_M(i)$. Background pixels of color i may increase $H_I(i)$ but the whole expression will be limited by $H_M(i)$. The background may only have an influence if the input image does not contain the model object. In this case, a background pixel may increase the contents of a bucket that is not already at the maximum value for the model object.

Schiele and Crowley (1996, 2000) have suggested the χ^2 divergence measure to compare multi-dimensional histograms. In contrast to histogram intersection, this measure does not require a sparse distribution of colors to distinguish between different objects. The χ_e^2-divergence is calculated as

$$\chi_e^2(H_I, H_M) = \sum_{\mathbf{c}} \frac{(H_I(\mathbf{c}) - H_M(\mathbf{c}))^2}{H_M(\mathbf{c})} \tag{13.5}$$

when the exact distribution of the model histogram H_M is known. If the distribution is not known exactly, then the following measure should be used.

$$\chi^2(H_I, H_M) = \sum_{\mathbf{c}} \frac{(H_I(\mathbf{c}) - H_M(\mathbf{c}))^2}{H_I(\mathbf{c}) + H_M(\mathbf{c})}. \tag{13.6}$$

Schiele and Crowley (1996) reported that the χ^2_e-divergence measure worked slightly better than the χ^2-divergence measure. However, since the exact distribution is not known in this context, it is appropriate to use the χ^2-divergence measure. Using color as a single cue limits the discriminatory power of the approach. However, other types of image information may also be included. Schiele and Crowley (1996, 2000) suggested the computing of multi-dimensional receptive field histograms using gradient magnitude and direction as an additional cue. Similar and extended approaches which include the computation of various invariant features are used by some image search engines (Siggelkow et al. 2001). Since our focus is the development of color constancy algorithms, we will be using color histograms only.

Comparison of color histograms does not necessarily have to be done in RGB space. Histograms can also be computed for a variety of color spaces or certain subspaces. Swain and Ballard computed the histograms using three opponent color axes: red-green (RG), blue-yellow (BY), and black-white (BW). Let $\mathbf{c} = [c_r, c_g, c_b]^T$ be the color of the input pixel, then the transformed pixel coordinates $\mathbf{c} = [c_{\text{rg}}, c_{\text{by}}, c_{\text{bw}}]^T$ are given as

$$c_{\text{rg}} = c_r - c_g \tag{13.7}$$

$$c_{\text{by}} = 2c_b - c_r - c_g \tag{13.8}$$

$$c_{\text{bw}} = c_r + c_g + c_b. \tag{13.9}$$

They used a quantization of 8 for the BW-axis and a quantization of 16 for the RG- and BY-axis. Thus, the finer quantization was used for the color carrying components. For a real time implementation of their system, they have used the RGB components with equal quantization of all channels. According to Swain and Ballard, the choice of color axis is not crucial.

Swain and Ballard worked with 2048 bins but also experimented with other histogram sizes. They tested histogram sizes from 64 ($8 \times 8 \times 4$) bins to 8125 ($40 \times 40 \times 20$) bins. Recognition rates improved for larger histograms. Histogram sizes of 512 bins and more resulted in a recognition rate of over 99.7 when tested on a database with 66 models.

13.2 Object Recognition under Changing Illumination

If object recognition is based on color histograms, it is very important to use the actual colors of the object. However, if the light reflected by the object is simply measured and then used for object recognition, one may have trouble in recognizing the objects. The object colors seem to change if the illuminant changes. Therefore, a color constancy algorithm has to be applied before the method can be used successfully in different lighting conditions. Instead of running a color constancy algorithm and then computing a color histogram, one may also define a measure that is invariant to changes of the illuminant. One such method is described by Healey and Slater (1994).

Healey and Slater proposed to use moments of color distributions for illumination invariant object recognition. In Section 3.5 we have seen that the influence of a uniform illuminant can be viewed as a linear transformation of the image colors. If we are given two images of the same scene under a different illuminant and compute a color histogram for both images, then the histograms are related by a linear transformation of the coordinates. The color distributions can be aligned by computing the covariance matrix of the color distributions. Let $\mathbf{C}$ be the covariance matrix. The covariance matrix can be decomposed into two transformations $\mathbf{T}$ using a Cholesky decomposition $\mathbf{C} = \mathbf{TT}^T$. If we are given two images, one test image and one model image, we can compute the transformation $\mathbf{T}$ for both images. For each image, the corresponding transformation is applied. As a result, the two image distributions are related by an orthogonal transformation. Healey and Slater suggest computing the eigenvector of centered moment matrices for the transformed color distributions and use the eigenvectors for object recognition.

Let H be the histogram of the colors of the input image where $H(\mathbf{c})$ denotes the probability that color $\mathbf{c}$ occurs in the image. The centered moment is computed as follows:

$$A(\alpha) = \int (\mathbf{c} - \bar{H})^\alpha H(\mathbf{c}) d\mathbf{c} \tag{13.10}$$

where $\bar{H} = \int \mathbf{c} H(\mathbf{c}) d\mathbf{c}$ denotes the mean of the histogram and $\mathbf{c}^\alpha$ denotes the monomial $\prod_i c_i^{\alpha_i}$. For their experiments, Healey and Slater compute the following moments:

$$\mathbf{A}(\alpha) = \int \mathbf{X}_{[2,2]} \left(\mathbf{c} - \bar{H} \right) H(\mathbf{c}) d\mathbf{c} \tag{13.11}$$

where $\mathbf{X}_{[2,2]}(\mathbf{c}) = \mathbf{X}_{[2]}(\mathbf{c})\mathbf{X}_{[2]}(\mathbf{c})^T$ and $\mathbf{X}_{[2]}(\mathbf{c}) = \left(\frac{1}{\sqrt{2}} c_r^2, c_r c_g, c_r c_b, \frac{1}{\sqrt{2}} c_g^2, c_g c_b, \frac{1}{\sqrt{2}} c_b^2 \right)^T$. The six eigenvalues of this matrix are used for recognition. A match between test image and model image is made based on the Euclidean distance between the eigenvectors computed for the two images. Obviously, the eigenvectors for the model images can be precomputed. Healey and Slater (1994) report results for a database of nine objects and report that their method is more accurate than color constant color indexing that was developed by Funt and Finlayson (1995). In order to distinguish a larger number of objects, the number of invariants would probably have to be increased.

Funt and Finlayson (1995) proposed histogramming of color ratios. They suggested computing the ratios between adjacent pixels. Let $\mathbf{c} = [c_r, c_g, c_b]^T$ and $\mathbf{c}' = [c'_r, c'_g, c'_b]^T$ be two colors that are obtained from two adjacent pixel positions. Assuming that the sensors are narrow band, the intensity measured by the sensor at position (x, y) is given by

$$I_i(x, y) = G(x, y) R_i(x, y) L_i(x, y) \tag{13.12}$$

where $G(x, y)$ is a factor that depends on the scene geometry, $L_i(x, y)$ is the irradiance at position (x, y) for wavelength λ_i, and R_i is the reflectance for wavelength λ_i. If we now assume that there is a linear relationship between the sensor's response and the pixel colors then we have $I_i = c_i$. For two adjacent pixels with colors c_i and c'_i, the illuminant factor L_i and the geometry factor G are almost identical. Thus, if we form the ratio between

EVALUATION OF ALGORITHMS

adjacent pixels with reflectances R_i and R'_i, we obtain

$$\frac{c_i}{c'_i} = \frac{GR_iL_i}{GR'_iL_i} = \frac{R_i}{R'_i} \tag{13.13}$$

which is a color constant descriptor.

Funt and Finlayson (1995) suggest computing such a color constant descriptor by first computing the logarithm of the input pixels and then applying the Laplace operator to the image. The Laplace operator essentially computes the average difference between adjacent pixels. Since we initially applied the logarithm, this leads to a color constant descriptor based on ratios. The Laplacian of an image with colors $\mathbf{p}(x, y) = [p_r(x, y), p_g(x, y), p_b(x, y)]^T$ is computed as

$$L(x, y) = -4\mathbf{p}(x, y) + \mathbf{p}(x - 1, y) + \mathbf{p}(x + 1, y) + \mathbf{p}(x, y - 1) + \mathbf{p}(x, y + 1). \tag{13.14}$$

If we have applied the logarithm, i.e. $p_i = \log c_i$ then we can rewrite the Laplacian as

$$L(x, y) = \log\frac{\mathbf{c}(x - 1, y)}{\mathbf{c}(x, y)} + \log\frac{\mathbf{c}(x + 1, y)}{\mathbf{c}(x, y)} + \log\frac{\mathbf{c}(x, y - 1)}{\mathbf{c}(x, y)} + \log\frac{\mathbf{c}(x, y + 1)}{\mathbf{c}(x, y)} \tag{13.15}$$

$$= \log\frac{\mathbf{R}(x - 1, y)}{\mathbf{R}(x, y)} + \log\frac{\mathbf{R}(x + 1, y)}{\mathbf{R}(x, y)} + \log\frac{\mathbf{R}(x, y - 1)}{\mathbf{R}(x, y)} + \log\frac{\mathbf{R}(x, y + 1)}{\mathbf{R}(x, y)} \tag{13.16}$$

where division is defined component-wise and the logarithm is also applied component-wise. Thus, we see that the Laplacian applied to the logarithm of an image is basically a color constant descriptor. Funt and Finlayson take the output of the Laplacian and compute a three-dimensional histogram from this data. Matching is done using histogram intersection. Note that care must be taken when forming histograms as the data is not uniformly distributed. Funt and Finlayson suggest the adjusting of the bin partitioning to reflect this fact.

Finlayson et al. (1995) have proposed yet another method for color constant object recognition. They suggested the viewing of the bands of the input image as vectors. If the sensors of the measuring device, which produced the image are very narrow shaped, then these vectors are only scaled by the illuminant. The length of these vectors vary when the illuminant changes. The orientation of the vectors remains the same. Finlayson et al. suggested the use of orientation between these vectors as an index into a database. The index consists of only three angles $(\theta_1, \theta_2, \theta_3)$. Let $\mathbf{c}_i$ be the entire color band i with $i \in \{r, g, b\}$ of the input image. Then the three angles are computed as follows:

$$\theta_1 = \cos^{-1}\left(\frac{\mathbf{c}_r \mathbf{c}_g}{|\mathbf{c}_r||\mathbf{c}_g|}\right) \tag{13.17}$$

$$\theta_2 = \cos^{-1}\left(\frac{\mathbf{c}_g \mathbf{c}_b}{|\mathbf{c}_g||\mathbf{c}_b|}\right) \tag{13.18}$$

$$\theta_3 = \cos^{-1}\left(\frac{\mathbf{c}_b \mathbf{c}_r}{|\mathbf{c}_b||\mathbf{c}_r|}\right). \tag{13.19}$$

Finlayson et al. (1995) report good results on a small database of 55 objects. The method seems to perform slightly worse than color constant color indexing that was developed by Funt and Finlayson (1995).

Berwick and Lee (1998) suggested the use of the following chromaticity space for object recognition. They simply choose one of the color channels and divide the remaining channels by the chosen color channel. Let $\mathbf{c} = [c_r, c_g, c_b]^T$ be the color of an image pixel. The chromaticity space is then defined as

$$(\xi_1, \xi_2) = [\log \frac{c_r}{c_g}, \log \frac{c_b}{c_g}]. \quad (13.20)$$

Dividing by one of the channels has the advantage that geometry information, which is constant for all three channels, is automatically removed. A nonwhite illuminant scales the three color channel by $\mathbf{S}(s_r, s_g, s_b)$ as discussed in Section 4.6, provided that the camera's sensors are sufficiently narrow band. Therefore, two chromaticity distributions ξ and ξ' of the same object in (ξ_1, ξ_2) chromaticity space are invariant up to a translation, i.e.

$$\xi' = \xi + \eta \quad \text{with} \quad \eta = [\log \frac{s_r}{s_g}, \log \frac{s_b}{s_g}]. \quad (13.21)$$

A match between model and test image can be established by choosing the best correlation between test and model distributions. Berwick and Lee (1998) suggest the use of phase matching of the Fourier transform of the object signature in ξ-space to establish a match between model and test image.

13.3 Evaluation on Object Recognition Tasks

Swain and Ballard (1991) have noted that histogram-based object recognition can also be used to evaluate the quality of color constancy algorithms. Object recognition based on color histograms relies on using actual object colors that should be constant irrespective of the illuminant. Finlayson et al. (1998) and Funt et al. (1998) have shown that object recognition improves if input images are color corrected first. Finlayson et al. (1998) demonstrated this using their method of comprehensive color normalization. Funt et al. (1998) showed that the object recognition performance is linearly related to the root mean squared error in color prediction. We will now be using color histograms to evaluate the quality of the different color constancy algorithms. Each algorithm is evaluated on a number of different image sets. The image sets are listed in Table 13.1.

Sets 1 through 5 were created by Barnard et al. (2002c). The first set contains only objects with minimal specularities, i.e. Lambertian reflectors. Set 2 contains objects with metallic specularities. Set 3 contains objects with nonnegligible dielectric specularities. Set 4 contains objects with at least one fluorescent surface. Set 5 differs from sets 1 through 4 in that the object was purposely moved whenever the illuminant was changed. Sets 6 through 9 were derived from set 1. Images of sets 6 and 7 contain a slowly changing illuminant. Images of sets 8 and 9 contain a sharply changing illuminant. Images from sets 10 through 13 show natural scenes with varying illumination within a single image. For each set, the number of different objects and the number of images per objects, i.e. number of different illuminants are shown. Barnard et al. used 11 different illuminants for each object but were forced to remove some images because of deficiencies in the calibration data. The total number of images in each set is also shown. The illuminants used by Barnard et al. are listed in Table 13.2.

Table 13.1: Set of Images. Image sets 1 through 5 were assembled by Barnard et al. (2002c). Image sets 6 through 9 were derived from set 1. Image sets 10 through 13 show scenes with varying illumination.

No.	Name	No. of Objects	Images per Objects	No. of Images	Alignment
1	Lambertian objects	21	2–11	212	aligned
2	Metallic objects	14	9–11	146	aligned
3	Specular objects	9	4–11	83	aligned
4	Fluorescent objects	6	9–11	59	aligned
5	Different objects	20	11	220	not
6	Smooth gradient (set 1)	21	2–10	202	aligned
7	Smooth circular gradient (set 1)	21	2–10	202	aligned
8	Sharp gradient (set 1)	21	2–10	202	aligned
9	Sharp circular gradient (set 1)	21	2–10	202	aligned
10	Natural scenes (CD-ROM)	4	13–27	73	aligned
11	Natural scenes (Kodak photo CD)	5	5–27	54	aligned
12	Natural scenes (Canon FS4000US)	9	4–28	130	aligned
13	Natural scenes (Canon 10D)	10	12–20	137	not

Table 13.2: Illuminants used by Barnard et al. (2002c).

No.	Type
1	Sylvania 50MR16Q (12VDC)
2	Sylvania 50MR16Q (12VDC) + Roscolux 3202 filter
3	Solux 3500K (12VDC)
4	Solux 3500K (12VDC) + Roscolux 3202 filter
5	Solux 4100K (12VDC)
6	Solux 4100K (12VDC) + Roscolux 3202 filter
7	Solux 4700K (12VDC)
8	Solux 4700K (12VDC) + Roscolux 3202 filter
9	Sylvania warm white fluorescent
10	Sylvania cool white fluorescent
11	Philips ultralume fluorescent

For the experiments which will be described in the next section, we require that the objects shown in different images are exactly aligned, i.e. it is important that neither the camera nor the object was moved when the illuminant was changed. We have aligned all images of an object with subpixel accuracy. Image sets 1 through 4 and 6 through 13 were aligned. Image set 5 was not aligned. This set was created by Barnard et al. to test object recognition algorithms. The objects were purposely moved whenever the illuminant was changed. Therefore, the images from this set could not be aligned.

The image alignment was made using the normal vector of each image channel. The normal vector was used because images had to be aligned even though the illuminant differed greatly between two images. Let I_1 be the first image of a particular object or scene. Let I_i be one of the remaining images. This image is shifted by (dx, dy). The shift is determined by minimizing the following expression:

$$\min_{dx,dy} \text{RMSE}_{x,y}(N(I_1(x, y)), N(I_i(x + dx, y + dy))) \qquad (13.22)$$

where RMSE denotes the root mean squared error and $N(I)$ denotes the image with the normal vectors. For a three band color image I, the image $N(I)$ contains nine bands. The root mean squared error was computed across all nine bands. Because of the alignment, some border pixels had to be cropped. After alignment, images from sets 1 through 4 were down-sampled to half the original size. The images then had approximately 317×233 pixels. Images from set 5 were also down-sampled to half the size.

We found that the third object of set 1 (a ball) could not be properly aligned. The ball was slightly moved when the illuminant was changed. Therefore, the ball was removed from set 1. Three images were removed from set 2, 15 images were removed from set 3. These images could not be aligned exactly. These image sets contained metallic or specular objects with highlights. The changing highlights made alignment more difficult than for the other sets.

Sets 6 through 9 were derived from set 1. These sets were created to test the ability of color constancy algorithms to cope with varying illumination within a single image. The images were created by merging two images of a single object with different illumination from set 1. Let $\mathbf{c}_1(x, y)$ be the color of the first image at pixel (x, y). Let $\mathbf{c}_2(x, y)$ be the color of the second image. Then the resulting image is computed as

$$\mathbf{c}(x, y) = \mathbf{c}_1(x, y) + s(x, y)\mathbf{c}_2(x, y) \qquad (13.23)$$

where $s(x, y)$ is simply a one band image that defines how the two images are merged. By varying $s(x, y)$, we can control how the illuminant changes over the image. The intensity of the second image was increased by a factor of three before merging. Before merging two images, they have to be properly aligned. This is another reason for the alignment of the images, which was described in the preceding text.

Set 6 was assembled by creating all possible combinations of illuminants for each object. In other words, if we have an object illuminated with n different illuminants, we created $n(n - 1)$ images. Since most objects of set 1 were illuminated by 11 different illuminants, this would give us 110 images per object. To limit the size of the database, we created only a maximum of 10 images per object. If the number of combinations exceeded 10, then 10 combinations were selected at random (without replacement) using the random generator of Matsumoto and Nishimura (1998). Sets 7, 8, and 9 were created similarly except that a different illumination gradient was used. Figure 13.3 shows the illumination gradients for sets 6 through 9. The first two illumination gradients change smoothly over the entire image. The first gradient was created using a sigmoidal function

$$f(x) = \frac{1}{1 + e^{-\frac{x}{\sigma}}} \qquad (13.24)$$

with $\sigma = 50$. The second gradient was created using the same function except that the radius from a center point was used as an index to the sigmoidal function. This illumination

EVALUATION OF ALGORITHMS 285

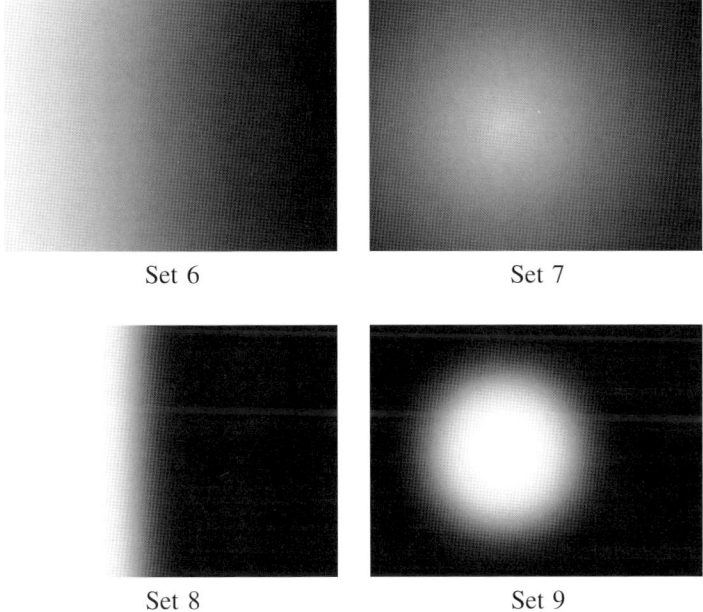

Figure 13.3 Illumination gradients that were used to create image sets 6 through 9.

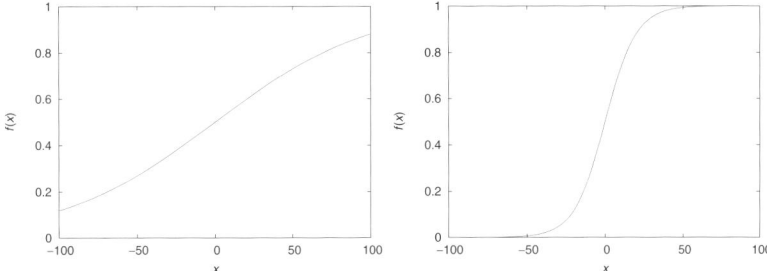

Figure 13.4 Two sigmoidal functions. The two functions were used to create the gradients of Figure 13.3.

gradient resembles a spotlight. Illumination gradients of sets 8 and 9 were created similarly except that we have used $\sigma = 10$. The two sigmoidal functions are shown in Figure 13.4. The first function creates a smoothly changing gradient while the second function creates a sharply changing gradient.

Sets 10, 11, 12, and 13 contain images with one or more illuminants. The images show large scenes, i.e. a section of a room or a cupboard. The scenes were illuminated with a variety of different light sources. Light sources included red, green, blue, and yellow light bulbs. These light bulbs are just ordinary light bulbs with a color coating. The light sources were placed to the left and right of the scene. Images were taken with either one or both lights turned on. If a desk lamp was present, the lamp was turned on for some of the images, creating a local illumination. Ceiling lamps and natural illumination was also present in some of the images. Images of sets 10, 11, and 12 were taken with an analog camera (Canon EOS 5). Although the camera was not moved when the images were taken, the digitized images had to be aligned with subpixel accuracy. Images of set 10 were digitized by ordering a CD-ROM along with the developed film. The original image sizes were 1500 × 980 pixels. Images of set 11 are from a Kodak Picture CD made from some of the negatives. The images were decoded with the program `hpcdtoppm` which is available through the Netpbm-Package (Henderson 2003). The size of the decoded images was 3072 × 2048 (16Base). Images of set 12 were scanned using a calibrated film scanner (Canon FS4000US). The images were scanned with 4000 dpi and then down-sampled to 2787 × 1850. Scanner and film were calibrated using IT8-targets. The images of set 13 were taken with a digital camera (Canon EOS 10D). The white balance was set to 6500 K. The images from the digital camera were not aligned because the camera remained stationary at all times. These images were of size 3072 × 2048. The images were saved using the sRGB color space. Sets 10, 11, and 12 can be used to evaluate the quality of color constancy algorithms in the context of different image digitization techniques while images of set 13 were taken with a digital camera. Due to the large size of the images of sets 10, 11, 12, and 13, the images were down-sampled to a range of around 360 × 240 pixel for further processing. The resulting images contained only small amounts of noise. Appendix D shows all objects or scenes from each image set under a single illuminant.

Histogram-based object recognition was used to evaluate the quality of the different color constancy algorithms. In the course of this work, we have discussed many of the color constancy algorithms that are available from the literature in Chapter 6 and Chapter 7. Algorithms based on local space average color were described in Chapter 11 and Chapter 12. All algorithms that were evaluated using histogram-based object recognition are summarized in Appendix E and Appendix F. Some of the algorithms have a number of different parameters that can be varied. The settings that were used for each algorithm are listed in the appendix.

The experimental setting is shown in Figure 13.5. A color constancy algorithm is applied to an input image. The resulting image is then used to match images from a database to which the same color constancy algorithm has been applied. Object recognition results for sets 1 through 5 are shown in Table 13.3. Each row shows the results of a single color constancy algorithm on all five image sets. Evaluation was done as follows. We

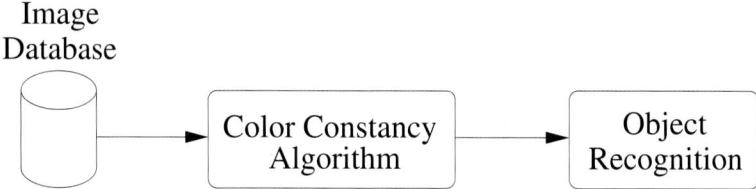

Figure 13.5 Experimental setting that is used to test the performance of color constancy algorithms. Each input image from the database is processed by a color constancy algorithm. The processed images are then used for object recognition.

EVALUATION OF ALGORITHMS

Table 13.3: Results for image sets 1 through 5. Histograms were computed in RGB space. For each image set, random performance is also shown.

Algorithm	1	2	3	4	5
Random recognition rate	0.048	0.071	0.111	0.167	0.050
Full range per band	0.562	0.524	0.764	0.702	0.483
White patch retinex	0.613	0.672	0.939	0.885	0.409
Gray world assumption	0.805	0.726	0.980	0.968	0.421
Simplified horn	0.462	0.451	0.744	0.617	0.182
Gamut constraint 3D	0.555	0.513	0.741	0.707	0.468
Gamut constraint 2D	0.404	0.421	0.634	0.618	0.309
Color cluster rotation	0.465	0.544	0.640	0.608	0.510
Comprehensive normalization	0.593	0.533	0.889	0.820	0.236
Risson (2003)	0.583	0.546	0.760	0.652	0.477
Horn (1974)/Blake (1985)	0.480	0.409	0.597	0.613	0.254
Moore et al. (1991) Retinex	0.907	0.810	0.767	0.812	0.504
Moore et al. (1991) Extended	0.430	0.526	0.643	0.623	0.288
Rahman et al. (1999)	0.849	0.830	0.793	0.832	0.488
Homomorphic filtering	0.305	0.425	0.559	0.578	0.209
Homomorphic filtering (HVS)	0.346	0.345	0.569	0.577	0.172
Intrinsic (min. entropy)	0.189	0.333	0.446	0.562	0.155
Local space average color (1)	**0.948**	**0.885**	**1.000**	**1.000**	0.545
Local space average color (2)	0.457	0.530	0.598	0.720	0.373
Local space average color (3)	0.571	0.623	0.721	0.765	0.551
Local space average color (4)	0.477	0.532	0.619	0.700	0.346
Local space average color (5)	0.598	0.652	0.752	0.785	0.513
Local space average color (6)	0.467	0.518	0.617	0.703	0.347
Local space average color (7)	0.583	0.646	0.747	0.780	0.528
Anisotropic L.S.A. color (1)	**0.948**	0.879	0.987	**1.000**	0.519
Anisotropic L.S.A. color (2)	0.474	0.520	0.619	0.693	0.388
Anisotropic L.S.A. color (3)	0.560	0.611	0.702	0.750	0.561
Anisotropic L.S.A. color (4)	0.472	0.529	0.633	0.693	0.353
Anisotropic L.S.A. color (5)	0.580	0.628	0.742	0.750	0.530
Anisotropic L.S.A. color (6)	0.461	0.524	0.624	0.692	0.351
Anisotropic L.S.A. color (7)	0.567	0.621	0.726	0.747	0.545
Combined WP/L.S.A. color	0.942	0.791	0.877	0.998	**0.769**
Comprehensive L.S.A. color	0.504	0.626	0.664	0.668	0.558

selected two different images for each object from the database. The images were chosen at random. This gave us two image sets where each set contains exactly one image per object as shown in Figure 13.6. The first set of images was our test set, i.e. it contained the objects to be recognized. The second set of images were our model images. RGB histograms were computed for each image. A quantization of 100 per channel was used. The χ^2-divergence measure was computed for every combination of test image and model

Figure 13.6 Two images were selected for each object. The two images differ in the type of illuminant used. A color constancy algorithm is evaluated by applying the algorithm to both image sets. A perfect color constancy algorithm would produce exactly the same images for both the test and the model image. The recognition rate is computed by establishing a match between the test and model images. (Original image data from "Data for Computer Vision and Computational Colour Science" made available through http://www.cs.sfu.ca/~colour/data/index.html. See Barnard K, Martin L, Funt B and Coath A 2002 A data set for color research, Color Research and Application, Wiley Periodicals, 27(3), 147–151. Reproduced by permission of Kobus Barnard.)

image. The lowest χ^2-divergence measure was our match. All correct matches between test images and model images were counted. This procedure was repeated 100 times. The number of correct matches was divided by the total number of matches. This is our recognition rate. At the top of the table, the random performance for each set is shown. The random performance for a single run is simply $\frac{1}{n}$ where n is the size of the database, i.e. the number of different objects or scenes in the image set. The best recognition rate for each image set was printed in bold face. Table 13.4 shows the results when images were transformed to RGB chromaticity space before matching.

When we look at Table 13.3, we see that there is no single algorithm with best performance on all image sets. Recall again that set 1 contains Lambertian reflectors, set 2 contains objects with metallic specularities, set 3 contains nonnegligible dielectric specularities, and set 4 contains objects with at least one fluorescent surface. Computing local space average color and then scaling the input pixels by twice the local space average color worked best for sets 1, 2, 3, and 4. Computing anisotropic local space average color performed similarly on set 1 and 4. Local space average color (1) resulted in a perfect recognition rate for sets 3 and 4. Anisotropic local space average color (1) resulted in a perfect recognition rate for set 4. Note that even though a single illuminant was used for sets 1 through 5, an illumination or intensity gradient may nevertheless be present in the image. An intensity gradient is present if the light source was attached at one side of the image. This explains the lower performance of using the gray world assumption on the entire image. Use of the combined WP/L.S.A. color algorithm (described in Section 11.5) produced best results on set 5.

In performing color-based object recognition, we are not interested in using the geometry information. The intensity can be removed by computing chromaticities. Note that if we compute chromaticities, we can no longer distinguish between black, gray, and white. Results for object recognition based on RGB chromaticities are shown in Table 13.4. The

Table 13.4: Results for image sets 1 through 5. Histograms were computed in RGB chromaticity space. For each image set, random performance is also shown.

Algorithm	1	2	3	4	5
Random recognition rate	0.048	0.071	0.111	0.167	0.050
Full range per band	0.574	0.482	0.723	0.767	0.357
White patch retinex	0.696	0.735	0.969	0.960	0.429
Gray world assumption	0.861	0.813	0.990	**1.000**	0.464
Simplified horn	0.463	0.422	0.624	0.637	0.251
Gamut constraint 3D	0.571	0.478	0.696	0.730	0.354
Gamut constraint 2D	0.440	0.517	0.632	0.688	0.254
Color cluster rotation	0.609	0.615	0.782	0.897	0.477
Comprehensive normalization	0.727	0.632	0.970	0.952	0.166
Risson (2003)	0.589	0.593	0.867	0.675	0.371
Horn (1974)/Blake (1985)	0.370	0.362	0.500	0.643	0.176
Moore et al. (1991) Retinex	0.940	0.924	0.987	0.985	0.562
Moore et al. (1991) Extended	0.397	0.488	0.640	0.660	0.284
Rahman et al. (1999)	**0.956**	0.893	**1.000**	**1.000**	0.499
Homomorphic filtering	0.289	0.355	0.450	0.532	0.199
Homomorphic filtering (HVS)	0.271	0.418	0.450	0.537	0.213
Intrinsic (min. entropy)	0.138	0.283	0.306	0.378	0.067
Local space average color (1)	0.944	0.916	0.992	**1.000**	0.579
Local space average color (2)	0.491	0.599	0.707	0.770	0.320
Local space average color (3)	0.831	0.791	0.911	0.968	0.455
Local space average color (4)	0.489	0.570	0.694	0.765	0.301
Local space average color (5)	0.824	0.790	0.913	0.970	0.441
Local space average color (6)	0.475	0.549	0.668	0.730	0.303
Local space average color (7)	0.823	0.764	0.902	0.940	0.444
Anisotropic L.S.A. color (1)	0.950	0.924	0.989	**1.000**	0.590
Anisotropic L.S.A. color (2)	0.522	0.613	0.720	0.755	0.309
Anisotropic L.S.A. color (3)	0.831	0.790	0.881	0.955	0.438
Anisotropic L.S.A. color (4)	0.523	0.593	0.702	0.753	0.287
Anisotropic L.S.A. color (5)	0.826	0.789	0.876	0.960	0.426
Anisotropic L.S.A. color (6)	0.507	0.589	0.677	0.745	0.289
Anisotropic L.S.A. color (7)	0.830	0.768	0.867	0.930	0.434
Combined WP/L.S.A. color	0.936	0.734	0.843	0.987	**0.764**
Comprehensive L.S.A. color	0.937	**0.929**	0.971	**1.000**	0.478

gray world assumption resulted in a perfect recognition rate on set 4. The algorithm of Rahman et al. (1999) gave best results for sets 1, 3 and 4 with a perfect recognition rate for sets 3 and 4. Local space average color (1) resulted in a perfect recognition rate on set 4. Anisotropic local space average color (1) also resulted in a perfect recognition rate on set 4. The combined WP/L.S.A. color algorithm (described in Section 11.5) gave best results on set 5. Comprehensive local space average color performed best on set 2 and also resulted in a perfect recognition rate on set 4. In total, five algorithms resulted in a perfect recognition rate on set 4.

When we look at Table 13.3 and Table 13.4 we note that the algorithms that achieve best results are all based on the computation of local space average color. We also see that the recognition rates on set 5 are much lower (across all algorithms) than the recognition rates for the other sets. This is understandable as the object recognition task using set 5 is more difficult. Set 5 is different from the other sets in that the objects were placed in a different position whenever the illuminant was changed.

Sets 6 through 13 were created in order to see how the algorithms cope with nonuniform illumination. Results for image sets 6 through 13 are shown in Table 13.5 and Table 13.6. Table 13.5 shows the results when histograms are computed in RGB space. Table 13.6 shows the results when histograms are computed in chromaticity space. Local space average color (1) gave best results on sets 8, 11, and 12. Anisotropic local space average color (1) performed best on sets 6, 7, 9, 10, and 11. Comprehensive local space average color gave best results on set 13.

When using RGB chromaticity space, comprehensive local space average color normalization performed best on all image sets. Again, irrespective of whether we compute the histograms in RGB color space or in RGB chromaticity space, algorithms based on the computation of local space average color gave best results.

13.4 Computation of Color Constant Descriptors

In the previous section, we have turned to histogram-based object recognition to evaluate the quality of the different color constancy algorithms. The results, however, may not only be influenced by the type of algorithm used but also by the particular method which was used to perform object recognition. The ultimate goal of a color constancy algorithm is that the computed color for an image pixel is constant irrespective of the illuminant used. But at the same time, the result has to be nontrivial. Otherwise, a gray image could be output by an algorithm. This would produce constant "colors" all the time. Thus, the result has to be nontrivial. An ideal scenario would be an optimal use of the available color space. Otherwise, constant colors would be produced for a subspace only. In this case, object recognition would be more difficult because less information is available about the objects. Therefore, an optimal algorithm would produce color constant descriptors that populate the entire color space.

Image sets were already described in Section 13.3. Image sets 1 through 4 and 6 through 12 were aligned with subpixel accuracy. Image set 13 was not aligned but the camera was not moved when taking the image. Using these image sets we can evaluate how constant the output of the different algorithms is. An ideal algorithm would produce the same output for a particular image pixel of an object or scene for all images which show this object or scene. We do not know what the output should be. Whatever it is, it should be constant

EVALUATION OF ALGORITHMS

Table 13.5: Results for image sets 6 through 13. Histograms were computed in RGB space. For each image set, random performance is also shown.

Algorithm	6	7	8	9	10	11	12	13
Random recognition rate	0.048	0.048	0.048	0.048	0.250	0.200	0.111	0.100
Full range per band	0.549	0.627	0.496	0.443	0.400	0.418	0.346	0.265
White patch retinex	0.675	0.739	0.557	0.576	0.370	0.332	0.369	0.290
Gray world assumption	0.700	0.783	0.559	0.599	0.393	0.312	0.370	0.329
Simplified horn	0.591	0.630	0.495	0.469	0.507	0.468	0.328	0.169
Gamut constraint 3D	0.543	0.613	0.489	0.429	0.398	0.370	0.340	0.263
Gamut constraint 2D	0.476	0.545	0.445	0.456	0.532	0.368	0.341	0.215
Color cluster rotation	0.552	0.635	0.466	0.511	0.430	0.338	0.276	0.186
Comprehensive normalization	0.637	0.728	0.510	0.607	0.410	0.386	0.373	0.367
Risson (2003)	0.587	0.613	0.430	0.554	0.347	0.294	0.348	0.243
Horn (1974)/Blake (1985)	0.621	0.617	0.564	0.587	0.435	0.438	0.304	0.152
Moore et al. (1991) Retinex	0.920	0.951	0.864	0.866	0.512	0.374	0.428	0.427
Moore et al. (1991) Extended	0.661	0.667	0.644	0.612	0.468	0.382	0.409	0.244
Rahman et al. (1999)	0.867	0.894	0.802	0.740	0.417	0.318	0.368	0.384
Homomorphic filtering	0.588	0.574	0.530	0.491	0.510	0.382	0.342	0.168
Homomorphic filtering (HVS)	0.572	0.605	0.476	0.501	0.435	0.394	0.350	0.263
Intrinsic (min. entropy)	0.320	0.349	0.262	0.259	0.350	0.248	0.306	0.177
Local space average color (1)	0.965	0.940	**0.970**	0.823	0.562	**0.482**	**0.532**	0.430
Local space average color (2)	0.672	0.666	0.591	0.630	0.450	0.374	0.362	0.283
Local space average color (3)	0.751	0.756	0.665	0.660	0.490	0.356	0.366	0.353
Local space average color (4)	0.651	0.667	0.570	0.610	0.497	0.400	0.352	0.330
Local space average color (5)	0.743	0.752	0.666	0.646	0.505	0.364	0.366	0.334
Local space average color (6)	0.642	0.670	0.558	0.606	0.487	0.390	0.350	0.332
Local space average color (7)	0.731	0.742	0.642	0.640	0.505	0.382	0.372	0.336
Anisotropic L.S.A. color (1)	**0.977**	**0.978**	0.961	**0.881**	**0.570**	**0.482**	0.513	0.393
Anisotropic L.S.A. color (2)	0.687	0.691	0.596	0.632	0.502	0.392	0.371	0.315
Anisotropic L.S.A. color (3)	0.765	0.774	0.660	0.695	0.485	0.380	0.384	0.373
Anisotropic L.S.A. color (4)	0.670	0.692	0.581	0.614	0.495	0.412	0.381	0.355
Anisotropic L.S.A. color (5)	0.752	0.766	0.665	0.682	0.487	0.382	0.391	0.352
Anisotropic L.S.A. color (6)	0.661	0.690	0.579	0.614	0.505	0.408	0.373	0.351
Anisotropic L.S.A. color (7)	0.742	0.764	0.648	0.680	0.495	0.396	0.383	0.356
Combined WP/L.S.A. color	0.872	0.866	0.770	0.719	0.522	0.390	0.436	0.383
Comprehensive L.S.A. color	0.702	0.753	0.645	0.678	0.557	0.442	0.450	**0.471**

over all images showing a single object or scene. For each image pixel, we can compute the standard deviation of the output from the mean. Let $\sigma_i(x, y)$ be the standard deviation computed for channel $i \in \{r, g, b\}$ for position (x, y) of the RGB chromaticity image. The average standard deviation for the entire image is then given by

$$\bar{\sigma} = \frac{1}{3n} \sum_{i \in \{r,g,b\}} \sum_{x,y} \sigma_i(x, y) \tag{13.25}$$

Table 13.6: Results for image sets 6 through 13. Histograms were computed in RGB chromaticity space. For each image set, random performance is also shown.

Algorithm	6	7	8	9	10	11	12	13
Random recognition rate	0.048	0.048	0.048	0.048	0.250	0.200	0.111	0.100
Full range per band	0.560	0.618	0.499	0.433	0.495	0.412	0.303	0.172
White patch retinex	0.688	0.735	0.575	0.563	0.443	0.312	0.330	0.202
Gray world assumption	0.729	0.817	0.593	0.608	0.480	0.360	0.320	0.243
Simplified horn	0.603	0.632	0.516	0.484	0.453	0.470	0.316	0.138
Gamut constraint 3D	0.552	0.597	0.479	0.427	0.507	0.340	0.300	0.159
Gamut constraint 2D	0.535	0.568	0.500	0.482	0.485	0.318	0.304	0.147
Color cluster rotation	0.539	0.632	0.471	0.460	0.445	0.310	0.309	0.184
Comprehensive normalization	0.697	0.805	0.537	0.636	0.490	0.404	0.348	0.269
Risson (2003)	0.664	0.684	0.504	0.600	0.500	0.408	0.308	0.221
Horn (1974)/Blake (1985)	0.530	0.552	0.453	0.445	0.395	0.410	0.294	0.139
Moore et al. (1991) Retinex	0.983	0.980	0.966	0.935	0.583	0.462	0.454	0.329
Moore et al. (1991) Extended	0.651	0.630	0.582	0.574	0.403	0.398	0.393	0.234
Rahman et al. (1999)	0.902	0.938	0.810	0.737	0.472	0.416	0.401	0.314
Homomorphic filtering	0.570	0.560	0.523	0.491	0.445	0.408	0.348	0.138
Homomorphic filtering (HVS)	0.553	0.554	0.498	0.476	0.385	0.354	0.316	0.168
Intrinsic (min. entropy)	0.231	0.236	0.198	0.209	0.375	0.216	0.257	0.104
Local space average color (1)	0.965	0.950	0.949	0.787	0.510	0.432	0.456	0.314
Local space average color (2)	0.736	0.724	0.696	0.620	0.475	0.380	0.379	0.249
Local space average color (3)	0.927	0.893	0.880	0.722	0.477	0.434	0.352	0.225
Local space average color (4)	0.718	0.710	0.667	0.588	0.485	0.390	0.376	0.251
Local space average color (5)	0.898	0.871	0.857	0.712	0.477	0.432	0.347	0.224
Local space average color (6)	0.690	0.683	0.640	0.577	0.477	0.388	0.366	0.257
Local space average color (7)	0.887	0.864	0.846	0.704	0.470	0.460	0.359	0.222
Anisotropic L.S.A. color (1)	0.981	0.980	0.958	0.877	0.517	0.462	0.449	0.324
Anisotropic L.S.A. color (2)	0.757	0.760	0.722	0.664	0.477	0.428	0.388	0.254
Anisotropic L.S.A. color (3)	0.952	0.931	0.908	0.803	0.502	0.454	0.373	0.224
Anisotropic L.S.A. color (4)	0.743	0.748	0.701	0.640	0.475	0.420	0.384	0.241
Anisotropic L.S.A. color (5)	0.932	0.919	0.881	0.785	0.482	0.452	0.372	0.220
Anisotropic L.S.A. color (6)	0.721	0.732	0.672	0.634	0.458	0.420	0.386	0.247
Anisotropic L.S.A. color (7)	0.926	0.908	0.879	0.773	0.485	0.486	0.370	0.215
Combined WP/L.S.A. color	0.827	0.850	0.719	0.651	0.453	0.418	0.402	0.313
Comprehensive L.S.A. color	**0.989**	**0.990**	**0.986**	**0.975**	**0.652**	**0.498**	**0.511**	**0.336**

where n is the number of image pixels. Let $\bar{\sigma}(j)$ be the average standard deviation for image j from the image set. The average standard deviation over the entire database DB is then defined as

$$\bar{\sigma}(\text{DB}) = \frac{1}{n_{\text{obj}}} \sum_{j=1}^{n_{\text{obj}}} \bar{\sigma}(j) \qquad (13.26)$$

Table 13.7: Average standard deviation computed per pixel per band for image sets 1 through 4.

Algorithm	1	2	3	4
Full range per band	0.041	0.078	0.049	0.046
White patch retinex	0.036	0.067	0.029	0.034
Gray world assumption	0.034	0.063	0.028	0.032
Simplified horn	0.042	0.074	0.039	0.039
Gamut constraint 3D	0.042	0.078	0.049	0.058
Gamut constraint 2D	0.084	0.086	0.054	0.074
Color cluster rotation	0.056	0.082	0.065	0.086
Comprehensive normalization	0.048	0.076	0.031	0.036
Risson (2003)	0.044	0.073	0.036	0.046
Horn (1974)/Blake (1985)	0.026	0.037	0.031	0.029
Moore et al. (1991) Retinex	**0.009**	0.016	**0.008**	**0.009**
Moore et al. (1991) Extended	0.022	0.025	0.019	0.019
Rahman et al. (1999)	**0.009**	0.016	0.009	**0.009**
Homomorphic filtering	0.066	0.086	0.058	0.060
Homomorphic filtering (HVS)	0.040	0.043	0.037	0.037
Intrinsic (min. entropy)	0.083	0.097	0.087	0.087
Local space average color (1)	0.016	0.026	0.015	0.016
Local space average color (2)	0.052	0.065	0.044	0.048
Local space average color (3)	0.019	0.029	0.018	0.019
Local space average color (4)	0.064	0.085	0.057	0.062
Local space average color (5)	0.029	0.045	0.028	0.031
Local space average color (6)	0.061	0.081	0.055	0.059
Local space average color (7)	0.026	0.041	0.025	0.028
Anisotropic L.S.A. Color (1)	0.015	0.026	0.016	0.017
Anisotropic L.S.A. Color (2)	0.043	0.060	0.039	0.042
Anisotropic L.S.A. Color (3)	0.018	0.028	0.017	0.019
Anisotropic L.S.A. Color (4)	0.056	0.079	0.052	0.056
Anisotropic L.S.A. Color (5)	0.028	0.044	0.028	0.030
Anisotropic L.S.A. Color (6)	0.052	0.072	0.049	0.051
Anisotropic L.S.A. Color (7)	0.023	0.036	0.025	0.026
Combined WP/L.S.A. Color	0.012	**0.013**	0.014	0.018
Comprehensive L.S.A. Color	0.014	0.024	0.014	0.014

where n_{obj} is the number of images stored in the database. This is our measure to evaluate the quality of color constancy algorithms. Results for image sets 1 through 4 using this measure are shown in Table 13.7. Results for image sets 6 through 13 are shown in Table 13.8. The smaller the measure, the better the algorithm. Again, best values are printed in bold face.

The retinex algorithm of Moore et al. (1991) clearly produces the most constant output. The algorithm of Rahman et al. (1999) also produced a small standard deviation in the output color. The algorithm described in Section 11.5, the combined WP/L.S.A.

Table 13.8: Average standard deviation computed per pixel per band for image sets 6 through 13.

Algorithm	6	7	8	9	10	11	12	13
Full range per band	0.047	0.039	0.062	0.066	0.161	0.111	0.106	0.148
White patch retinex	0.040	0.032	0.057	0.059	0.148	0.113	0.088	0.120
Gray world assumption	0.036	0.028	0.050	0.048	0.130	0.088	0.073	0.115
Simplified horn	0.031	0.028	0.039	0.041	0.108	0.089	0.065	0.122
Gamut constraint 3D	0.047	0.042	0.064	0.066	0.163	0.098	0.106	0.149
Gamut constraint 2D	0.073	0.061	0.081	0.080	0.171	0.127	0.127	0.228
Color cluster rotation	0.072	0.055	0.070	0.065	0.154	0.135	0.113	0.097
Comprehensive normalization	0.042	0.035	0.054	0.050	0.122	0.084	0.069	0.145
Risson (2003)	0.045	0.037	0.061	0.053	0.152	0.105	0.080	0.170
Horn (1974)/Blake (1985)	0.023	0.020	0.028	0.024	0.073	0.083	0.063	0.088
Moore et al. (1991) Retinex	**0.007**	**0.006**	**0.007**	**0.008**	**0.021**	**0.021**	**0.014**	**0.020**
Moore et al. (1991) Extended	0.015	0.014	0.017	0.018	0.035	0.028	0.027	0.042
Rahman et al. (1999)	**0.007**	0.007	0.008	0.009	0.025	0.024	0.016	0.022
Homomorphic filtering	0.043	0.040	0.050	0.052	0.132	0.076	0.097	0.175
Homomorphic filtering (HVS)	0.031	0.029	0.035	0.034	0.085	0.062	0.069	0.096
Intrinsic (min. entropy)	0.062	0.058	0.068	0.067	0.167	0.103	0.121	0.211
Local space average color (1)	0.014	0.013	0.017	0.020	0.081	0.053	0.045	0.062
Local space average color (2)	0.038	0.036	0.043	0.044	0.102	0.078	0.072	0.107
Local space average color (3)	0.015	0.015	0.018	0.021	0.065	0.051	0.039	0.042
Local space average color (4)	0.046	0.044	0.053	0.055	0.117	0.102	0.091	0.130
Local space average color (5)	0.023	0.023	0.029	0.033	0.100	0.078	0.060	0.064
Local space average color (6)	0.045	0.044	0.050	0.053	0.116	0.097	0.089	0.128
Local space average color (7)	0.021	0.022	0.026	0.030	0.095	0.069	0.056	0.059
Anisotropic L.S.A. color (1)	0.014	0.013	0.016	0.017	0.073	0.048	0.042	0.056
Anisotropic L.S.A. color (2)	0.033	0.031	0.037	0.037	0.091	0.069	0.066	0.096
Anisotropic L.S.A. color (3)	0.015	0.014	0.017	0.017	0.058	0.046	0.035	0.038
Anisotropic L.S.A. color (4)	0.042	0.040	0.047	0.048	0.109	0.092	0.085	0.120
Anisotropic L.S.A. color (5)	0.023	0.022	0.026	0.028	0.089	0.071	0.056	0.058
Anisotropic L.S.A. color (6)	0.039	0.038	0.043	0.045	0.107	0.085	0.082	0.116
Anisotropic L.S.A. color (7)	0.020	0.020	0.022	0.024	0.085	0.062	0.051	0.053
Combined WP/L.S.A. color	0.016	0.015	0.020	0.022	0.071	0.054	0.054	0.063
Comprehensive L.S.A. color	0.012	0.011	0.014	0.014	0.062	0.043	0.037	0.050

color algorithm gave best results on set 2. Compared to other algorithms like gamut-constraint algorithms or color cluster rotation, the algorithms local space average color (1) and anisotropic local space average color (1) also had a small pixel standard deviation.

Apart from producing a constant output irrespective of the illuminant, the entire color space should be used. We can compute the standard deviation of the pixels of a single image in RGB chromaticity space and average the standard deviation over all images of the image set. This will measure how well the available color space is used. The color

Table 13.9: Average standard deviation of image pixels for image sets 1 through 4.

Algorithm	1	2	3	4
Full range per band	0.132	0.179	0.149	0.145
White patch retinex	0.129	0.183	0.150	0.147
Gray world assumption	0.129	0.185	0.152	0.150
Simplified horn	0.097	0.152	0.105	0.105
Gamut constraint 3D	0.130	0.179	0.148	0.144
Gamut constraint 2D	0.128	0.180	0.150	0.149
Color cluster rotation	**0.170**	0.170	**0.227**	**0.218**
Comprehensive normalization	0.146	**0.201**	0.157	0.157
Risson (2003)	0.131	0.182	0.151	0.150
Horn (1974)/Blake (1985)	0.055	0.058	0.068	0.064
Moore et al. (1991) Retinex	0.034	0.046	0.036	0.035
Moore et al. (1991) Extended	0.045	0.055	0.050	0.046
Rahman et al. (1999)	0.032	0.043	0.037	0.034
Homomorphic filtering	0.104	0.144	0.118	0.112
Homomorphic filtering (HVS)	0.057	0.071	0.066	0.068
Intrinsic (min. entropy)	0.066	0.084	0.068	0.084
Local space average color (1)	0.073	0.084	0.077	0.080
Local space average color (2)	0.115	0.130	0.109	0.113
Local space average color (3)	0.073	0.084	0.075	0.076
Local space average color (4)	0.145	0.169	0.143	0.148
Local space average color (5)	0.114	0.130	0.119	0.122
Local space average color (6)	0.137	0.157	0.132	0.141
Local space average color (7)	0.103	0.115	0.105	0.111
Anisotropic L.S.A. color (1)	0.066	0.078	0.069	0.073
Anisotropic L.S.A. color (2)	0.101	0.120	0.099	0.101
Anisotropic L.S.A. color (3)	0.065	0.077	0.067	0.068
Anisotropic L.S.A. color (4)	0.130	0.159	0.130	0.135
Anisotropic L.S.A. color (5)	0.103	0.121	0.107	0.110
Anisotropic L.S.A. color (6)	0.119	0.141	0.118	0.124
Anisotropic L.S.A. color (7)	0.089	0.100	0.093	0.096
Combined WP/L.S.A. color	0.075	0.040	0.049	0.091
Comprehensive L.S.A. color	0.062	0.077	0.063	0.068

space measure is shown in Table 13.9 and Table 13.10. A large measure indicates that a large part of the color space is used.

Color cluster rotation produces the largest standard deviation on most of the image sets. This is of no surprise, since color cluster rotation explicitly rescales the data points to fill most of the color cube. Comprehensive normalization produces the largest standard deviation on image set 2 and 13. The algorithm of Risson produces the largest standard deviation on image set 10.

Table 13.10: Average standard deviation of image pixels for image sets 6 through 13.

Algorithm	6	7	8	9	10	11	12	13
Full range per band	0.133	0.130	0.137	0.135	0.183	0.155	0.109	0.123
White patch retinex	0.130	0.128	0.133	0.132	0.189	0.134	0.115	0.146
Gray world assumption	0.131	0.128	0.135	0.134	0.202	0.147	0.120	0.158
Simplified horn	0.089	0.088	0.092	0.094	0.152	0.103	0.068	0.094
Gamut constraint 3D	0.131	0.128	0.135	0.133	0.185	0.151	0.107	0.121
Gamut constraint 2D	0.124	0.126	0.125	0.128	0.195	0.144	0.113	0.122
Color cluster rotation	**0.180**	**0.176**	**0.161**	**0.153**	0.203	**0.194**	**0.199**	0.177
Comprehensive normalization	0.141	0.138	0.147	0.146	0.202	0.152	0.121	**0.186**
Risson (2003)	0.134	0.130	0.138	0.136	**0.206**	0.147	0.123	0.172
Horn (1974)/Blake (1985)	0.054	0.055	0.054	0.054	0.082	0.091	0.076	0.086
Moore et al. (1991) Retinex	0.032	0.033	0.032	0.033	0.039	0.048	0.033	0.040
Moore et al. (1991) Extended	0.044	0.045	0.044	0.045	0.049	0.049	0.041	0.050
Rahman et al. (1999)	0.030	0.031	0.030	0.031	0.045	0.049	0.036	0.041
Homomorphic filtering	0.102	0.100	0.105	0.106	0.137	0.125	0.097	0.103
Homomorphic filtering (HVS)	0.063	0.060	0.065	0.061	0.097	0.083	0.077	0.088
Intrinsic (min. entropy)	0.064	0.063	0.070	0.066	0.105	0.058	0.054	0.043
Local space average color (1)	0.080	0.080	0.080	0.079	0.135	0.108	0.092	0.105
Local space average color (2)	0.122	0.123	0.122	0.123	0.156	0.146	0.148	0.145
Local space average color (3)	0.079	0.080	0.079	0.079	0.103	0.103	0.077	0.070
Local space average color (4)	0.149	0.150	0.149	0.151	0.168	0.185	0.168	0.177
Local space average color (5)	0.123	0.124	0.122	0.122	0.149	0.146	0.116	0.103
Local space average color (6)	0.139	0.142	0.137	0.142	0.167	0.177	0.165	0.174
Local space average color (7)	0.108	0.112	0.106	0.108	0.144	0.135	0.107	0.097
Anisotropic L.S.A. color (1)	0.074	0.073	0.073	0.072	0.125	0.102	0.086	0.098
Anisotropic L.S.A. color (2)	0.107	0.108	0.107	0.108	0.144	0.137	0.138	0.138
Anisotropic L.S.A. color (3)	0.071	0.072	0.071	0.071	0.095	0.097	0.071	0.065
Anisotropic L.S.A. color (4)	0.135	0.137	0.135	0.137	0.160	0.175	0.160	0.169
Anisotropic L.S.A. color (5)	0.112	0.114	0.111	0.112	0.138	0.139	0.109	0.097
Anisotropic L.S.A. color (6)	0.121	0.125	0.120	0.124	0.157	0.165	0.155	0.164
Anisotropic L.S.A. color (7)	0.094	0.098	0.092	0.094	0.132	0.126	0.100	0.089
Combined WP/L.S.A. color	0.074	0.077	0.070	0.071	0.109	0.102	0.109	0.104
Comprehensive L.S.A. color	0.067	0.067	0.067	0.065	0.103	0.096	0.077	0.092

Let $\bar{\sigma}_1$ be the constancy measure and let $\bar{\sigma}_2$ be the color space measure. We can combine both measures into a color constancy measure m and compute the fraction

$$m = \frac{\bar{\sigma}_2}{\bar{\sigma}_1}. \tag{13.27}$$

The results for this color constancy measure are shown in Table 13.11 and Table 13.12. The largest value for each set is printed in bold face.

In order to see if the combined measure correlates with the performance of the algorithms, we can enter both data into a common graph. The performance is used as the x

Table 13.11: Combined color constancy measure for image sets 1 through 4.

Algorithm	1	2	3	4
Full range per band	3.203	2.287	3.067	3.134
White patch retinex	3.591	2.750	5.142	4.293
Gray world assumption	3.818	2.926	**5.338**	4.691
Simplified horn	2.335	2.063	2.660	2.730
Gamut constraint 3D	3.078	2.290	3.034	2.485
Gamut constraint 2D	1.525	2.099	2.784	2.028
Color cluster rotation	3.054	2.079	3.484	2.541
Comprehensive normalization	3.042	2.652	5.014	4.337
Risson (2003)	2.975	2.505	4.159	3.225
Horn (1974)/Blake (1985)	2.095	1.579	2.190	2.231
Moore et al. (1991) Retinex	3.980	2.851	4.347	4.064
Moore et al. (1991) Extended	2.105	2.230	2.593	2.426
Rahman et al. (1999)	3.625	2.722	4.137	3.950
Homomorphic filtering	1.567	1.678	2.044	1.862
Homomorphic filtering (HVS)	1.418	1.665	1.775	1.842
Intrinsic (min. entropy)	0.786	0.866	0.781	0.967
Local space average color (1)	4.615	3.172	5.044	**4.964**
Local space average color (2)	2.220	1.989	2.517	2.385
Local space average color (3)	3.847	2.865	4.153	3.927
Local space average color (4)	2.271	1.990	2.484	2.386
Local space average color (5)	3.915	2.879	4.243	3.943
Local space average color (6)	2.233	1.946	2.413	2.375
Local space average color (7)	3.981	2.822	4.134	3.974
Anisotropic L.S.A. color (1)	4.286	3.012	4.399	4.291
Anisotropic L.S.A. color (2)	2.316	2.005	2.562	2.428
Anisotropic L.S.A. color (3)	3.662	2.753	3.843	3.660
Anisotropic L.S.A. color (4)	2.312	2.000	2.490	2.397
Anisotropic L.S.A. color (5)	3.715	2.757	3.890	3.666
Anisotropic L.S.A. color (6)	2.307	1.962	2.427	2.424
Anisotropic L.S.A. color (7)	3.867	2.745	3.775	3.716
Combined WP/L.S.A. color	**6.269**	2.972	3.626	4.940
Comprehensive L.S.A. color	4.476	**3.183**	4.506	4.780

coordinate. The color constancy measure m is used as the y-coordinate. Figure 13.7 shows the results for sets 1 through 4, sets 6 through 9, and sets 10 through 13. The data points clearly form a line for image sets 1 through 4 and image sets 6 through 9. This tells us that the color constancy measure correlates with the performance of the algorithms. Object recognition becomes better the more constant the computed output is. For color-based object recognition, it also helps if the entire color space is used.

The correlation is not very good for sets 10 through 13. Sets 10 through 13 are different from sets 1 through 9 in that the images of sets 1 through 9 were created using a camera

Table 13.12: Combined color constancy measure for image sets 6 through 13.

Algorithm	6	7	8	9	10	11	12	13
Full range per band	2.836	3.324	2.199	2.063	1.137	1.400	1.028	0.831
White patch retinex	3.247	3.958	2.330	2.234	1.281	1.179	1.316	1.222
Gray world assumption	3.677	4.574	2.691	2.801	1.559	1.663	1.650	1.366
Simplified horn	2.875	3.193	2.368	2.311	1.412	1.156	1.047	0.773
Gamut constraint 3D	2.785	3.075	2.095	2.020	1.137	1.537	1.010	0.815
Gamut constraint 2D	1.682	2.053	1.542	1.602	1.137	1.133	0.891	0.536
Color cluster rotation	2.516	3.198	2.302	2.347	1.322	1.440	1.758	1.824
Comprehensive normalization	3.379	3.991	2.712	2.908	1.651	1.817	1.755	1.278
Risson (2003)	2.937	3.537	2.284	2.579	1.357	1.397	1.535	1.010
Horn (1974)/Blake (1985)	2.319	2.727	1.970	2.263	1.131	1.101	1.213	0.976
Moore et al. (1991) Retinex	4.990	5.204	4.350	4.068	**1.871**	**2.332**	**2.306**	**1.979**
Moore et al. (1991) Extended	2.960	3.143	2.612	2.549	1.381	1.754	1.548	1.171
Rahman et al. (1999)	4.445	4.725	3.700	3.373	1.777	2.052	2.227	1.872
Homomorphic filtering	2.359	2.505	2.097	2.045	1.040	1.635	1.000	0.590
Homomorphic filtering (HVS)	2.015	2.105	1.835	1.782	1.139	1.347	1.112	0.914
Intrinsic (min. entropy)	1.028	1.076	1.026	0.992	0.629	0.567	0.446	0.202
Local space average color (1)	**5.849**	**5.946**	4.741	3.964	1.665	2.042	2.031	1.692
Local space average color (2)	3.231	3.378	2.864	2.792	1.535	1.879	2.046	1.361
Local space average color (3)	5.246	5.362	4.293	3.751	1.576	2.006	1.984	1.670
Local space average color (4)	3.251	3.382	2.847	2.742	1.439	1.811	1.840	1.361
Local space average color (5)	5.328	5.377	4.285	3.668	1.500	1.883	1.910	1.617
Local space average color (6)	3.101	3.260	2.737	2.679	1.444	1.823	1.855	1.358
Local space average color (7)	5.024	5.166	4.073	3.562	1.508	1.945	1.913	1.632
Anisotropic L.S.A. color (1)	5.351	5.582	4.651	4.333	1.699	2.095	2.052	1.744
Anisotropic L.S.A. color (2)	3.272	3.465	2.895	2.900	1.589	2.003	2.102	1.427
Anisotropic L.S.A. color (3)	4.865	5.074	4.203	4.033	1.637	2.088	2.019	1.717
Anisotropic L.S.A. color (4)	3.224	3.416	2.853	2.832	1.467	1.898	1.887	1.411
Anisotropic L.S.A. color (5)	4.914	5.099	4.217	3.992	1.550	1.962	1.949	1.660
Anisotropic L.S.A. color (6)	3.149	3.327	2.790	2.773	1.477	1.934	1.905	1.411
Anisotropic L.S.A. color (7)	4.807	5.020	4.135	3.920	1.564	2.037	1.935	1.684
Combined WP/L.S.A. color	4.724	5.341	3.565	3.265	1.544	1.887	2.030	1.647
Comprehensive L.S.A. color	5.416	5.787	**4.779**	**4.761**	1.682	2.250	2.082	1.839

with linear sensors (Barnard et al. 2002c) whereas the images from sets 10 through 13 were created with several different methods. Also, when creating the image sets 1 through 9, it was ensured that there is no camera offset (response to no light) for sets 1 through 5. The unknown gamma factor may be a cause for the generally lower performance obtained on sets 10 through 13. Another cause may be that the color sensors of the camera used for sets 1 through 5 are fairly narrow band which may not be the case for image sets 10 through 13.

EVALUATION OF ALGORITHMS

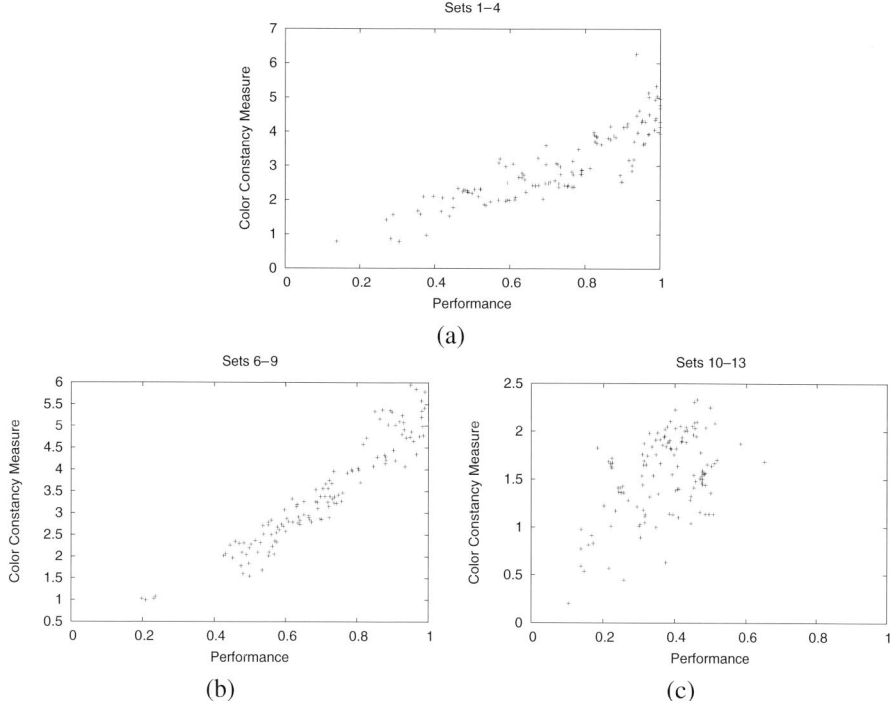

Figure 13.7 Color constancy measure as a function of the performance of the algorithms. The data for sets 1 through 4 are shown in graph (a). The data for sets 6 through 9 and sets 10 through 13 are shown in the two graphs (b) and (c).

13.5 Comparison to Ground Truth Data

An in-depth comparison of computational color constancy algorithms was performed by Barnard et al. (2002a,b). Experiments were conducted with synthesized data as well as real image data. Barnard et al. (2002c) measured the ground truth data for the illuminants for the image sets. Thus, we can perform the same analysis here. Barnard et al. (2002a,b) measured the angular error between the actual and the estimated illuminant. They also measured the distance between the actual and the estimated illuminant in RG chromaticity space. Let $\mathbf{L}_E = [L_{E,r}, L_{E,g}, L_{E,b}]^T$ be the estimated illuminant and let $\mathbf{L}_T = [L_{T,r}, L_{T,g}, L_{T,b}]^T$ be the actual illuminant. The angular error E_α is then given by

$$E_\alpha = \cos^{-1} \frac{\mathbf{L}_E \mathbf{L}_T}{|\mathbf{L}_E||\mathbf{L}_T|}. \tag{13.28}$$

Let $\hat{\mathbf{L}}_E = \frac{1}{L_{E,r}+L_{E,g}+L_{E,b}}[L_{E,r}, L_{E,g}, L_{E,b}]^T$ be the chromaticity of the estimated illuminant and let $\hat{\mathbf{L}}_T = \frac{1}{L_{T,r}+L_{T,g}+L_{T,b}}[L_{T,r}, L_{T,g}, L_{T,b}]^T$ be the chromaticity of the actual illuminant. Then the distance E_d between the estimated and the actual illuminant is given by

$$E_d = \sqrt{(\hat{L}_{E,r} - \hat{L}_{T,r})^2 + (\hat{L}_{E,g} - \hat{L}_{T,g})^2}. \tag{13.29}$$

Table 13.13: Comparison to ground truth data for image sets 1 through 5.

Algorithm	1		2		3		4		5	
	E_α	E_d	E_α	E_d	E_α	E_d	E_α	E_d	E_α	E_d
Full range per band	**8.6**	**0.057**	**13.1**	**0.094**	12.5	0.088	**10.3**	**0.087**	10.2	0.068
White patch retinex	9.1	0.064	18.7	0.137	13.8	0.106	15.5	0.136	15.4	0.120
Gray world assumption	11.0	0.088	18.9	0.141	19.9	0.167	19.4	0.173	10.8	0.081
Gamut constraint 3D	**8.6**	**0.057**	**13.1**	**0.094**	12.3	0.086	13.7	0.119	10.6	0.073
Gamut constraint 2D	18.2	0.132	21.0	0.157	22.3	0.159	18.5	0.146	16.7	0.120
Risson (2003)	9.8	0.070	15.1	0.110	**10.4**	**0.070**	14.4	0.142	12.7	0.086
Local space average color	10.7	0.084	18.3	0.134	19.4	0.158	19.0	0.168	10.8	0.079
Anisotropic L.S.A. color	10.7	0.084	18.4	0.134	19.0	0.154	19.0	0.167	10.9	0.079

Table 13.14: Comparison to ground truth data for image sets 6 through 9.

Algorithm	6		7		8		9	
	E_α	E_d	E_α	E_d	E_α	E_d	E_α	E_d
Local space average color	**3.4**	**0.305**	**3.4**	**0.301**	3.5	**0.308**	**3.4**	**0.303**
Anisotropic L.S.A. color	3.5	0.310	**3.4**	0.306	**3.5**	0.310	**3.4**	0.304

Barnard et al. (2002a,b) report the root mean squared error of these error measures for synthetic as well as real image data. For some algorithms, they also report illuminant RGB error and the brightness error $(R + G + B)$. Both are not considered here. They report results for the basic gray world assumption (described in Section 6.2), the white patch retinex algorithm, various variants of the gamut-constraint algorithm originally developed by Forsyth (1988, 1992) (described in Section 6.4), Finlayson's (1996) color in perspective method (described in Section 6.5), a neural net approach by Funt et al. (1996) (described in Section 8.2) as well as the algorithm color by correlation (Finlayson et al. 1997). The algorithms color cluster rotation (described in Section 6.6) and comprehensive normalization (described in Section 6.7), were not tested. The algorithm by Risson (2003) (described in Section 6.8) which appeared at a later time could also not be included in the study. Algorithms that estimate the illuminant locally for each image pixel were excluded from the study.

We evaluated all algorithms with respect to the same error measures as introduced by Barnard et al. (2002a,b). Like Barnard et al., we excluded pixels less than 2 from the error measure. Results for image sets 1 through 5 are shown in Table 13.13. Results for image sets 6 through 9 are shown in Table 13.14.

Table 13.13 shows the results for algorithms that assume a constant illuminant as well as the results for algorithms that assume a nonuniform illuminant across the image. For algorithms that estimate the illuminant locally, we simply computed the global average of the locally estimated illuminant in order to compare it to the ground truth data. Simply estimating the illuminant by considering the maximum and the minimum of the measured data as done by the algorithm "full range per band", produces best results on sets 1, 2, and 4. This is of no surprise, as the image sets 1 through 5 contain no images whose maximum

EVALUATION OF ALGORITHMS

values have been clipped. Also, these sets all have no camera offset (black level of zero). The three-dimensional gamut-constraint algorithm also performed very well. Barnard et al. (2002b) report excellent results for the three-dimensional gamut-constraint algorithm. As expected, the algorithm of Risson, which was specifically designed for scenes with specular surfaces, performed best on set 3.

Table 13.14 shows the results when either isotropic local space average color or anisotropic local space average color is used to estimate the illuminant. Here, the root mean squared error measure was computed for all pixels of the image. The data shows that isotropic local space average color can be used just as well as anisotropic local space average color. However, in order to use isotropic local space average color, we need to know the size of the smoothing parameter. In contrast, if the correct lines of constant illumination were known, the smoothing parameter could be set to a sufficiently large value.

14

Agreement with Data from Experimental Psychology

We will now investigate if the algorithms have any relation to human color perception. In order to compare the algorithms to human color perception, we need to look at data from experimental psychology, i.e. experiments with human subjects.

14.1 Perceived Color of Gray Samples When Viewed under Colored Light

Helson (1938) experimented with gray patches illuminated with colored light. He constructed a light-tight booth for his experiments. The booth was lined with white and gray cardboard. Samples placed on a shelf located in a corner of the booth were shown to a subject. Light could enter the booth through two openings of equal size on the top of the booth. One opening was used to illuminate the booth with filtered light and the other was used to illuminate it with white light. The composition of the light could be varied by placing filters and screens in front of the two openings. A 500-W lamp placed above one of the openings served as a single light source. The opening below the light source could be equipped with red, green, blue or yellow filters. The other opening did not contain any filter. Instead, a mirror was placed above it. This mirror was used to deflect the light from the lamp such that the booth could also be illuminated with white light. A glass screen was located below the openings to diffuse the light inside the booth. This ensured that the samples that were shown to the subjects were uniformly illuminated. The subjects were trained before they entered the booth. They were given a Munsell book of colors. The training included recognizing and reporting on differences in color perception. Once the subject entered the booth, the results were accepted without question. Helson notes that in order to obtain reproducible results both the stimulus and the subject have to be standardized.

A total of 16 different samples were used as stimuli. The size of the samples was 4×5 cm^2. All samples were achromatic when viewed under daylight illumination. The

---

Color Constancy M. Ebner
© 2007 John Wiley & Sons, Ltd

Table 14.1: Color perceived by a single subject (data from Helson (1938)). The table shows the perceived color for each sample with reflectance R. The perceived color depends on the color of the illuminant and the type of background. The experiment was carried out with four different illuminants and three different backgrounds.

R	Red Illumination			Green Illumination		
	White Background	Gray Background	Black Background	White Background	Gray Background	Black Background
0.80	R	R	R	YG	YG	YG
0.52	R	R	R	YG	YG	YG
0.39	A	bR	R	A	YG	YG
0.34	A	bR	R	A	YG	YG
0.27	B	RB	bR	BR	YG	yG
0.23	B	RB	bR	BR	YG	yG
0.22	B	RB	bR	BR	A	yG
0.17	B	RB	bR	BR	A	yG
0.16	B	RB	bR	BR	A	yG
0.15	B	A	bR	BR	A	yG
0.13	B	A	bR	BR	A	yG
0.13	B	A	bR	BR	A	yG
0.11	B	A	bR	BR	A	yG
0.10	B	bG	bR	BR	RB	yG
0.07	B	bG	bR	BR	RB	yG
0.07	B	bG	bR	BR	RB	yG
0.05	B	bG	bR	BR	RB	yG
0.03	gB	bG	bR	BR	RB	yG
0.03	BG	BG	A	BR	RB	A

R	Blue Illumination			Yellow Illumination		
	White Background	Gray Background	Black Background	White Background	Gray Background	Black Background
0.80	B	B	B	gY	Y	Y
0.52	B	B	B	gY	Y	Y
0.39	A	B	B	A	Y	Y
0.34	rB	B	B	A	Y	Y
0.27	rB	rB	B	RB	Y	Y
0.23	Y	rB	B	RB	Y	Y
0.22	Y	rB	rB	RB	A	Y
0.17	Y	A	rB	RB	A	Y
0.16	Y	A	rB	RB	A	Y
0.15	Y	A	rB	RB	A	Y
0.13	Y	A	rB	RB	A	Y
0.13	Y	A	rB	RB	A	Y
0.11	Y	A	rB	RB	RB	Y
0.10	Y	Y	rB	RB	RB	A
0.07	Y	Y	rB	RB	RB	A
0.07	Y	Y	RB	RB	RB	A
0.05	Y	Y	RB	RB	RB	RB
0.03	Y	RY	A	RB	RB	RB
0.03	Y	RY	A	rB	B	A

samples were illuminated with colored light and the task of the subjects was to report anything they saw. Apart from the illuminant, the background on which the samples were presented was also varied. Three different backgrounds (white, gray, and black) were used. Helson tabulated the perceived hue, saturation, and lightness of each sample for each subject. The experiment was carried out with 27 subjects. Table 14.1 shows the results of this experiment for a single subject. The table shows the perceived hue of each sample for a given illuminant and a particular background on which the samples were presented. Four primary colors were used to classify the perceived hue: red (R), green (G), blue (B), and yellow (Y). The primary colors were denoted with a single letter and other colors were denoted with two letters. If both components were equally dominant, two upper case letters were used. If one component was more dominant than the other, then the dominant color was written with an upper case letter and the other with a lower case letter, i.e. yG corresponds to a yellowish-green. If no color could be perceived, the sample was said to be achromatic (A). The data in Table 14.1 shows that the perceived color depends on not only the color of the illuminant but also the color of the background.

Helson (1938) drew the following main conclusion from his experiments. If the illumination is constant, then the perceived color of a patch depends on the brightness of the patch as well as the color of the background. On a gray background, a bright patch will have the color of the illuminant and a dark patch will have the complementary color of the illuminant. Patches of intermediate reflectance will appear achromatic. In other words, the background will have an impact on the perceived color. The brighter the background, the brighter the patch must be for it to appear as having the color of the illuminant.

We now investigate how the algorithms, which were described in the preceding text, perform on this task. The algorithms can be analyzed either theoretically or by supplying sample images that correspond to the stimuli used in Helson's experiments.

14.2 Theoretical Analysis of Color Constancy Algorithms

We start with a theoretical analysis of the algorithms. Before we address each algorithm in turn, let us define a suitable stimulus. We assume that a sample is placed on a uniform background. Let R_p be the reflectance of the sample and let R_{bg} be the reflectance of the background. Let $\mathbf{L}$ be the vector that describes the intensity of the illuminant for the three primaries red, green, and blue. Then, the color $\mathbf{c}_p = [c_r, c_g, c_b]^T$ of the sample (as measured by the sensor) is given by

$$\mathbf{c}_p = R_p \mathbf{L}. \tag{14.1}$$

Similarly, the color of the background $\mathbf{c}_{bg}$ is

$$\mathbf{c}_{bg} = R_{bg} \mathbf{L}. \tag{14.2}$$

These are the two colors measured by the image sensor. In practice, intermediate colors also occur at the boundary of the patch; however, this shall not be our concern here. This input stimulus is processed by a color constancy algorithm. In the following text, we will calculate the output color for each algorithm in turn.

White Patch Retinex

The white patch retinex algorithm described in Section 6.1 divides all color channels by the maximum of each channel over the entire image. We consider three input stimuli where the reflectance of the sample is either higher than the background reflectance, equivalent to the background reflectance, or lower than the reflectance of the background. If the reflectance of the sample is higher than the background reflectance, then the white patch retinex algorithm will estimate the illuminant as follows:

$$\mathbf{L}_{\max} = \mathbf{c}_p. \tag{14.3}$$

In this case, the output color $\mathbf{o} = [o_r, o_g, o_b]^T$ of the sample will be

$$\mathbf{o} = \frac{\mathbf{c}_p}{\mathbf{c}_p} = [1, 1, 1]^T = \mathbf{1} \tag{14.4}$$

where division is defined componentwise. Therefore, the output will be achromatic if the reflectance of the patch is higher than that of the background. In this case, the output color does not depend on the color of the illuminant.

If the reflectance of the sample is equivalent to the reflectance of the background, then all input pixels will be equivalent. If we divide the pixel color by the maximum, then we again obtain an output value of $[1, 1, 1]^T$. Therefore, the output will be achromatic.

If the reflectance of the sample is lower than the reflectance of the background, then the color of the illuminant will be assumed to be

$$\mathbf{L}_{\max} = \mathbf{c}_{\text{bg}}. \tag{14.5}$$

Given an estimate of the illuminant, the color of the sample is computed as

$$\mathbf{o} = \frac{\mathbf{c}_p}{\mathbf{c}_{\text{bg}}}. \tag{14.6}$$

With the derivation of the color of the sample and the color of the background, we obtain

$$\mathbf{o} = \frac{R_p \mathbf{L}}{R_{\text{bg}} \mathbf{L}} = \frac{R_p}{R_{\text{bg}}} \mathbf{1}. \tag{14.7}$$

Again, the color of the output pixel is independent of the illuminant. Note that both the sample and the background are achromatic. Therefore, the sample will appear to be achromatic because all color channels will have the same value. Hence, the white patch retinex algorithm is not in agreement with the results obtained by Helson.

The Gray World Assumption

The gray world assumption was discussed in detail in Section 6.2. Algorithms based on the gray world assumption first compute global space average color and then divide the input pixel by the global space average color. Let $\mathbf{a}$ be the global space average color. The space average color is a mixture of the sample color and the color of the background. It depends on the size of the patch relative to the background. Let α be the fraction of the size of the patch relative to the size of the background. Then, global space average color is given by

$$\mathbf{a} = \alpha \mathbf{c}_p + (1 - \alpha) \mathbf{c}_{\text{bg}}. \tag{14.8}$$

AGREEMENT WITH DATA FROM EXPERIMENTAL PSYCHOLOGY

If we apply the gray world assumption, then the output color **o** of the sample is given by

$$\mathbf{o} = \frac{\mathbf{c}_p}{f\mathbf{a}} = \frac{R_p \mathbf{L}}{f\alpha R_p \mathbf{L} + f(1-\alpha)R_{bg}\mathbf{L}} \quad (14.9)$$

where f is a constant scaling factor. Again, the illuminant **L** drops out of the equation.

$$\mathbf{o} = \frac{R_p}{f\alpha R_p + f(1-\alpha)R_{bg}} \cdot \mathbf{1} \quad (14.10)$$

Thus, the output color is independent of the light that illuminates the sample. The output color will be achromatic for all combinations of illuminants and backgrounds. Thus, the gray world assumption is not in agreement with the results obtained by Helson.

Variant of Horn's Algorithm

The variant of Horn's algorithm (Horn 1974, 1986) that assumes a uniform illumination just eliminates the following steps: computation of Laplacian, thresholding the result, and reintegrating. Here, since a uniform illumination is used, the result will be equivalent to the original formulation of Horn, which is discussed in the subsequent text.

Gamut-Constraint Methods

The three-dimensional gamut-constraint method assumes that a canonical illuminant exists. The method first computes the convex hull $\mathcal{H}_c$ of the canonical illuminant. The points of the convex hull are then scaled using the set of image pixels. Here, the convex hull would be rescaled by the inverse of the two pixel colors $\mathbf{c}_p$ and $\mathbf{c}_{bg}$. The resulting hulls are then intersected and a vertex with the largest trace is selected from the hull. The following result would be obtained for the intersection of the maps $\mathcal{M}_\cap$:

$$\mathcal{M}_\cap = \begin{cases} \mathcal{H}_c/\mathbf{c}_p & \text{if } R_p > R_{bg} \\ \mathcal{H}_c/\mathbf{c}_p & \text{if } R_p = R_{bg} \\ \mathcal{H}_c/\mathbf{c}_{bg} & \text{if } R_p < R_{bg}. \end{cases} \quad (14.11)$$

If we now assume that the vertex that describes white under the canonical illuminant is the vertex with the largest trace, then the map **m** that will be chosen from the intersection of the maps $\mathcal{M}_\cap$ will be given by

$$\mathbf{m} = \begin{cases} 1/\mathbf{c}_p & \text{if } R_p > R_{bg} \\ 1/\mathbf{c}_p & \text{if } R_p = R_{bg} \\ 1/\mathbf{c}_{bg} & \text{if } R_p < R_{bg}. \end{cases} \quad (14.12)$$

If this map is applied to the color measured by the sensor in the center of the image, we obtain the same result as we obtained for the white patch retinex algorithm. The output color will be achromatic for all three cases.

$$\mathbf{o} = \begin{cases} [1,1,1]^T & \text{if } R_p > R_{bg} \\ [1,1,1]^T & \text{if } R_p = R_{bg} \\ \frac{\mathbf{c}_p}{\mathbf{c}_{bg}} & \text{if } R_p < R_{bg} \end{cases} \quad (14.13)$$

Therefore, the three-dimensional gamut-constraint method is not in agreement with the results obtained by Helson.

The two-dimensional gamut-constraint algorithm first projects all colors **c** onto the plane at $b = 1$. Because both the sample and the background are achromatic, the results will be independent of the reflectance R of the sample.

$$\left[\frac{c_r}{c_b}, \frac{c_g}{c_b}, 1\right]^T = \left[\frac{RL_r}{RL_b}, \frac{RL_g}{RL_b}, 1\right]^T = \left[\frac{L_r}{L_b}, \frac{L_g}{L_b}, 1\right]^T \qquad (14.14)$$

If no constraints are imposed on the illuminant, then the map obtained by intersecting the convex hulls obtained from this data will be equivalent to

$$\mathbf{m} = \left[\frac{L_b}{L_r}, \frac{L_b}{L_g}\right]^T \mathbf{m}_c \qquad (14.15)$$

where $\mathbf{m}_c = [m_x, m_y]^T$ is the map with the largest trace obtained from the canonical gamut. If this map is applied to the input color, we obtain

$$\mathbf{o} = [m_x, m_y, 1]^T \qquad (14.16)$$

which is independent of the color of the illuminant. If the illuminant is constrained to lie on the curve of the black-body radiator, then the input image will be scaled by the same illuminant in all three cases. Similar to the three-dimensional gamut-constraint algorithm, this algorithm does not correspond to the results obtained by Helson.

Color Cluster Rotation

Color cluster rotation, which is described in Section 6.6, views pixels of the input image as a cloud of points. A principal component analysis is done to determine the main axis of this cloud of points. The main axis is rotated onto the gray vector. For the input data in Helson's experiments, there are only two different colors sensed by the sensor. The two colors line up along the axis $\mathbf{e}_1$, which is defined by the illuminant.

$$\mathbf{e}_1 = [L_r, L_g, L_b]^T \qquad (14.17)$$

After rotating this axis onto the gray vector and also applying a shift such that the center of the two points is located at the center of the RGB cube, both colors, the color measured for the sample and the color measured for the background will be located on the gray vector. Owing to clipping, some color may be present in the computed image. Again, the results do not match the observations made by Helson.

Comprehensive Color Normalization

Comprehensive color normalization, developed by Finlayson et al. (1998), which is described in detail in Section 6.7, interleaves color normalization and normalization based on the gray world assumption. First, the image pixels are normalized by dividing each channel by the sum over all channels. This algorithm is denoted by $\mathcal{A}_{\text{norm}}$. Next, the image pixels are normalized by dividing each channel independently by the sum over all pixels.

This algorithm is denoted by $\mathcal{A}_{\text{gray}}$. Algorithm $\mathcal{A}_{\text{norm}}$ computes the following output color $\mathbf{o}'$ for the sample:

$$\mathbf{o}' = \frac{\mathbf{c}}{\sum_{i \in \{r,g,b\}} c_i} = \frac{R\mathbf{L}}{RL_r + RL_g + RL_b} = \frac{\mathbf{L}}{L_r + L_g + L_b}. \qquad (14.18)$$

Thus, after the first step of the algorithm, the reflectance of either the sample or the background drops out of the equation. Now all pixels have the same color. If we now calculate global average color $\mathbf{a}$, we get

$$\mathbf{a} = \frac{\mathbf{L}}{L_r + L_g + L_b}. \qquad (14.19)$$

In other words, the average will have the color of the illuminant.

The second step of the algorithm of Finlayson et al ($\mathcal{A}_{\text{gray}}$) actually divides all channels by three times the global average color. The factor of 3 was used so that the sum over all channels of a single pixel is equal to $\frac{1}{3}$ and the sum over all pixels of a single channel is equal to $\frac{n}{3}$. Thus, after we have applied algorithm $\mathcal{A}_{\text{gray}}$, we obtain

$$\mathbf{o} = \frac{\mathbf{o}'}{3\mathbf{a}} = \frac{1}{3}. \qquad (14.20)$$

After this step, the output will no longer change. The whole image will appear to be achromatic. The patch can no longer be distinguished from the background because the pixel values of the patch and the background will be equivalent.

Note that the result does not change if we change the order of the two normalization steps. If we perform $\mathcal{A}_{\text{gray}}$ first and then $\mathcal{A}_{\text{norm}}$, we can use the results from our analysis of the algorithm based on the gray world assumption. In this case, all pixels will appear to be achromatic right after the first step. The second normalization step results in $\mathbf{o} = [1/3, 1/3, 1/3]^T$ as before. Again, we see that this algorithm does not agree with the results obtained by Helson.

Algorithms Based on the Dichromatic Reflection Model

Algorithms based on the dichromatic reflection model are not applicable as the samples are considered to be matte.

The Retinex Theory of Color Vision

Land's alternative retinex theory of color vision (Land 1986a), which was described in Section 7.1, calculates the logarithm of the ratio of the color of the current pixel to the average color over an extended field. The average is computed using sensors that are nonuniformly distributed over the surrounding area. The density of the sensors varies as $\frac{1}{r^2}$, where r is the distance from the current pixel. The sensors that are located above the patch will measure the color of the patch and the sensors that are located above the background will measure the color of the background. The resulting average will be a linear combination of the two colors. It will depend on the size of the patch relative to the background. Let α be a factor that depends on the distribution of the sensors and the size of the patch relative to the size of the background. Then, the average color **a** computed from the sensors will be given by

$$\mathbf{a} = \alpha \mathbf{c}_p + (1-\alpha)\mathbf{c}_{bg} \tag{14.21}$$

for some $\alpha \in [0, 1]$. Land's retinex theory calculates the following output color **o** for a pixel color **c**:

$$\mathbf{o} = \log(\mathbf{c}) - \log(\mathbf{a}) = \log\left(\frac{\mathbf{c}}{\mathbf{a}}\right). \tag{14.22}$$

The pixel color measured by the sensor in the center will be equivalent to the color of the sample.

$$\mathbf{o} = \log\left(\frac{\mathbf{c}_p}{\mathbf{a}}\right) \tag{14.23}$$

$$= \log\left(\frac{\mathbf{c}_p}{\alpha \mathbf{c}_p + (1-\alpha)\mathbf{c}_{bg}}\right) \tag{14.24}$$

$$= \log\left(\frac{R_p \mathbf{L}}{\alpha R_p \mathbf{L} + (1-\alpha) R_{bg} \mathbf{L}}\right) \tag{14.25}$$

We now see that the illuminant **L** drops out of the equation.

$$\mathbf{o} = \log\left(\frac{R_p}{\alpha R_p + (1-\alpha) R_{bg}}\right) \mathbf{1} \tag{14.26}$$

The computed output color is independent of the illuminant. The retinex theory of color vision was also analyzed in depth by Brainard and Wandell (1992). Brainard and Wandell experimented with color Mondrians consisting of nine colored patches. They found that the retinex algorithm does not correctly predict color appearance when the composition of the Mondrian is varied. The results of Brainard and Wandell could be due to the use of infinite path lengths for the retinex algorithm. The effect of the path length on the retinex algorithm is discussed by (Ciurea and Funt 2004).

Computation of Lightness and Color

Horn's algorithm (described in Section 7.2) is just an alternative formulation of Land and McCann's retinex theory. Therefore, the following discussion also applies to Land and McCann's retinex theory (Land and McCann 1971) and the improvements made by Blake (1985). Horn's algorithm (Horn 1974, 1986) is based on the assumption that smooth changes of intensity are due to a change of the illuminant. Sharp edges are associated with a change of reflectance. An edge detection operation is used to locate sharp changes of the illuminant. Edge detection is performed by applying the Laplace operator. Next, a threshold is used to separate the smooth changes from the sharp changes of the illuminant. Only sharp edges are retained and then the result is spatially integrated. Finally, all the color channels are normalized to the range [0, 1].

If we apply this algorithm to an image that consists of a gray sample lying on a uniform gray background, then the only edges that will be detected will be the edges around the sample. The output of the Laplace operator at the border between the sample and the background is given by

$$\log\left(\frac{\mathbf{c}_{bg}}{\mathbf{c}_p}\right) = \log\left(\frac{R_{bg}\mathbf{L}}{R_p\mathbf{L}}\right) = \log\left(\frac{R_{bg}}{R_p}\mathbf{1}\right) \tag{14.27}$$

which is again independent of the color of the illuminant. The output will be the same constant over all three channels because both the sample and the background are achromatic. Thus, if we assume a uniform illumination, then the image obtained after the integration step will be equivalent to the reflectance image except for a constant offset. The normalization operation will transform all channels to the range [0,1]. Therefore, Horn's algorithm will also output a constant value for all the color channels at the position of the sample. Since the output will be achromatic irrespective of the stimulus, the algorithm is not in agreement with Helson's results.

The real-time neural system developed by Moore et al. (1991) is also based on the retinex theory. This holds for the original implementation and the extension proposed by Moore et al. In the extended version, a smoothed version of the derivative of the input image is used to modulate the blurred version of the input image before it is subtracted from the input image.

$$\mathbf{o} = \log(\mathbf{c}) - \left(\log(\mathbf{c}) \otimes e^{-\frac{|\mathbf{r}|}{\sigma}}\right)\left(|\partial \log \mathbf{c}| \otimes e^{-\frac{|\mathbf{r}|}{\sigma}}\right) \tag{14.28}$$

The scaling is done uniformly over all three bands: red, green, and blue. Thus, we only need to consider the simple version, which is another variant of Land's retinex theory.

$$\mathbf{o} = \log(\mathbf{c}) - \left(\log(\mathbf{c}) \otimes e^{-\frac{|\mathbf{r}|}{\sigma}}\right) \tag{14.29}$$

Let us assume that the scaling factor σ is chosen such that the blurring occurs over a very large area, i.e. the entire image. In this case, the second term computes the geometric mean of the pixel values. Let

$$\mathbf{a} = \alpha \log(\mathbf{c}_p) + (1-\alpha)\log(\mathbf{c}_{bg}) \tag{14.30}$$

be the computed average. The expression for the output color of the sample is then given by

$$\mathbf{o} = \log(\mathbf{c}) - \alpha \log(\mathbf{c}_p) - (1-\alpha)\log(\mathbf{c}_{bg}) \tag{14.31}$$

$$= \log\left(\frac{\mathbf{c}_p}{\mathbf{c}_p^{\alpha}\mathbf{c}_{bg}^{(1-\alpha)}}\right). \tag{14.32}$$

We now use the assumption that pixel values are proportional to reflectance and the intensity of the illuminant for a given wavelength. This gives us

$$\mathbf{o} = \log\left(\frac{R_p\mathbf{L}}{R_p^{\alpha}\mathbf{L}^{\alpha} R_{bg}^{(1-\alpha)}\mathbf{L}^{(1-\alpha)}}\right) = \log\left(\frac{R_p}{R_p^{\alpha} R_{bg}^{(1-\alpha)}}\right) \tag{14.33}$$

which is independent of the color of the illuminant.

The algorithm of Rahman et al. (1999) is another extension of Land's retinex theory (Land 1986a). The adjustment is done for a number of different scales. The results are added together using a set of weights. Since each intermediate output is independent of the color of the illuminant, the combined output computed by the algorithm of Rahman et al. will also be independent of the color of the illuminant.

Homomorphic Filtering

Homomorphic filtering as described in Section 7.5 can be used to reduce or even eliminate the influence of a slowly varying illuminant. First the logarithm is applied to the input image. Then the image is transformed to Fourier space where a high emphasis filter is applied. This is supposed to remove or suppress slowly varying components of the image while maintain the high-frequency components. If the sample has a higher or lower reflectance than the background, the high-frequency components will be caused by the border of the sample. After transforming the image back to the spatial domain, the image will be essentially equivalent (except for rescaling) to the input image. If the sample has the same reflectance as the background, then each band of the input image will be constant. The low-frequency components are only attenuated and not removed entirely and the output will be equivalent to the input (except for rescaling). In any case, the output obtained using homomorphic filtering will not produce the results described by Helson.

Intrinsic Images

The intrinsic images obtained by the method of Finlayson and Hordley (2001a,b) only depend on the reflectance of the object points provided the illuminant can be approximated by a black-body radiator. Since a green illuminant cannot be approximated by a black-body radiator, the result will not be an intrinsic image. That aside, here, achromatic patches are illuminated by four different illuminants. Finlayson et al. compute the logarithm for each channel and subtract the geometric mean. The resulting coordinates are denoted by ρ_i for $i \in \{r, g, b\}$:

$$\rho_i = \log(c_i) - \log(c_M) \tag{14.34}$$

with $c_M = \sqrt[3]{c_r c_g c_b}$. Thus, we obtain

$$\rho_i = \log\left(\frac{c_i}{c_M}\right) = \log\left(\frac{RL_i}{R\sqrt[3]{L_r L_g L_b}}\right) = \log\left(\frac{L_i}{\sqrt[3]{L_r L_g L_b}}\right) \tag{14.35}$$

The coordinates are proportional to the color of the illuminant. Therefore, the output will not match the results obtained by Helson.

14.3 Theoretical Analysis of Algorithms Based on Local Space Average Color

Let us now address the algorithms of Section 11, all of which compute output color on the basis of local space average color. All these algorithms first compute local space average color as described in Section 10. Here, we assume that the local space average color at the sample will be intermediate between the two theoretical colors of the background and the sample. In other words, we assume that local space average color **a** is given by

$$\mathbf{a} = \alpha \mathbf{c}_p + (1 - \alpha) \mathbf{c}_{\text{bg}} \tag{14.36}$$

with $\alpha \in [0, 1]$. The parameter α depends on the size of the sample relative to the size of the background. In the following text we refer to the individual algorithms using local space average color by the same number as specified in appendix F.

Scaling Input Values (No. 1)

The first algorithm that is described in Section 11.1 is based on the gray world assumption. Each channel is divided by the product of the local space average and a constant factor. From our analysis in the preceding text, we have already seen that the output will be constant for all color channels in this case. Thus, even though the algorithm is based on local space average color and is therefore able to cope with scenes of varying illumination, it is not in agreement with the results obtained by Helson.

Color Shifts (No. 2)

The algorithm described in Section 11.2 subtracts the component of local space average color that is orthogonal to the gray vector from the color of the input pixel. Let $\mathbf{c}_p$ be the color of the input pixel and let $\mathbf{a}$ be the local space average color. Let $\bar{c} = \frac{1}{3}[c_r, c_g, c_b]$ be the average value of the components of a pixel, i.e. $\bar{a}$ denotes the average value of the components of local space average color. Then the color of the output pixel $\mathbf{o}$ is given by

$$o_i = c_i - a_i + \bar{a} \tag{14.37}$$

$$= R_p L_i - \left(\alpha R_p L_i + (1-\alpha) R_{bg} L_i\right) + \left(\alpha R_p \bar{L} + (1-\alpha) R_{bg} \bar{L}\right) \tag{14.38}$$

$$= (1-\alpha) R_p L_i - (1-\alpha) R_{bg} L_i + \left(\alpha R_p \bar{L} + (1-\alpha) R_{bg} \bar{L}\right) \tag{14.39}$$

$$= (1-\alpha)(R_p - R_{bg}) L_i + \left(\alpha R_p \bar{L} + (1-\alpha) R_{bg} \bar{L}\right) \tag{14.40}$$

with $i \in \{r, g, b\}$ and $\bar{L} = \frac{1}{3}[L_r, L_g, L_b]^T$ is the average of the components of the illuminant. This equation has the form

$$o_i = a(R_p - R_{bg}) L_i + b \tag{14.41}$$

with $a = (1-\alpha)$ and $b = (\alpha R_p \bar{L} + (1-\alpha) R_{bg} \bar{L})$. If $R_p = R_{bg}$, then the output will be $o_i = b$. In other words, in this case, the color of the output pixel will be achromatic. If the reflectance of the sample is higher than the reflectance of the background, then a vector that is proportional to the color of the illuminant will be added to the constant vector $[b, b, b]^T$. The output color will appear to have the color of the illuminant. If the reflectance of the sample is smaller than the reflectance of the background, then a vector that is proportional to the color of the illuminant will be subtracted from the constant vector $[b, b, b]^T$. In this case, the color of the sample will appear to have the complementary color of the illuminant. Thus, this algorithm produces an output color that corresponds to the experimental data obtained by Helson (1938).

Normalized Color Shifts (No. 3)

The algorithm described in Section 11.3 normalizes both the color of the input pixel and local space average color before subtracting local space average color from the color of the input pixel. The resulting color is then scaled to have the same intensity as the input pixel. Let $\mathbf{c}_p$ be the color of the input pixel and let $\mathbf{a}$ be the local space average color. Then, the color of the output pixel o is given by

$$o_i = c_i - \frac{\bar{c}}{\bar{a}}(a_i - \bar{a}) \tag{14.42}$$

$$= c_i - \bar{c}\frac{a_i}{\bar{a}} + \bar{c} \tag{14.43}$$

with $i \in \{r, g, b\}$. Let $\bar{L} = \frac{1}{3}(L_r + L_g + L_b)$ be the average of the components of the illuminant. Local space average color is again assumed to be intermediate between the measured colors of the sample and the measured color of the background. This gives us

$$o_i = R_p L_i - R_p \bar{L} \left(\frac{\alpha R_p L_i + (1-\alpha) R_{bg} L_i}{\alpha R_p \bar{L} + (1-\alpha) R_{bg} \bar{L}} \right) + R_p \bar{L} \tag{14.44}$$

$$= R_p L_i - R_p L_i + R_p \bar{L} \tag{14.45}$$

$$= R_p \bar{L}. \tag{14.46}$$

Again, the output is constant over all the color channels. Therefore, the sample will appear to be achromatic irrespective of the color of the illuminant. Thus, this algorithm is not in agreement with the results obtained by Helson.

Color Shifts and Normalized Color Shifts in Combination with an Increase of Saturation (Nos 4-7)

In Section 11.4 it was discussed that the saturation of colors produced by algorithm Nos 2 and 3 may be low. Saturation may be increased by a scaling factor that depends on the size of the color shift. This rescaling is done in a direction perpendicular to the gray vector. Therefore, the results obtained with algorithm Nos 4 and 6 will be the same as for algorithm No. 2. Similarly, the results obtained with algorithm Nos 5 and 7 will be the same as the results obtained with algorithm No. 3.

Anisotropic Local Space Average Color

Computation of anisotropic local space average color along the line of iso-illumination was discussed in Chapter 12. Since only two different colors are present in the input image, local space average color **a** computed along the line of constant illumination will again be given by

$$\mathbf{a} = \alpha \mathbf{c}_p + (1-\alpha)\mathbf{c}_{bg} \tag{14.47}$$

for some $\alpha \in [0, 1]$. Thus, the results obtained will be equivalent to the previously discussed methods, which are based on simple local space average color.

Combined White Patch Retinex and Local Space Average Color

The algorithm described in Section 11.5 subtracts local space average color from the color of the input pixel. Each channel is then scaled by a factor that depends on the maximum

AGREEMENT WITH DATA FROM EXPERIMENTAL PSYCHOLOGY

deviation between the color at the input pixel and local space average color. Finally, the result is fed through a sigmoidal output. After we subtract local space average color from the color of a pixel located on the sample, we obtain

$$\mathbf{o}'' = \mathbf{c}_p - \mathbf{a} \tag{14.48}$$
$$= R_p \mathbf{L} - \left(\alpha R_p \mathbf{L} + (1-\alpha) R_{bg} \mathbf{L}\right) \tag{14.49}$$
$$= (1-\alpha) R_p \mathbf{L} - (1-\alpha) R_{bg} \mathbf{L} \tag{14.50}$$
$$= (1-\alpha) \mathbf{L} \left(R_p - R_{bg}\right). \tag{14.51}$$

Thus, we see that the computed value is proportional to the color of the illuminant. The reflectance of the sample and the reflectance of the background determine the sign of proportionality. If both are equal, then all channels are zero. The computed value is divided by the maximum deviation between the color of the input pixel and local space average color. The maximum deviation is simply

$$d_i = \left| (1-\alpha) L_i \left(R_p - R_{bg}\right) \right|. \tag{14.52}$$

This value is used to scale o_i'' before it is sent through a sigmoidal activation function.

$$o_i' = \frac{c_{p,i} - a_i}{d_i} = \begin{cases} 1 & \text{if } R_p > R_{bg} \\ 0 & \text{if } R_p = R_{bg} \\ -1 & \text{if } R_p < R_{bg} \end{cases} \tag{14.53}$$

Therefore, if the computed value for a channel is greater than zero, then it will be scaled to one. If the computed value for a channel is lesser than zero, it will be scaled to minus one. If both reflectances are equivalent, then the output will be zero. The output is constant over all channels. The sigmoidal activation function does not change the computed color. In any case, the output is constant over all channels. Therefore, the patch will appear to be achromatic if this algorithm is used. The results obtained by this algorithm do not agree with the results obtained by Helson.

Comprehensive Local Space Average Color

Comprehensive color normalization based on local space average color was discussed in Section 6.7. As we have already seen earlier, after the first step of the algorithm, when the colors are normalized, the reflectance of either the sample or the background drops out of the equation. After normalization, we obtain the output $\mathbf{o}'$:

$$\mathbf{o}' = \frac{\mathbf{c}}{\sum_{i \in \{r,g,b\}} c_i} = \frac{R\mathbf{L}}{RL_r + RL_g + RL_b} = \frac{\mathbf{L}}{L_r + L_g + L_b}. \tag{14.54}$$

Now all pixels have the same color. Computation of local space average color instead of global space average color will not make a difference. Since comprehensive color normalization does not correspond to the results obtained by Helson, neither will comprehensive color normalization based on local space average color.

14.4 Performance of Algorithms on Simulated Stimuli

As an additional test, we ran all algorithms on input images that were similar to the stimuli used in Helson's experiments. This is particularly important as some of the algorithms operate on a grid of processing elements and the output may not be uniform over the entire sample. Also, some simplifying assumptions had to be made in the theoretical analysis. We will see in a moment that the calculated output colors correspond to the colors that were theoretically computed.

As input images, we created small 32×32 images. The image is filled with a reflectance of 0.5, which gives us a gray background. The center 12×12 area of the image is filled with the reflectance of the current sample. Only three types of samples are used: a sample with a reflectance higher than that of the background (0.9), a sample with a reflectance equivalent to that of the background (0.5), and a sample with a reflectance lower than that of the background (0.1). After the sample has been placed on the background, it is illuminated with colored light. Four types of light sources are used:

$$\mathbf{L}_1 = [1.0, 0.1, 0.1]^T \quad \text{(red)} \tag{14.55}$$

$$\mathbf{L}_2 = [0.1, 1.0, 0.1]^T \quad \text{(green)} \tag{14.56}$$

$$\mathbf{L}_3 = [0.1, 0.1, 1.0]^T \quad \text{(blue)} \tag{14.57}$$

$$\mathbf{L}_4 = [1.0, 1.0, 0.1]^T \quad \text{(yellow)} \tag{14.58}$$

The color of each input pixel c is obtained by multiplying the reflectance at each pixel with the light given off by the virtual light source:

$$\mathbf{c}(x, y) = R(x, y)\mathbf{L}_i \tag{14.59}$$

where $R(x, y)$ is the reflectance at position (x, y) and i is the number of the light source. Three samples and four illuminants give us a total of 12 input images.

The perceived color of the 12 input stimuli are shown in Table 14.2, i.e. if the sample has a higher reflectance than the background, then to a human observer the sample appears to have the color of the illuminant. If the sample has the same reflectance as the background, then the sample appears to be achromatic. If the sample has a lower reflectance than the background, then the sample appears to have the complementary color of the illuminant. Table 14.3 shows the performance of the algorithms discussed in Chapters 6 and 7 on the 12 input images. Table 14.4 shows the performance of the algorithms discussed in Chapter 11

Table 14.2: Theoretical classification of input stimuli according to the result of Helson (1938). Three different samples are illuminated with four different light sources. The reflectance of the background was 0.5 for all three samples.

Illuminant	Red			Green			Blue			Yellow		
Reflectance of sample	0.9	0.5	0.1	0.9	0.5	0.1	0.9	0.5	0.1	0.9	0.5	0.1
Stimulus												
Theory	R	–	C	G	–	M	B	–	Y	Y	–	B

AGREEMENT WITH DATA FROM EXPERIMENTAL PSYCHOLOGY

Table 14.3: Each algorithm is tested with four different illuminants and three different patches. One patch has a higher reflectance than the background, the second has the same reflectance as the background and the third has a lower reflectance than the background. Algorithms that have the same performance as a human observer are marked with ✓.

Illuminant	Red			Green			Blue			Yellow			Agreement
Patch reflectance	0.9	0.5	0.1	0.9	0.5	0.1	0.9	0.5	0.1	0.9	0.5	0.1	
Stimulus													
Full range per band	−	−	−	−	−	−	−	−	−	−	−	−	
White patch retinex													
Gray world assumption													
Simplified Horn		−	−		−	−		−	−		−	−	
Gamut constraint 3D		−	−		−	−		−	−		−	−	
Gamut constraint 2D	Rg	Rg	Rg	Gr	Gr	Gr	Bg	Bg	Bg	Rg	Rg	Rg	
Color cluster rotation	−	−	−	−	−	−	−	−	−	−	−	−	
Comprehensive normalization	−	−	−	−	−	−	−	−	−	−	−	−	
Horn (1974)/Blake (1985)	−	−	−	−	−	−	−	−	−	−	−	−	
Moore et al. (1991) retinex	−	−	C	−	−	M	−	−	Y	−	−	B	
Moore et al. (1991) extended	−	R	C	−	G	M	−	B	Y	−	Y	B	
Rahman et al. (1999)	−	−	C	−	−	M	−	−	Y	−	−	B	
Homomorphic filtering	R	R	R	G	G	G	B	B	B	Y	Y	Y	
Homomorphic filtering (HVS)	R	R	R	M	M	M	Y	Y	Y	Y	Y	Y	
Intrinsic (minimum entropy)	Gr	Gr	Gr	G	G	Gr	B	B	B	Y	Y	Gr	

318 AGREEMENT WITH DATA FROM EXPERIMENTAL PSYCHOLOGY

Table 14.4: Results for algorithms described in Chapter 10 through 12.

Illuminant	Red			Green			Blue			Yellow			Agreement
Patch reflectance	0.9	0.5	0.1	0.9	0.5	0.1	0.9	0.5	0.1	0.9	0.5	0.1	
Stimulus													
Local space average color (1)	–	–	–	–	–	–	–	–	–	–	–	–	
Local space average color (2)	R	–	C	G	–	M	B	–	Y	Y	–	B	✓
Local space average color (3)	–	–	–	–	–	–	–	–	–	–	–	–	
Local space average color (4)	R	–	C	G	–	M	B	–	Y	Y	–	B	✓
Local space average color (5)	–	–	C	–	–	M	–	–	Y	–	–	–	
Local space average color (6)	R	–	C	G	–	M	B	–	Y	Y	–	B	✓
Local space average color (7)	–	–	C	–	–	M	–	–	Y	–	–	–	
Anisotropic L.S.A. color (1)	–	–	–	–	–	–	–	–	–	–	–	–	
Anisotropic L.S.A. color (2)	R	–	C	G	–	M	B	–	Y	Y	–	B	✓
Anisotropic L.S.A. color (3)	–	–	–	–	–	–	–	–	–	–	–	–	
Anisotropic L.S.A. color (4)	R	–	C	G	–	M	B	–	Y	Y	–	B	✓
Anisotropic L.S.A. color (5)	–	–	C	–	–	M	–	–	Y	–	–	–	
Anisotropic L.S.A. color (6)	R	–	C	G	–	M	B	–	Y	Y	–	B	✓
Anisotropic L.S.A. color (7)	–	–	C	–	–	M	–	–	Y	–	–	–	
Combined White Patch (WP)/L.S.A. color	–	–	–	–	–	–	–	–	–	–	–	–	
Comprehensive L.S.A. color	–	–	C	–	–	–	–	–	Y	–	–	B	

AGREEMENT WITH DATA FROM EXPERIMENTAL PSYCHOLOGY 319

and Chapter 12. Some of the algorithms have a number of different parameters that can be varied. The settings used for the experiments are summarized in Appendix F.

In order to compare Helson's results with the output created by our algorithms, we need to classify the possible output colors. The calculated output color is first normalized by dividing each color channel by the sum of the channels, i.e. we compute the chromaticities. The color is said to be achromatic if the computed color is within a distance of 0.03 from the gray point $[1/3, 1/3, 1/3]^T$. If the smallest two channels differ by less than 0.1 and the difference between the largest and the second largest channel is greater than 0.1, then we classify the output color as red (R), green (G), or blue (B) depending on which channel is the largest. If the largest two channels differ by less than 0.1 and the difference between the second largest and the third largest channel is larger than 0.1, then we classify the output color as yellow (Y), magenta (M), or cyan (C) depending on which channels are the largest. Otherwise we will report the two largest channels with the major and minor components, i.e. Gr if green is the largest channel and red is the second largest.

The first row for each algorithm shows the classification of the average color of the patch. The second row shows the actual output images. If we compare the results for each of the algorithms from Table 14.3 with the correct theoretical output from Table 14.2, we see that none of the algorithms have the same response as a human observer. Table 14.4 shows the results for algorithms based on local space average color, which were discussed in Chapter 10 through Chapter 12. We see that the output computation that uses local space average color to shift the given color toward the gray vector (discussed in Section 11.2) does indeed have the same response as a human observer.

14.5 Detailed Analysis of Color Shifts

We will now have a detailed look at the performance of the algorithm described in Section 11.2 on the entire set of experiments. Table 14.5 shows the results for four different illuminants and three different backgrounds. The data is shown in the same format as Table 14.1 in order to compare the theoretical results with the experimental results obtained by Helson better. If we compare the two tables, we see that the results correspond exactly to the data obtained by Helson. If the reflectance of the patch is higher than the reflectance of the background, then the patch appears to have the color of the illuminant. If the reflectance of the patch is equivalent to the reflectance of the background, then the patch appears to be achromatic. If the reflectance of the patch is lower than the reflectance of the background, then the patch appears to have the complementary color of the illuminant.

Instead of classifying the color of the patch, we can also plot the RGB values and the chromaticities over all reflectances. The result is shown in Figure 14.1 for a red illuminant $[1.0, 0.1, 0.1]^T$. The three graphs (a), (c), and (e) in Figure 14.1 show the RGB values. The three graphs (b), (d), and (f) show the chromaticities. In each case, the results are shown for three different background reflectances: 0.1, 0.5 and 0.9. In order to further confirm if the algorithm of Section 11.2 is actually used by the human visual system, one would have to look for a neuron with a similar response characteristic as the data shown in Figure 14.1. This, however, is outside the scope of this research.

Table 14.5: Detailed results for the algorithm described in Section 11.2.

R	Red Illumination			Green Illumination		
	White Background	Gray Background	Black Background	White Background	Gray Background	Black Background
1.000000	R	R	R	G	G	G
0.800000	C	R	R	M	G	G
0.500000	C	–	R	M	–	G
0.450000	C	C	R	M	M	G
0.400000	C	C	R	M	M	G
0.350000	C	C	R	M	M	G
0.300000	C	C	R	M	M	G
0.250000	C	C	R	M	M	G
0.200000	C	C	R	M	M	G
0.150000	C	C	R	M	M	G
0.100000	C	C	–	M	M	–
0.050000	C	C	C	M	M	M
0.030000	C	C	C	M	M	M

R	Blue Illumination			Yellow Illumination		
	White Background	Gray Background	Black Background	White Background	Gray Background	Black Background
1.000000	B	B	B	–	Y	Y
0.800000	Y	B	B	B	Y	Y
0.500000	Y	–	B	B	–	Y
0.450000	Y	Y	B	B	B	Y
0.400000	Y	Y	B	B	B	Y
0.350000	Y	Y	B	B	B	Y
0.300000	Y	Y	B	B	B	Y
0.250000	Y	Y	B	B	B	Y
0.200000	Y	Y	B	B	B	Y
0.150000	Y	Y	B	B	B	Y
0.100000	Y	Y	–	B	B	–
0.050000	Y	Y	Y	B	B	B
0.030000	Y	Y	Y	B	B	B

14.6 Theoretical Models for Color Conversion

The simple color shift, described in Section 11.2, is not the only theory that fits the data of Helson. Theoretical models are also given by Judd (1940) and Richards and Parks (1971). Judd (1940) describes the hue of a sample as the orientation of a vector that points from the achromatic point $[n_r, n_g, n_b]^T$ to the given color $[c_r, c_g, c_b]^T$.

$$\text{hue} = \tan^{-1}\left(\frac{c_r - n_r}{c_g - n_g}\right) \qquad (14.60)$$

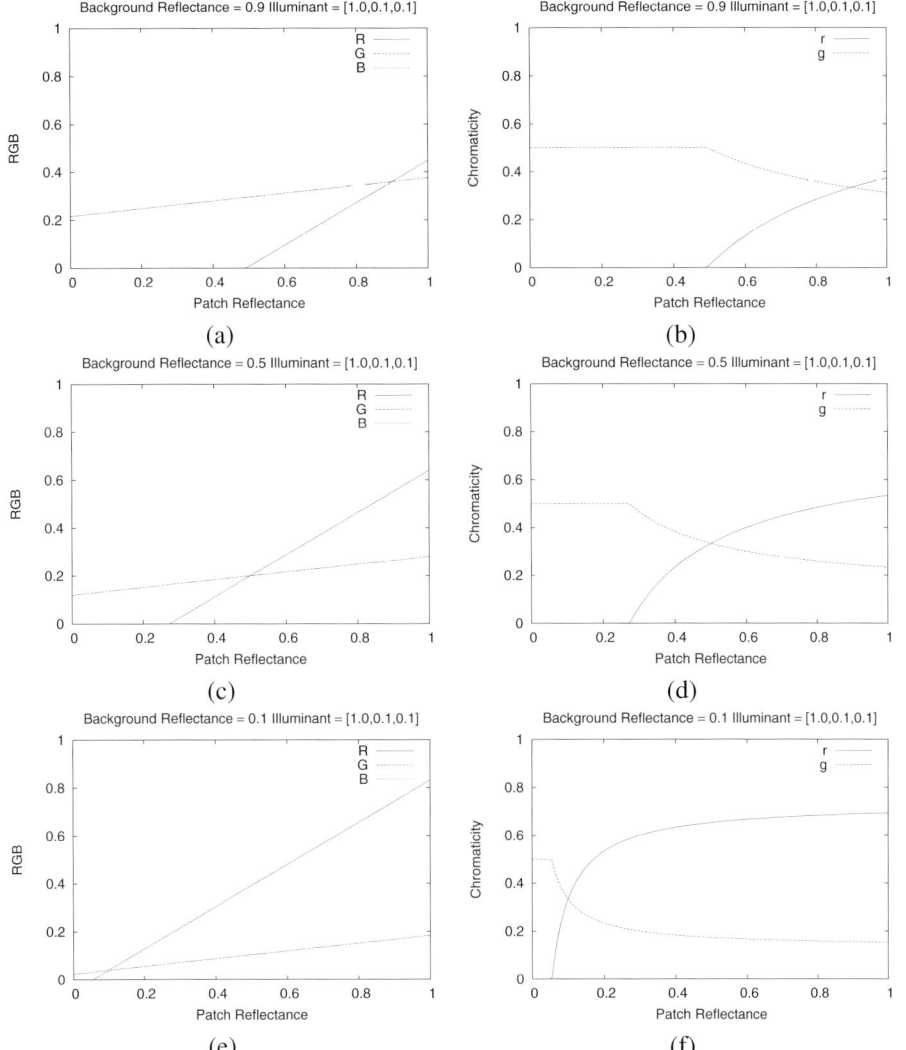

Figure 14.1 RGB values and chromaticities computed by the algorithm described in Section 11.2 over all patch reflectances. A red illuminant $[1.0, 0.1, 0.1]^T$ is assumed. The graphs (a), (c), and (e) show the RGB results, and the graphs (b), (d), and (f) show the chromaticities. A background reflectance of 0.9 was assumed for the two graphs (a) and (b), a background reflectance of 0.5 was assumed for the two graphs (c) and (d), and a background reflectance of 0.1 was assumed for the two graphs (e) and (f).

All colors are represented as trilinear coordinates where the amounts of the three primary stimuli are expressed as fractions of their total, i.e. the colors are normalized. In the model, the achromatic point is a function of the illuminant, the particular sample fixated, and the samples near the fixation point in the immediate past:

$$n_r = L_r - D\left[0.1L'(L_r - 0.360) - 0.018L_b\bar{R}L'^2 \log_{10} 2000I\right] \tag{14.61}$$

$$n_g = L_g - D\left[0.1L'(L_g - 0.300) - 0.030\right] \tag{14.62}$$

where $\mathbf{L} = [L_r, L_g, L_b]^T$ denotes the tristimulus values of the illuminant, D describes the distance from the daylight color to the color of the illuminant

$$D = \sqrt{(L_r - 0.44)^2 + (L_g - 0.47)^2 + (L_b - 0.09)^2}, \tag{14.63}$$

and L' is an adjusted lightness, which in turn is computed from the apparent reflectance R of the sample:

$$L' = 10\frac{R}{R + \bar{R}} - 3 \tag{14.64}$$

where $\bar{R}$ denotes the average apparent reflectance of the samples and background weighted according to proximity in past time and in space to the central part of the visual field, and I denotes the illuminance of the sample plane in foot-candles. A simpler version is also given. The simpler version sets the white point at an intermediate position between the color of the illuminant $\mathbf{L}$ and the original white point denoted by $[w_r, w_g, w_b]^T$:

$$n_r = (1 - k)L_r + kw_r \tag{14.65}$$

$$n_g = (1 - k)L_g + kw_g \tag{14.66}$$

with $k \in [0, 1]$. Table 14.6 shows the results computed using the equations in the preceding text for three hypothetical nonselective samples for both light and dark backgrounds, four illuminants, and two observing situations.

Note that, in order to explain the experimental data, Judd also incorporated factors into his equations that depend on the size of the samples relative to the background and the amount of time for which the samples were fixated. The equations given by Judd are quite complicated. In fact, Land (1960) criticized the formulation of Judd for being wavelength-rich and time-dependent. According to Land, the formulation should be nearly independent of wavelength and fully independent of time.

Compared to the formulation of Judd, our use of simple color shifts is much more elegant. A temporal effect is introduced when we assume that the local averaging, as described in Chapter 10, takes a finite amount of time. It takes some time until the process converges. Of course, intermediate results can still be obtained at any point in time. The computed output color would then depend on outdated information about local space average color, which is stored in the parallel network. This would explain why afterimages occur when the focus of fixation is suddenly moved to a different location.

A completely different model is given by Richards and Parks (1971). It is based on modified von Kries coefficients. If we assume that the sensor response functions have the shape of delta functions, then it is possible to transform a given color of a patch taken under one illuminant to the color of the patch when viewed under a different illuminant by multiplying the colors using three constant factors for the three channels. These factors are known as *von Kries coefficients*, as described in Section 4.6. The von Kries coefficients are defined as

$$K_i = \frac{I_i}{I'_i} \tag{14.67}$$

Table 14.6: Computed hue of nonselective samples on light and dark backgrounds for four different illuminants (data from Judd 1940). The color of the samples is specified using the Munsell notation.

Illuminant	Background Reflectance	Sample	Momentary Fixation of Sample	Fixation of Sample for Several Seconds
Red	0.80	N 2/0	B	B
Red	0.80	N 5/0	RB	R
Red	0.80	N 9/0	YR	YR
Red	0.10	N 2/0	B	B
Red	0.10	N 5/0	YR	YR
Red	0.10	N 9/0	YR	YR
Green	0.80	N 2/0	RB	RB
Green	0.80	N 5/0	RB	RB
Green	0.80	N 9/0	YG	YG
Green	0.10	N 2/0	RB	RB
Green	0.10	N 5/0	YG	YG
Green	0.10	N 9/0	YG	YG
Blue	0.80	N 2/0	RB	RB
Blue	0.80	N 5/0	RB	RB
Blue	0.80	N 9/0	B	B
Blue	0.10	N 2/0	RB	RB
Blue	0.10	N 5/0	B	B
Blue	0.10	N 9/0	B	B
Yellow	0.80	N 2/0	B	B
Yellow	0.80	N 5/0	RB	R
Yellow	0.80	N 9/0	Y	Y
Yellow	0.10	N 2/0	RB	RB
Yellow	0.10	N 5/0	YR	Y
Yellow	0.10	N 9/0	Y	Y

where I_i and I'_i denote the measurements made by a sensor at a particular image position for two different illuminants. The effect described by Helson is modeled by introducing an exponent P_i that is applied to the von Kries coefficient K_i. This results in a modified von Kries coefficient k_i of the form

$$k_i = K_i^{P_i}. \tag{14.68}$$

The modified von Kries coefficients are then applied to the measured colors c_i to obtain corresponding colors o_i under another possibly neutral illuminant.

$$o_i = k_i c_i \tag{14.69}$$

The exponent P_i is defined as

$$P_i = \left(\frac{I_{\text{avg}}}{I_p}\right)^{E_i} \tag{14.70}$$

where I_p is the luminance of the sample and I_{avg} is the average luminance. The exponents E_i are set to $E = [\frac{1}{8}, \frac{1}{4}, \frac{1}{4}]$ for the three channels red, green, and blue. By introducing a dependence on the luminance of the sample in relation to the average luminance, the desired effect is obtained. If the sample is brighter than the background, then the sample seems to have the color of the illuminant ($P_i \to 0$). If the sample has the same luminance as the average luminance, then we have $P_i = 1$, which gives normal von Kries color constancy. If the sample has a lower luminance than the average, we obtain $P_i > 1$. In this case, the original color is factored out and the color is pushed toward the inverse of the illuminant.

Both models, the one given by Judd and the model of Richards and Parks, are psychophysical models of color perception. No explanation is given on how or why the results would depend on either the average apparent reflectance or the average luminance. They are phenomenological descriptions of color vision and not computational theories of color vision (Marr 1982). In this respect, the parallel model described in Chapter 10 in combination with local color shifts described in Section 11.2 is much more satisfying.

14.7 Human Color Constancy

One of the still unsolved questions is, what color constancy algorithm is actually used by the human visual system? Among the algorithms that would lend themselves to a biological realization are the parallel algorithms of Land and McCann (1971), Land (1986a), Horn (1974), Blake (1985), and Ebner (2002, 2003a, 2004b,c, 2006). Each of those algorithms could, in principle, be realized on the massively parallel hardware of the human visual system. Linnell and Foster (1997) suggest that observers are mainly using space average color to estimate the color of the illuminant in Mondrian scenes. Apart from space average color, the color of the highest luminance patch may also be used to a certain extent. McCann (1997) argues that the human visual system basically determines color by normalizing each color channel independently to the maximum in the field of view.

How the human visual system computes color constant descriptors is as yet unknown. What is known, however, is that color constant cells have been found inside visual area V4 (Zeki and Marini 1998). Visual area V4 contains cells with a very large receptive field, which may be exactly the cells that respond to either local or global space average color. Even though the experiments done by Helson point to the use of color shifts, it may also be that the subjects of Helson's experiments have used the wavelength sensitive cells of V1 when describing the color of the patches. The cells of V1 respond to light of different wavelengths.

It would be interesting to know how exactly the cells of V4 are wired up. Hurlbert (1986) noted that the integration, i.e. the computation of local space average color, could either be done in space or in time as is done by the algorithms of Horn (1974), Blake (1985), and Ebner (2002, 2003a, 2004b,c, 2006). In this case, only recurrent neurons that are also connected to their nearest neighbors are required. Instead of computing local space average color iteratively, it could also be computed by consecutively applying a Gaussian blur. This method of computation would resemble the algorithm of Rahman et al. (1999). If such an algorithm is used by the visual system, the neurons would form a hierarchy of increasing receptive fields. The first neuron at the bottom of the hierarchy would have a very small receptive field. The second neuron would have a slightly bigger receptive field. It would see a slightly blurred part of the image. The final neuron at the top of the

AGREEMENT WITH DATA FROM EXPERIMENTAL PSYCHOLOGY 325

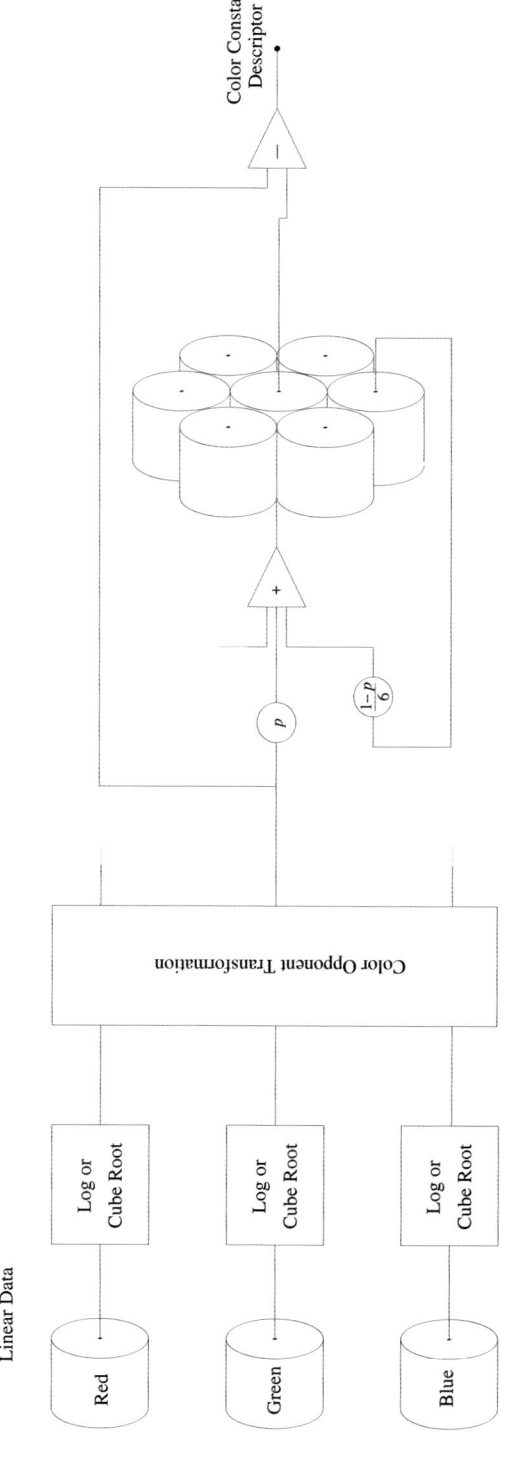

Figure 14.2 Likely description of the color constancy algorithm used by the human visual system to compute color constant descriptors. The parameter p is simply a small percentage as described in Section 10.2.

hierarchy would see a completely blurred image, essentially global space average color. Another suggestion made by Hurlbert (1986) is that rods in the periphery of the retina are used to take a spatial average over the image boundary. It has also been suggested that color constancy is due to an adaptation mechanism (D'Zmura and Lennie 1986). As the eye, head, and body move, a single region of the retina is exposed to different parts of the scene. The adapted state at any point of the retina would be a function of the space average color viewed by this point in the course of time. The experiments of Land and McCann (1971) using short exposure time with no change in the ability to perceive colors as constant suggest that color constancy is a result of computations that can be done by perceiving an image for only a fraction of a second.

Given that V4 is the first visual area where color constant cells have been found, it seems most likely that the algorithm of Land (1986a), Horn (1974), or Ebner (2003a, 2004b,c) is used by the visual system. The computation of local space average color using a hierarchy of neurons would require a much larger neural architecture. It would be possible that local differencing is used to implement a Laplace operator. Since the response of the photoreceptors is logarithmic (or nearly logarithmic), the result would be color constant descriptors (Herault 1996). The integration of the thresholded output would then be done in V4 using resistive coupled neurons. This would be the color constancy algorithm of Horn (1974). The second alternative would be that the actual color signals are averaged as described by Ebner (2002, 2004b,c). This has the advantage that no threshold has to be set. Local space average color would then be subtracted from the color of the given pixel. This type of architecture is illustrated in Figure 14.2.

The cones of the retina measure the light for the three color bands red, green, and blue. There is some dispute about whether the relationship between lightness and reflectance is logarithmic, as proposed by Faugeras (1979), or follows a cube root or square root relationship (Hunt 1957). In any case, as we have already discussed in Section 2.4 either one is a possible approximation. Note that the CIE $L^*a^*b^*$ color space (Colourware Ltd 2001; Glasser et al. 1958; International Commission on Illumination 1996) also uses a cube root transformation. Thus, the first processing step can probably be described by application of a logarithmic or other closely related function. Next, a coordinate transformation due to the color opponent cells follows. Now, the color space is described by a red–green, a blue–yellow, and a black–white axis. In this color space, local space average color is computed using interconnected neurons. Finally, at the last processing stage, local space average color is subtracted from the color at the current pixel. In my view, this is the most likely architecture that best describes human color constancy. At this point, however, it is still an educated guess. Future advances in brain research will hopefully tell us one day what mechanism is actually used by the human visual system.

15

Conclusion

Obtaining a solution to the problem of color constancy is highly important for object recognition. In particular, color-based object recognition can be improved by accurately estimating the object reflectances. Accurate color reproduction is also desirable for consumer photography. By understanding how human color constancy works, we can devise better algorithms for automatic color correction. In essence, we are interested in developing a computational theory of color constancy. A merely phenomenological description of color vision is not sufficient. In developing a computational theory of color constancy, care must be taken that the algorithm can be mapped to the neural hardware. A computational theory of color constancy should also be able to reproduce all known phenomenological results.

Starting with the visual system, some background on the theory of color information and color reproduction, we have looked at many different color constancy algorithms. At first sight, the algorithms appear to be highly diverse. However, quite frequently, simple assumptions are incorporated one way or another. For instance, the white patch retinex algorithm assumes that somewhere in the image there is a white patch that reflects all incident light. Another frequently made assumption is that, on average, the world is gray or that the illuminant can be approximated by a black-body radiator. The algorithms also differ in their computational complexity. Some algorithms can be reduced to a convolution followed by a point operation for each image pixel, while other algorithms require complex geometric computations. Apart from the basic color constancy algorithms, we have also covered learning color constancy as well as shadow removal and shadow attenuation. The method for computing intrinsic images based on the assumption that the illuminant can be approximated by a black-body radiator is particularly intriguing. The result is a single band image based only on the reflectance of the image pixels which is free from shadows.

Our own algorithms are based on the computation of local space average color, which is used to estimate the illuminant for each image pixel. This can be done by simply using a convolution. However, it also lends itself to a hardware implementation using either a grid of processing elements or even simpler, a resistive grid. A resistive grid can readily be mapped to a neural hardware. Once we have an estimate of the illuminant, we can use it to compute a color-corrected output image. The simplest method, dividing the output pixels by twice the local space average color, produces very good results when our goal is machine-based object recognition. It also automatically increases the brightness of pixels in dark areas of the image. If we seek to imitate human color perception, then algorithms that are based on local color shifts seem to be more appropriate.

We also considered anisotropic computation of local space average color. This method can be used if we have strong nonlinear illumination gradient in the image. Most algorithms require a number of parameters that have to be set before the algorithm can be used. The main parameter to be used for the computation of local space average color is the extent over which the image is to be blurred. If we are able to obtain a rough but qualitatively accurate estimate of the illuminant, then the dependence on the scaling parameter used for the image blur is reduced. Thus, computation of anisotropic local space average color is also a step toward a parameter free algorithm for color constancy based on local space average color.

The performance of the algorithms was evaluated using color-based object recognition. Algorithms that are based on the computation of local space average color performed much better than many of the other, often quite complicated, algorithms. Obtaining an accurate estimate of object reflectances is important from a machine vision point of view. If we are able to accurately estimate the actual color of objects, object recognition results improve considerably. However, another goal of color constancy algorithms is to mimic human color constancy. An important question in this respect is, which algorithm is used by the human visual system? In comparing the output computed by the algorithms with data from experimental psychology, it seems that the algorithm used by the human visual system is based on color shifts.

We evaluated all algorithms on a common data set from a variety of sources. The interested reader should have no problem choosing the color constancy algorithm that suits his or her needs best. No single color constancy algorithm gave best results on all image sets. However, we did see that algorithms based on local space average color and a simple scaling of the input pixels by local space average color produced very good results. Given the simplicity of the approach, use of these algorithms may be helpful in improving many machine vision tasks. The parallel algorithms may even be integrated directly into the imaging device.

Possible roads for future work include creating an algorithm that performs best on all image sets. Such an algorithm could be constructed by first analyzing the image data. Are there any specular surfaces? Are the objects mostly Lambertian reflectors? The problem is how to integrate all algorithms into one. A simple method would be to first analyze the image and then choose one of the algorithms and apply it to the image. However, an image may be composed of several different objects, i.e. a Lambertian reflector and a specular object. In this case, we would have to do the analysis locally and also estimate the illuminant locally. It would also be interesting to see if we can improve on the rough estimate of the illuminant which is used to compute anisotropic local space average color. It may also be possible to devise a way to adaptively find the scaling parameter that is optimal for the given image.

In summary, with this book, we have made a step toward a better understanding of a computational theory of color constancy. The main question, which algorithm is used by the human vision system, however, is still unsolved. In the future, noninvasive techniques that help to visualize the workings of the human brain in combination with additional psychophysical experiments may one day lead to the actual algorithm used by the visual system.

Appendix A

Dirac Delta Function

The Dirac delta function is not really a function because its value is not defined for all x. The delta function $\delta(x - x_0)$ can be considered to be the limit of the unit pulse with width 2ϵ and height $\frac{1}{2\epsilon}$ at the position $x = x_0$ as $\epsilon \to 0$ (Bronstein et al. 2001; Horn 1986).

$$\delta_\epsilon(x) = \begin{cases} \frac{1}{2\epsilon} & \text{for } |x| \leq \epsilon \\ 0 & \text{for } |x| > 0 \end{cases} \quad (A.1)$$

This sequence is illustrated in Figure A.1. The Dirac delta function can also be considered as the derivative of the unit step function $u(x)$.

$$u(x) = \begin{cases} 1 & \text{for } x > 0 \\ \frac{1}{2} & \text{for } x = 0 \\ 0 & \text{for } x < 0 \end{cases} \quad (A.2)$$

Consider the unit step function as the limit of the sequence $\{u_\epsilon(x)\}$.

$$u_\epsilon(x) = \begin{cases} 1 & \text{for } x > \epsilon \\ \frac{1}{2}\left(1 + \frac{x}{\epsilon}\right) & \text{for } |x| \leq \epsilon \\ 0 & \text{for } x < \epsilon \end{cases} \quad (A.3)$$

The derivative of this sequence is given by

$$\delta_\epsilon(x) = \frac{d}{dx} u_\epsilon(x) = \begin{cases} \frac{1}{2\epsilon} & \text{for } x \leq \epsilon \\ 0 & \text{for } |x| > \epsilon. \end{cases} \quad (A.4)$$

The area under the unit pulse is 1. Hence, we have

$$\int_{-\infty}^{\infty} \delta_\epsilon(x)\, dx = 1. \quad (A.5)$$

This sequence of functions $\{\delta_\epsilon\}$ defines the unit impulse or Dirac delta function.

$$\delta(x) = \lim_{\epsilon \to 0} \delta_\epsilon(x) \quad (A.6)$$

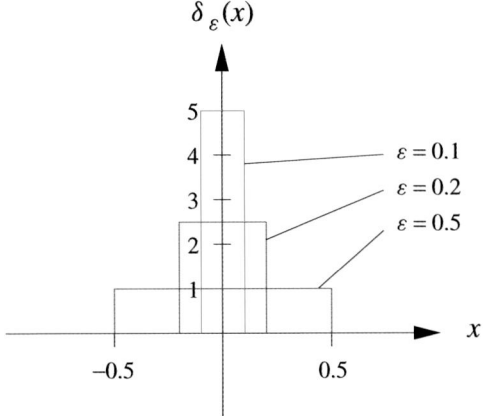

Figure A.1 Sequence of pulses.

The Dirac delta function is zero everywhere except for $x = 0$. We have,

$$\delta(x) = \lim_{\epsilon \to 0} \delta_\epsilon(x) = 0 \qquad \forall x \neq 0. \tag{A.7}$$

Let $f(x)$ be a continuous function. Then, we obtain

$$\int_{-\infty}^{\infty} f(x)\delta(x - x_0)\,dx = f(x_0). \tag{A.8}$$

Instead of using the unit step function, we can also view the delta function as the limit of a normal distribution.

$$\delta(x) = \lim_{\epsilon \to 0} \delta_\epsilon(x) = \lim_{\epsilon \to 0} \frac{1}{\epsilon\sqrt{2\pi}} e^{-\frac{x^2}{2\epsilon^2}} \tag{A.9}$$

Appendix B

Units of Radiometry and Photometry

A summary of symbols and units that are used in radiometry is shown in Table B.1. A summary of symbols and units that are used in photometry is shown in Table B.2. The photometric quantities use the same symbol except that the subscript v is added. A good introduction to the theory of radiometry is given by Poynton (2003), Jähne (2002), Horn (1986), and Levi (1993). See also International Commission on Illumination (1983). The radiometric units refer to physical quantities whereas the photometric units refer to perceptual quantities. In photometry, the physical quantities are weighted by the spectral sensitivity of the human visual system. Let $Q(\lambda)$ be the energy at wavelength λ. Then, the corresponding photometric quantity is given by

$$Q_v = 683.002 \frac{\text{lm}}{\text{W}} \int_{380 \text{ nm}}^{780 \text{ nm}} Q(\lambda) V(\lambda) \, d\lambda$$

where $V(\lambda)$ is the sensitivity function of photopic vision. Illuminance is obtained from irradiance by

$$E_v = 683.002 \frac{\text{lm}}{\text{W}} \int_{380 \text{ nm}}^{780 \text{ nm}} E(\lambda) V(\lambda) \, d\lambda$$

where $E(\lambda)$ denotes the irradiance at wavelength λ. Luminous exitance is given by

$$M_v = 683.002 \frac{\text{lm}}{\text{W}} \int_{380 \text{ nm}}^{780 \text{ nm}} M(\lambda) V(\lambda) \, d\lambda$$

where $M(\lambda)$ denotes the radiant exitance at wavelength λ. Luminous intensity is given by

$$I_v = 683.002 \frac{\text{lm}}{\text{W}} \int_{380 \text{ nm}}^{780 \text{ nm}} I(\lambda) V(\lambda) \, d\lambda$$

where $I(\lambda)$ denotes the radiant intensity at wavelength λ. Luminance is given by

$$L_v = 683.002 \frac{\text{lm}}{\text{W}} \int_{380 \text{ nm}}^{780 \text{ nm}} L(\lambda) V(\lambda) \, d\lambda$$

Color Constancy M. Ebner
© 2007 John Wiley & Sons, Ltd

UNITS OF RADIOMETRY AND PHOTOMETRY

Table B.1: Summary of radiometric units.

Name	Symbol	Unit
Radiant energy	Q	J
Radiant flux or power	$\Phi = P = \frac{dQ}{dt}$	W
Irradiance	$E = \frac{d\Phi}{dA}$	$\frac{W}{m^2}$
Radiant exitance	$M = \frac{d\Phi}{dA}$	$\frac{W}{m^2}$
Radiant intensity	$I = \frac{\Phi}{d\Omega}$	$\frac{W}{sr}$
Radiance	$L = \frac{d^2\Phi}{d\Omega\, dA \cos\theta}$	$\frac{W}{m^2\, sr}$

Table B.2: Summary of photometric units.

Name	Symbol	Unit
Luminous energy	Q_v	lm · s
Luminous flux	$\Phi_v = P_v = \frac{dQ_v}{dt}$	lumen = lm
Illuminance	$E_v = \frac{d\Phi_v}{dA}$	$\frac{lm}{m^2}$ = lux = lx
Luminous exitance	$M_v = \frac{d\Phi_v}{dA}$	$\frac{lm}{m^2}$ = lux = lx
Luminous intensity	$I_v = \frac{\Phi_v}{d\Omega}$	$\frac{lm}{sr}$ = candela = cd
Luminance	$L_v = Y = \frac{d^2\Phi_v}{d\Omega\, dA \cos\theta}$	$\frac{cd}{m^2}$ = nit

where $L(\lambda)$ denotes the radiance at wavelength λ.

Appendix C

Sample Output from Algorithms

The original input images and all the output images produced by the different algorithms are shown here for easy comparison. The reader should be aware that the printed colors that are displayed in this book may not be identical to the colors one would experience on a calibrated monitor. We have already discussed the difficulty of accurate color reproduction in depth in Chapter 4. When looking at the images in this chapter, please keep this in mind. Images may be slightly less saturated or slightly more saturated or the color balance may be slightly different from a calibrated output. Thus, the printed output should only be taken as indicative about the general performance of the algorithm but should not be taken as the absolute performance of the algorithm.

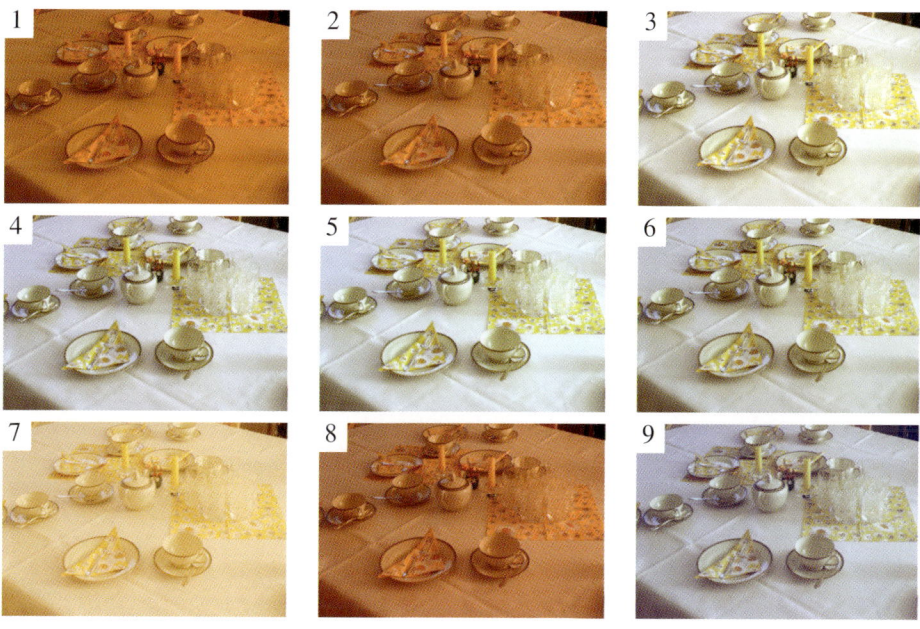

Color Constancy M. Ebner
© 2007 John Wiley & Sons, Ltd

SAMPLE OUTPUT FROM ALGORITHMS

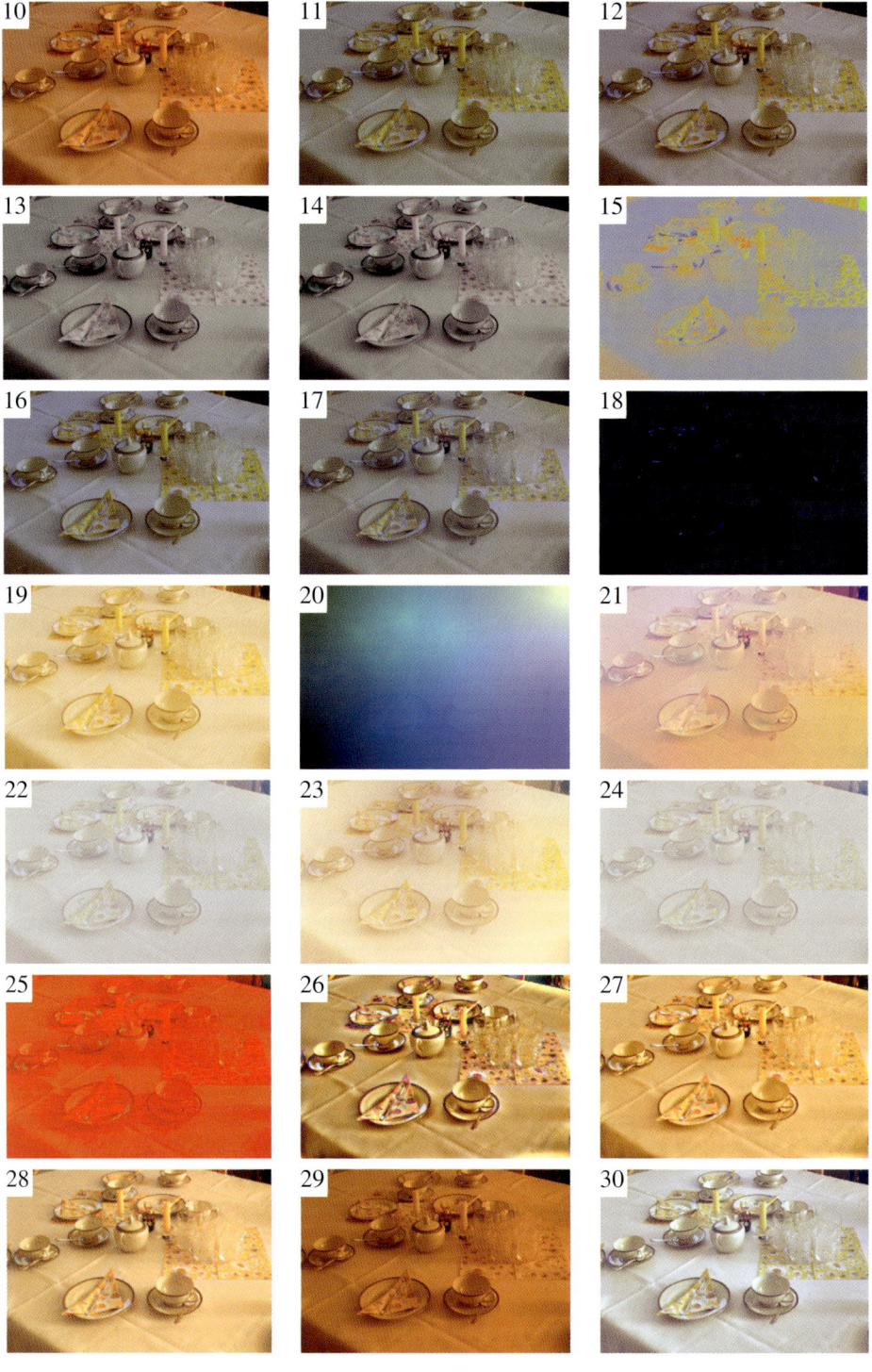

SAMPLE OUTPUT FROM ALGORITHMS

SAMPLE OUTPUT FROM ALGORITHMS

SAMPLE OUTPUT FROM ALGORITHMS

Legend

1. Original image.
2. White patch retinex algorithm, maximum value per channel.
3. White patch retinex algorithm, histogram.
4. Gray world assumption, averaging over all pixels.
5. Gray world assumption, averaging over segmented regions.
6. Gray world assumption, averaging over non-zero histogram buckets.
7. Variant of Horn's algorithm, normalization per band.
8. Variant of Horn's algorithm, white patch normalization.
9. Variant of Horn's algorithm, gray world normalization.
10. Gamut-constraint algorithm 3D.
11. Gamut-constraint algorithm 2D, no constraint placed on the illuminant, maximum trace selected.
12. Gamut-constraint algorithm 2D, illuminant assumed to be a black-body radiator.
13. Color cluster rotation, original average intensity is maintained.
14. Color cluster rotation, color cloud is centered and then rescaled to fill color cube.

15. Comprehensive color normalization, normalized colors.
16. Comprehensive color normalization, colors with original lightness.
17. Comprehensive color normalization using local space average color.
18. Color constancy using a dichromatic reflection model.
19. Horn's algorithm with threshold 0.
20. Horn's algorithm with threshold 6.
21. Blake's algorithm with threshold 3.
22. Moore et al.'s hardware implementation.
23. Moore et al.'s extended algorithm.
24. Color correction on multiple scales, no gain factor.
25. Color correction on multiple scales, gain factor = 5.
26. Homomorphic filtering, Parker's filter.
27. Homomorphic filtering, own filter.
28. Homomorphic filtering, perceptual color space.
29. Intrinsic color image, minimum entropy.
30. Local space average color, scaling of color channels.
31. Local space average color, color shifts.
32. Local space average color, normalized color shifts.
33. Local space average color, color shifts, increased saturation.
34. Local space average color, normalized color shifts, increased saturation.
35. Local space average color, combined white patch retinex and gray world assumption, linear output.
36. Local space average color, combined white patch retinex and gray world assumption, sigmoidal output.
37. Anisotropic local space average color, scaling of color channels.
38. Anisotropic local space average color, color shifts.
39. Anisotropic local space average color, normalized color shifts.

Appendix D

Image Sets

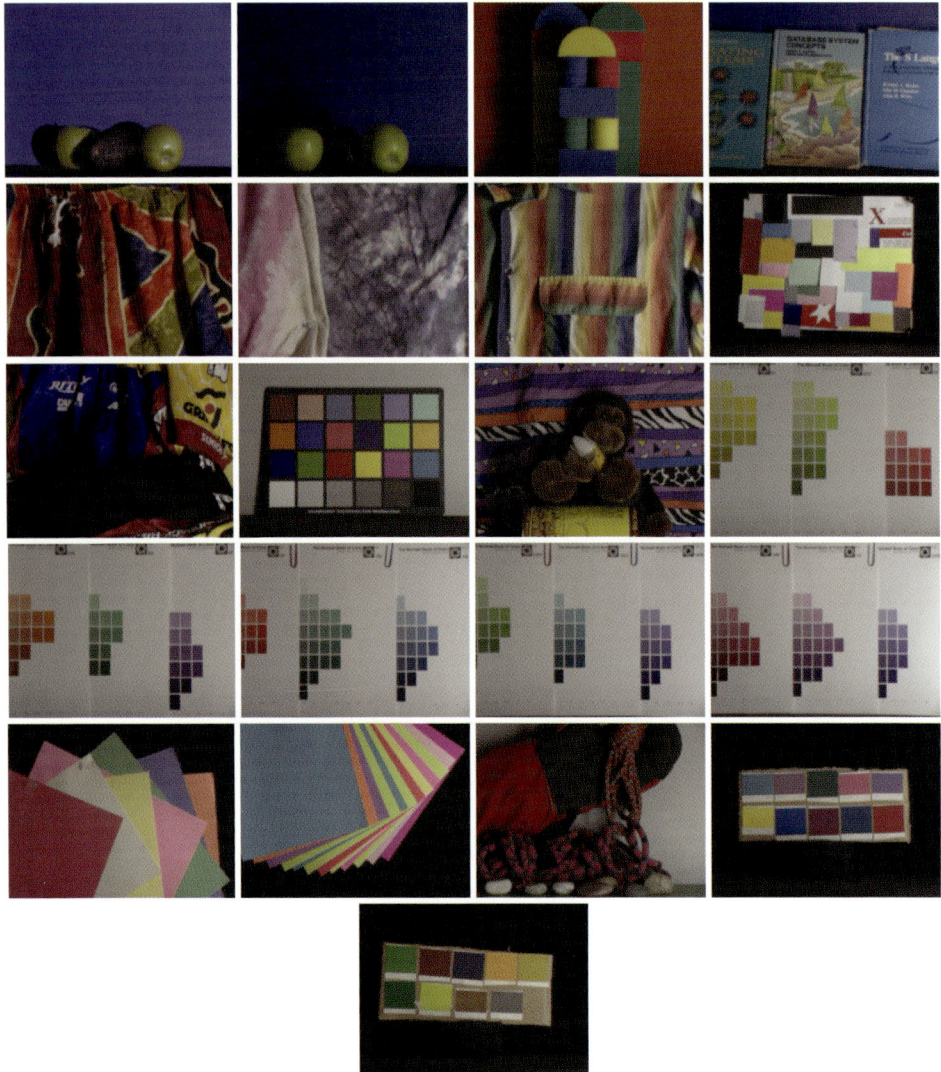

Figure D.1 Image set 1 (lambertian objects). (Original image data from "Data for Computer Vision and Computational Colour Science" made available through http://www.cs.sfu.ca/~colour/data/index.html. See Barnard K, Martin L, Funt B and Coath A 2002 A data set for color research, Color Research and Application, Wiley Periodicals, 27(3), 147–151. Reproduced by permission of Kobus Barnard.)

Figure D.2 Image set 2 (metallic objects). (Original image data from "Data for Computer Vision and Computational Colour Science" made available through http://www.cs.sfu.ca/~colour/data/index.html. See Barnard K, Martin L, Funt B and Coath A 2002 A data set for color research, Color Research and Application, Wiley Periodicals, 27(3), 147–151. Reproduced by permission of Kobus Barnard.)

Figure D.3 Image set 3 (specular objects). (Original image data from "Data for Computer Vision and Computational Colour Science" made available through http://www.cs.sfu.ca/~colour/data/index.html. See Barnard K, Martin L, Funt B and Coath A 2002 A data set for color research, Color Research and Application, Wiley Periodicals, 27(3), 147–151. Reproduced by permission of Kobus Barnard.)

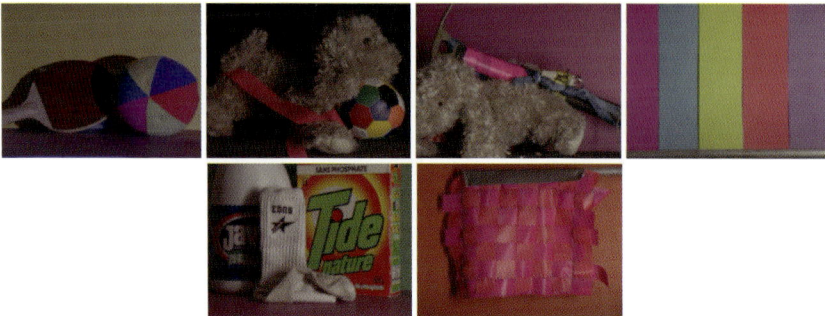

Figure D.4 Image set 4 (fluorescent objects). (Original image data from "Data for Computer Vision and Computational Colour Science" made available through http://www.cs.sfu.ca/~colour/data/index.html. See Barnard K, Martin L, Funt B and Coath A 2002 A data set for color research, Color Research and Application, Wiley Periodicals, 27(3), 147–151. Reproduced by permission of Kobus Barnard.)

Figure D.5 Image set 5 (objects in different positions). (Original image data from "Data for Computer Vision and Computational Colour Science" made available through http://www.cs.sfu.ca/~colour/data/index.html. See Barnard K, Martin L, Funt B and Coath A 2002 A data set for color research, Color Research and Application, Wiley Periodicals, 27(3), 147–151. Reproduced by permission of Kobus Barnard.)

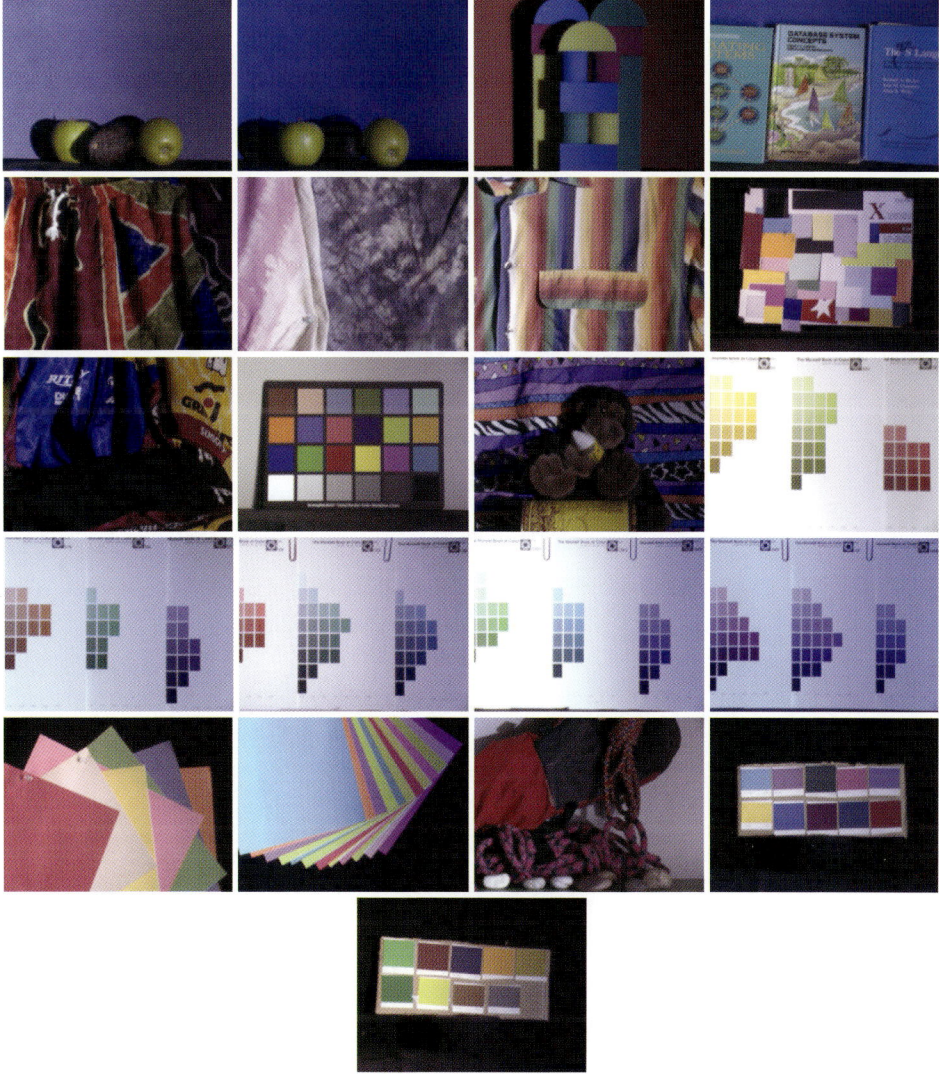

Figure D.6 Image set 6 (smooth illumination gradient). These images are derived from image set 1.

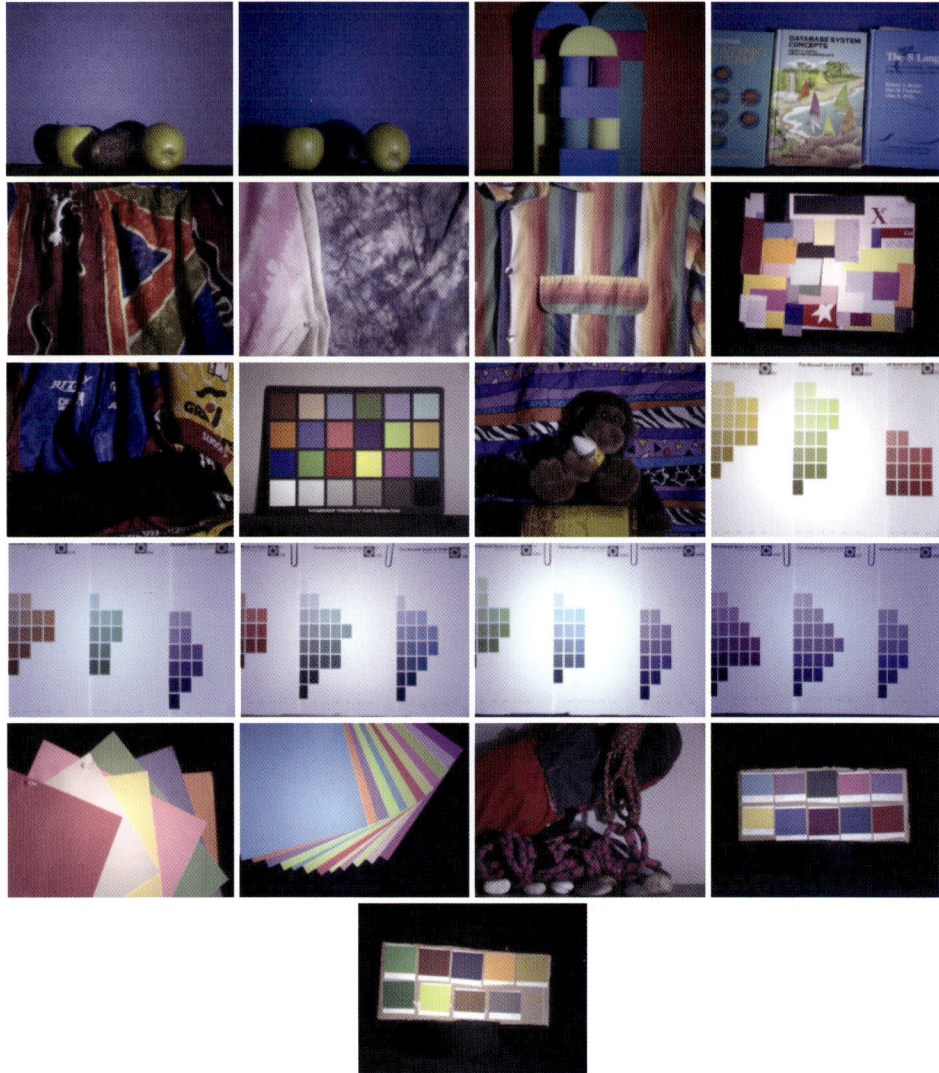

Figure D.7 Image set 7 (smooth circular illumination gradient). These images are derived from image set 1.

IMAGE SETS

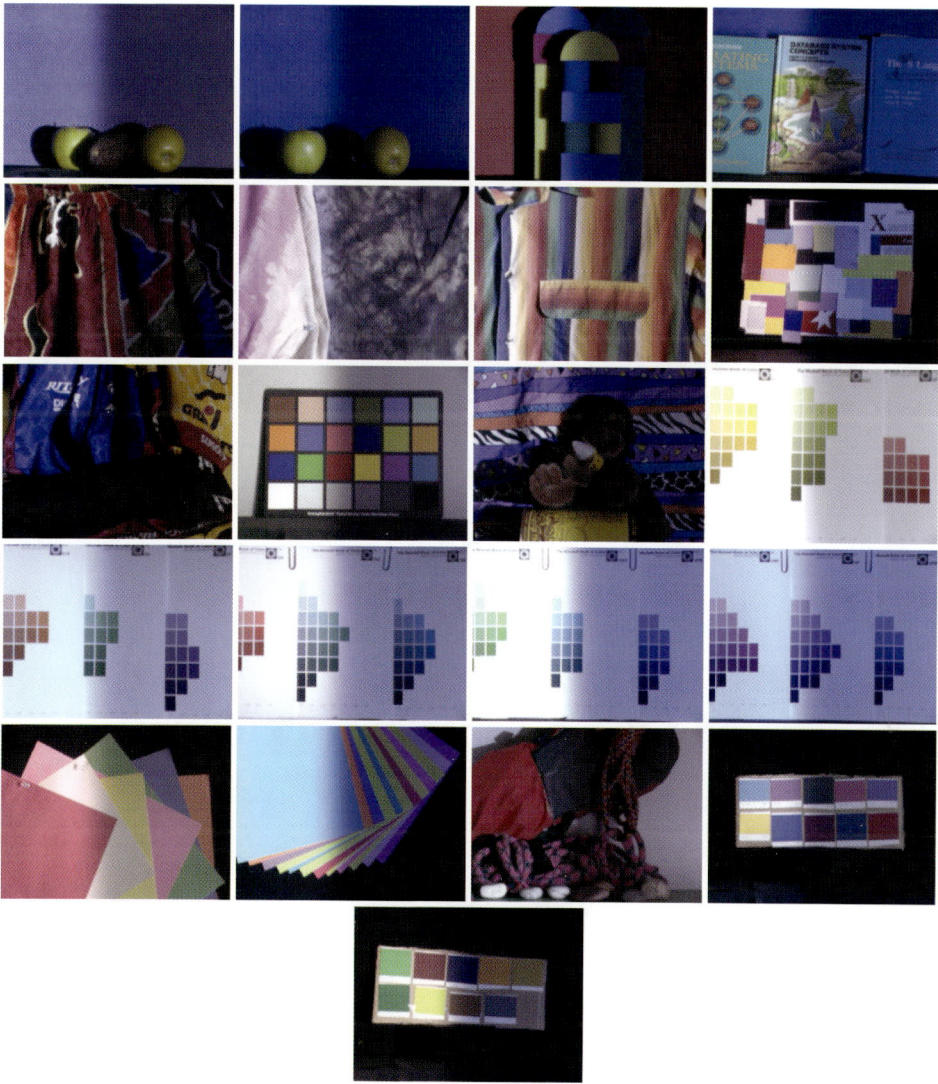

Figure D.8 Image set 8 (sharp illumination gradient). These images are derived from image set 1.

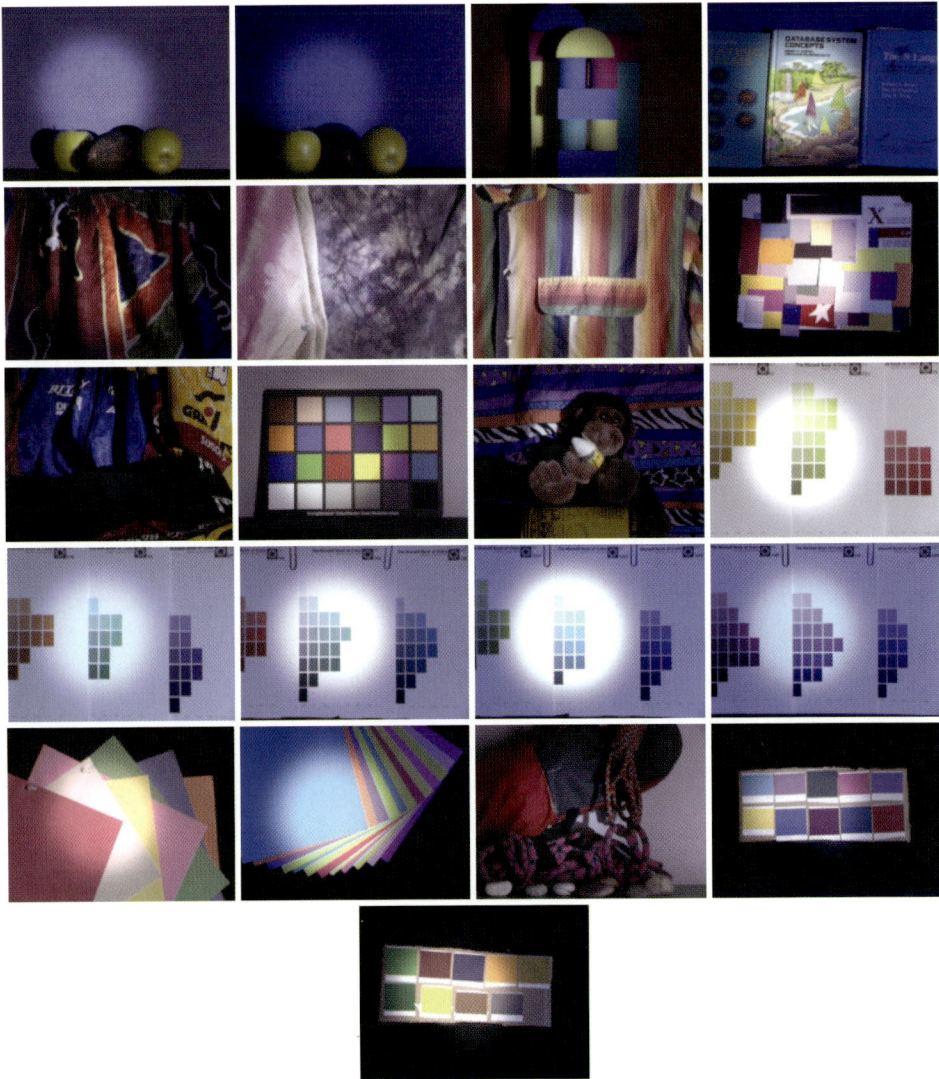

Figure D.9 Image set 9 (sharp circular illumination gradient). These images are derived from image set 1.

IMAGE SETS

Figure D.10 Image set 10 (natural scenes with illumination gradients). The images were taken with an analog camera and were digitized when the film was developed.

Figure D.11 Image set 11 (natural scenes with illumination gradients). The images were taken with an analog camera. A Kodak Photo CD was made from the negatives.

Figure D.12 Image set 12 (natural scenes with illumination gradients). The images were taken with an analog camera and digitized using a calibrated Canon FS4000US film scanner.

Figure D.13 Image set 13 (natural scenes with illumination gradients). The images were taken with a digital camera (Canon EOS 10D).

Appendix E

Program Code

All of the algorithms in the subsequent text are described in pseudocode. Note that this code will not compile as stated. It is meant to illustrate how the algorithms work. In order to get a code that will compile using C++, Java, or any other programming language, you will have to fill in the necessary details. The code shown here is meant to be as simple as possible in order to show what the algorithms do and how they work. Operators such as addition (+), subtraction (−), multiplication (*), or division (/) are understood to work with the individual components of the image. When the code is OutputImage=InputImage/maxValue and maxValue is a scalar value, it is meant that

$$\forall_{x,y,b} \text{OutputImage[x,y,b]=InputImage[x,y,b]/maxValue} \quad (E.1)$$

where the parameters x and y refer to image coordinates and the parameter b refers to the image band. When the code is OutputImage=InputImage*m and m is a three-element vector, it is meant that

$$\forall_{x,y,b} \text{OutputImage[x,y,b]=InputImage[x,y,b]*m[b]}. \quad (E.2)$$

Also, some optimizations can be made in order to speed up the code. For instance, instead of sorting all image values as is done in the pseudocode of the white patch retinex algorithm or the pseudocode of the gray world assumption, one may also compute a histogram and then compute the clip values from the histogram. In order to simplify the code further, some variables as well as functions are assumed to be common to all algorithms. Error checking has been omitted.

Image values are taken to be from the range [0, 1], i.e. we have

$$0 \leq \text{InputImage[x,y,b]} \leq 1. \quad (E.3)$$

Image values are assumed to be linear. If you process image data where the image is stored using the sRGB color space, you will need to linearize the data as described in Chapter 5. Similarly, the computed output image OutputImage will also contain linear

Color Constancy M. Ebner
© 2007 John Wiley & Sons, Ltd

values. These values have to be converted to the appropriate color space, i.e. sRGB, before display. After the output image is computed, output values may be outside the range [0, 1]. If this is the case, such values are simply clipped to the range [0, 1]. In some cases, the color of the illuminant will be estimated from the input image. This estimate will be denoted by `EstIll`.

The table in the subsequent text shows a list of symbols that are common to all algorithms.

Symbol	Meaning
`InputImage`	Original image data
`nPixel`	Number of pixels in `InputImage`
`EstIll`	Estimated illuminant
`OutputImage`	Output image
`Image[b]`	Refers to band b of image `Image`
`c[b]`	Refers to band b of color c

Definition of Functions

```
// transform image values to range [min,max]
function Transform(Image,min,max) {
  min=minimum over { Image[x,y,b] }
  max=maximum over { Image[x,y,b] }
  for all (x,y,b) do {
    Image[x,y,b]=(Image[x,y,b]-min)/(max-min)
  }
  return Image
}

// transform image values to full range per band
function FullRangePerBand(Image) {
  for all (b ∈ {r, g, b}) do {
    minValue=get minimum value of Image[b]
    maxValue=get maximum value of Image[b]
    Image[b]=Transform(Image[b],minValue,maxValue)
  }
  return Image
}

// copy lightness from original image to color constant image
function CopyLightness(Image,InputImage) {
  wr=0.2125; wg=0.7154; wb=0.0721
  for all (x,y) do {
    Lold=wr*Image[x,y,0]+wg*Image[x,y,1]+wb*Image[x,y,2]
```

```
    Lnew=wr*InputImage[x,y,0]+wg*InputImage[x,y,1]
        +wb*InputImage[x,y,2]
    OutputImage[x,y]=(Lnew/Lold)*Image[x,y]
  }
  return OutputImage
}

// normalize colors
function NormalizeColors(Image) {
  for all (x,y) do {
    Image[x,y]=Image[x,y]/
              (Image[x,y,0]+Image[x,y,1]+Image[x,y,2])
  }
  return Image
}

// convolve image with Gaussian using scale sigma
// G(x,y)=exp(-(x*x+y*y)/(2*sigma*sigma))
function ConvolveImageWithGaussian(Image,sigma) {
  for all (x,y) do {
    k(x,y)=1/\(∫∫b(x, y) dx dy)
                where b(x,y)=G(x,y) if (x,y) inside image
                      b(x,y)=0 if (x,y) outside image
    a[x,y]=k(x, y) ∫∫Image[x',y']G(x − x', y − y') dx' dy'
  }
  return a
}
```

Full Range per Band

Parameter	Description
—	—

```
for all (b ∈ {r, g, b}) do {
  minValue=get minimum value of InputImage[b]
  maxValue=get maximum value of InputImage[b]
  EstIll[b]=maxValue-minValue
  OutputImage[b]=Transform(InputImage[b],minValue,maxValue)
}
```

White Patch Retinex

Parameter	Description
pBlack	Percentage from below
pWhite	Percentage from above

```
for all (b ∈ {r, g, b}) do {
  SortedValues=sort values of InputImage[b] in ascending order
  minValue=SortedValues[nPixel*pBlack]
  maxValue=SortedValues[nPixel*(1-pWhite)]
  EstIll[b]=maxValue-minValue
  OutputImage[b]=Transform(InputImage[b],minValue,maxValue)
}
```

Gray World Assumption

Parameter	Description
pWhite	Percentage from above

```
a=∑_{x,y} InputImage[x,y]
EstIll=a
OutputImage=InputImage/a
SortedValues=sort values of OutputImage in ascending order
maxValue=SortedValues[3*nPixel*(1-pWhite)]
OutputImage=OutputImage/maxValue
```

Simplified Horn

Parameter	Description
—	—

```
TmpImage=log(InputImage*255+1)
TmpImage=FullRangePerBand(TmpImage)
TmpImage=exp(TmpImage)
OutputImage=FullRangePerBand(TmpImage)
```

Gamut Constraint 3D

Parameter	Description
CalibImage	Image of a calibration target under a canonical illuminant

```
// Let CalibImage be an image showing a calibration target
// under a canonical illuminant
if (CalibImage exists)
  CanonicalGamut=convex hull of colors in CalibImage
else
  CanonicalGamut=Unit Cube
ObservedGamut=convex hull of colors in InputImage
MapHull=∩_{v∈ObservedGamut} M(v)
       where M(v) = {v_c/v | v_c ∈ CanonicalGamut}
```

```
m=select map from MapHull, e.g. map with maximum trace
for all (x,y) do {
  OutputImage[x,y]=InputImage[x,y]*m
}
```

Gamut Constraint 2D

Parameter	Description
CalibImage	Image of a calibration target under a canonical illuminant

```
// Let CalibImage be an image showing a calibration target
// under a canonical illuminant
// Colors are projected onto the plane b=1
if (CalibImage exists)
  CanonicalGamut=2D convex hull of projected colors
                 from CalibImage
else
  CanonicalGamut=2D convex hull of projected unit cube
ObservedGamut=convex hull of projected colors from InputImage
MapHull=⋂v∈ObservedGamut M(v)
```

$$\text{where} \mathcal{M}(v) = \{v_c/v | v_c \in \text{CanonicalGamut}\}$$

```
m=select map from MapHull which lies on the curve of the
  black body radiator and has maximum trace
for all (x,y) do {
  c=InputImage[x,y]
  OutputImage[x,y]=[c[0]*m[0],c[1]*m[1],c[2]]
}
OutputImage=CopyLightness(OutputImage,InputImage)
```

Color Cluster Rotation

Parameter	Description
—	—

```
a=∑x,y InputImage[x,y]
```
$C = E\left[(c-a)(c-a)^T\right]$ // compute covariance matrix

```
lambda=compute largest eigenvalue of C
e=compute largest eigenvector of C
theta=compute angle between e and white vector
axis=compute axis of rotation
     (cross product between e and white vector)
for all (x,y) do {
  c=InputImage[x,y]
```

```
// Rot(axis,theta,c) rotates vector c around axis by theta
  c=Rot(axis,theta,c-a)/sqrt(lambda)+[0.5,0.5,0.5]
  OutputImage[x,y]=c
}
```

Comprehensive Color Normalization

Parameter	Description
—	—

```
TmpImage=InputImage
while (not yet converged) do {
  TmpImage=NormalizeColors(TmpImage)
```
$a=\sum_{x,y}$ `TmpImage[x,y]`
```
  TmpImage=TmpImage/(3*a)
}
OutputImage=CopyLightness(TmpImage,InputImage)
```

Risson (2003)

Parameter	Description
threshold	Threshold for segmentation
nLimit	Minimum number of pixels per region
minSaturation	Minimum saturation for average of region

```
Regions=segment InputImage into regions using threshold
Regions=Regions except for those regions which have less than
        nLimit pixels or which have less than minSaturation
Lines={}
for all (R ∈ Regions) do {
  // compute covariance matrix
  a=average of pixels in R
```
$C=E\left[(\hat{\mathbf{c}}-\hat{\mathbf{a}})(\hat{\mathbf{c}}-\hat{\mathbf{a}})^T\right]$ with $\mathbf{c} \in R$
```
  e=compute largest eigenvector of C
  Lines=Lines∪ e
}
SetOfIntersections={ (x,y) |
              lines e1, e2 ∈ Lines intersect at (x,y) }
p=[ median of SetOfIntersections[0],
    median of SetOfIntersections[1] ]
EstIll=[p[0],p[1],1-p[0]-p[1]]
OutputImage=InputImage/EstIll
OutputImage=Transform(OutputImage,0,max)
          where max is maximum over all bands
```

Horn (1974)/Blake (1985)

Parameter	Description
threshold	Threshold used to decide whether a change in pixel value is due to a change in reflectance or due to a change of the illuminant

```
InputImage=255*InputImage+1
TmpImage=log(InputImage)
// compute local differences and apply threshold
for all (x,y) do {
  // check for boundary conditions has been omitted here
  dxImage[x,y]=TmpImage[x+1,y]-TmpImage[x,y]
  dyImage[x,y]=TmpImage[x,y+1]-TmpImage[x,y]
  for all (b ∈ {r,g,b}) do {
    if (|InputImage[x+1,y,b]-InputImage[x,y,b]|<threshold)
      dxImage[x,y,b]=0
    if (|InputImage[x,y+1,b]-InputImage[x,y,b]|<threshold)
      dyImage[x,y,b]=0
  }
}
// compute laplacian from differences
for all (x,y) do {
  Laplacian[x,y]=dxImage[x,y]-dxImage[x-1,y]
              +dyImage[x,y]-dyImage[x,y-1]
}
// solve poisson equation
TmpImage=0
while (not yet converged) do {
  for all (x,y) do {
    TmpImage[x,y]=(TmpImage[x-1,y]+TmpImage[x+1,y]
                +TmpImage[x,y-1]+TmpImage[x+1,y+1]
                -Laplacian[x,y])/4
  }
}
TmpImage=FullRangePerBand(TmpImage)
TmpImage=exp(TmpImage)
OutputImage=FullRangePerBand(TmpImage)
```

Moore *et al.* (1991) Retinex

Parameter	Description
sigma	Scale of Gaussian

```
InputImage=255*InputImage+1
InputImage=log(InputImage)
a=ConvolveWithGaussian(InputImage,sigma)
```

```
OutputImage=InputImage-a
OutputImage=Transform(OutputImage,min,max) where min and max
         is the minimum and maximum over all bands
```

Moore *et al.* (1991) Extended

Parameter	Description
sigma	Scale of Gaussian

```
InputImage=255*InputImage+1
InputImage=log(InputImage)
a=ConvolveWithGaussian(InputImage,sigma)
for all (x,y) do {
  EdgeImage[x,y]=(|InputImage[x,y]-InputImage[x-1,y]|+
                  |InputImage[x,y]-InputImage[x,y-1]|+
                  |InputImage[x,y]-InputImage[x+1,y]|+
                  |InputImage[x,y]-InputImage[x,y-1]|)/4
}
EdgeImage=ConvolveWithGaussian(EdgeImage,sigma)
EdgeImage=Transform(OutputImage,0,max)
       where max is maximum over all bands
OutputImage=InputImage-(a*EdgeImage)
OutputImage=Transform(OutputImage,min,max) where min and max
         is the minimum and maximum over all bands
```

Rahman et al. (1999)

Parameter	Description
sigmaList[]	Scale of Gaussians

```
InputImage=255*InputImage+1
n=0
for all (sigma∈sigmaList) do {
  a=ConvolveWithGaussian(InputImage,sigma)
  OutputImage=log(InputImage)-log(a)
  n++
}
OutputImage/=n
OutputImage=Transform(OutputImage,min,max) where min and max
         is the minimum and maximum over all bands
```

Homomorphic Filtering

Parameter	Description
pWhite	Percentage from above

PROGRAM CODE

```
InputImage=255*InputImage+1
InputImage=log(InputImage)
```
Image=$\frac{1}{\sqrt{nPixel}} \sum_x \sum_y$ InputImage$[x,y]e^{-2\pi i \omega x} e^{-2\pi i v y}$
```
for all (x,y,b) do {
  Image[x,y,b]=filter(x,y,b,Image[x,y,b]) // filter Image
}
```
OutputImage=$\frac{1}{\sqrt{nPixel}} \sum_x \sum_y$ Image$[x,y]e^{2\pi i \omega x} e^{2\pi i v y}$
```
OutputImage=exp(OutputImage)
SortedValues=sort values of OutputImage in ascending order
maxValue=SortedValues[3*nPixel*(1-pWhite)]
OutputImage=OutputImage/maxValue
```

Homomorphic Filtering (HVS)

Parameter	Description
pWhite	Percentage from above

```
InputImage=255*InputImage+1
InputImage=log(InputImage)
```

$$\text{InputImage} = \begin{pmatrix} 0.612 & 0.369 & 0.019 \\ 1 & -1 & 0 \\ 1 & 0 & -1 \end{pmatrix} * \text{InputImage}$$

Image=$\frac{1}{\sqrt{nPixel}} \sum_x \sum_y$ InputImage$[x,y]e^{-2\pi i \omega x} e^{-2\pi i v y}$
```
for all (x,y) do {
  Image[x,y,1]=filter(x,y,1,Image[x,y,1]) // filter Image
  Image[x,y,2]=filter(x,y,2,Image[x,y,2]) // filter Image
}
```
OutputImage=$\frac{1}{\sqrt{nPixel}} \sum_x \sum_y$ Image$[x,y]e^{2\pi i \omega x} e^{2\pi i v y}$

$$\text{OutputImage} = \begin{pmatrix} 1 & 0.369 & 0.019 \\ 1 & -0.631 & 0.019 \\ 1 & 0.369 & -0.981 \end{pmatrix} * \text{OutputImage}$$

```
OutputImage=exp(OutputImage)
SortedValues=sort values of OutputImage in ascending order
maxValue=SortedValues[3*nPixel*(1-pWhite)]
OutputImage=OutputImage/maxValue
```

Intrinsic (Minimal Entropy)

Parameter	Description
—	—

```
InputImage=255*InputImage+1
for all (x,y) do {
  c=InputImage[x,y]
  gm=pow(c[0]*c[1]*c[2],1.0/3.0)
```
$$\text{ChiImage}[x,y] = \begin{pmatrix} \sqrt{\frac{2}{3}} & -\sqrt{\frac{1}{6}} & -\sqrt{\frac{1}{6}} \\ 0 & -\sqrt{\frac{1}{2}} & \sqrt{\frac{1}{2}} \end{pmatrix} \begin{bmatrix} \log(c[0]/gm) \\ \log(c[1]/gm) \\ \log(c[2]/gm) \end{bmatrix}$$
```
}
minEntropy=infinity
for all (angle ∈ [0, 180]) do {
  IntrinsicImage=dotProduct(ChiImage,[cos(angle),sin(angle)])
  entropy=−∑_g p(g) ln p(g)  where p(g) is the probability
          that g occurs in IntrinsicImage
  if (entropy<minEntropy) do {
    minEntropy=entropy
    minAngle=angle
  }
}

IntrinsicImage=
        dotProduct(ChiImage,[cos(minAngle),sin(minAngle)])
offset=compute extra light shift using data from ChiImage
x0=offset*sin(minAngle)
y0=-offset*cos(minAngle)
for all (x,y) do {
  [chi1,chi2]=ChiImage[x,y]
  chi=chi1*cos(minAngle)+chi2*sin(minAngle)
  chi1=x0+chi*cos(minAngle)
  chi2=y0+chi*sin(minAngle)
```
$$\text{OutputImage}[x,y] = \exp\left(\begin{pmatrix} \sqrt{\frac{2}{3}} & 0 \\ -\sqrt{\frac{1}{6}} & -\sqrt{\frac{1}{2}} \\ -\sqrt{\frac{1}{6}} & \sqrt{\frac{1}{2}} \end{pmatrix} \begin{bmatrix} chi1 \\ chi2 \end{bmatrix} \right)$$
```
}
NormalizeColors(OutputImage)
OutputImage=CopyLightness(OutputImage,InputImage)
```

PROGRAM CODE

Local Space Average Color (Convolution)

Parameter	Description
sigma	Scale of Gaussian

```
a=ConvolveWithGaussian(InputImage,sigma)
OutputImage=InputImage/(2*a)
```

Local Space Average Color (Parallel Version)

Parameter	Description
p	Small percentage

```
// xSize and ySize is the size of the image
while (not yet converged) do {
  for all (x,y) do {
    c=InputImage[x,y]
    avg=a[x,y]
    n=1
    if (x>0)       { avg+=a[x-1,y]; n++ }
    if (x<xSize-1) { avg+=a[x+1,y]; n++ }
    if (y>0)       { avg+=a[x,y-1]; n++ }
    if (y<ySize-1) { avg+=a[x,y+1]; n++ }
    avg/=n
    a[x,y]=c*p+avg*(1-p)
    OutputImage[x,y]=c/(2*avg)
  }
}
```

Anisotropic Local Space Average Color

Parameter	Description
pIsotropic	Small percentage for isotropic local space average color
pAnisotropic	Small percentage for anisotropic local space average color
q	Parameter that adjusts between isotropic and anisotropic averaging

```
while (not yet converged) do {
  for all (x,y) do {
    // compute isotropic local space average color
    // check for boundary conditions has been omitted here
    // see parallel algorithm for isotropic local
    // space average color
    c=InputImage[x,y]
    avg=(1/5)(IsoImage[x,y]+ IsoImage[x-1,y]+IsoImage[x+1,y]+
                             IsoImage[x,y-1]+IsoImage[x,y+1])
```

```
    IsoImage[x,y]=c*pIsotropic+avg*(1-pIsotropic)

    // compute estimate of illuminant
    F=IsoImage[0]+IsoImage[1]+IsoImage[2]

    // compute curvature
```
$\text{dx}= F_x = \frac{\partial}{\partial x} F$
$\text{dy}= F_y = \frac{\partial}{\partial x} F$
$F_{xy} = \frac{\partial}{\partial x} F_y$
$F_{yx} = \frac{\partial}{\partial y} F_x$
$F_{xx} = \frac{\partial}{\partial x} F_x$
$F_{yy} = \frac{\partial}{\partial y} F_y$

$$r = \left| \frac{(F_x^2+F_y^2)^{3/2}}{F_x F_{xy} F_y + F_x F_{yx} F_y - F_x^2 F_{yy} - F_y^2 F_{xx}} \right|$$

$[\text{dx}, \text{dy}] = [\text{dx}, \text{dy}] / (\sqrt{\text{dx}^2 + \text{dy}^2})$

```
       dxLeft=dxRight=
```
$\frac{1}{2r}$
```
    dyRight=sqrt(1-dxRight*dxRight)
    dyLeft=-dyRight
    rotate (dxLeft,dyLeft) and (dxRight,dyRight) in direction
       of gradient depending on direction of curvature

    // compute anisotropic local space average color
    // check for boundary conditions has been omitted here
    avg=AnIsoImage[x,y]
    avgFront=AnIsoImage[x+dx,y+dy]
    avgBack =AnIsoImage[x-dx,y-dy]
    avgLeft =AnIsoImage[x+dxLeft,y+dyLeft]
    avgRight=AnIsoImage[x+dxRight,y+dyRight]
    avg=(0.5-q)*avgLeft+(0.5-q)*avgRight+q*avgFront+q*avgBack
    avg=(1/3)*AnIsoImage[x,y]+(2/3)*a
    avg=c*pAnisotropic+a*(1-pAnisotropic)
    AnIsoImage[x,y]=avg

    OutputImage[x,y]=InputImage[x,y]/(2*avg)
  }
}
```

Combined White Patch Retinex and Local Space Average Color

Parameter	Description
p	Small percentage for isotropic local space average color
pd	Small percentage for attenuation of difference
s	Scaling of sigmoidal activation function

```
while (not yet converged) do {
  for all (x,y) do {
    // check for boundary conditions has been omitted here
    c=InputImage[x,y]
    avg=(1/5)(a[x,y]+a[x-1,y]+a[x+1,y]+a[x,y-1]+a[x,y+1])
    avg=c*p+avg*(1-p)
    a[x,y]=avg
    d=max {|c-avg|, DiffImage[x-1,y], DiffImage[x+1,y],
                    DiffImage[x,y-1], DiffImage[x,y+1] }
    d=(1-pd)*d
    DiffImage[x,y]=d
    o=(c-avg)/d
    OutputImage[x,y]=1/(1+exp(-o/s))
  }
}
```

Comprehensive Local Space Average Color

Parameter	Description
sigma	Scale of Gaussian

```
TmpImage=InputImage
while (not yet converged) do {
  TmpImage=NormalizeColors(TmpImage)
  a=ConvolveWithGaussian(TmpImage,sigma)
  TmpImage=TmpImage/(3*a)
}
OutputImage=CopyLightness(TmpImage,InputImage)
```

Appendix F

Parameter Settings

Algorithm	Parameter Settings
Full range per band	Simply transforms each color band to the range [0, 1].
White patch retinex	Described in Section 6.1. The white point is set at 4% per channel. A histogram quantization of 256 per channel is used. (pBlack=0, pWhite=0.04)
Gray world assumption	Described in Section 6.2. No segmentation is used. The factor f was estimated from the image by rescaling all the channels such that 0.02 of the pixels are clipped. (pWhite=0.02)
Simplified Horn	Described in Section 6.3. Zeros were removed from the input by transforming each channel with data in the range [0, 1] according to $y = (255x + 1)/256$. The logarithm is applied and the color bands are rescaled to the range [0, 1]. Then, the exponential function is applied and the result is again rescaled to the range [0, 1].
Gamut constraint 3D	Described in Section 6.4. If an empty intersection is created, the convex hulls are increased by a small amount. This process is repeated until a nonempty intersection is achieved. A histogram quantization of 256 per channel is used. (CalibImage={})
Gamut constraint 2D	Described in Section 6.5. If an empty intersection is created, the convex hulls are increased by a small amount. This process is repeated until a nonempty intersection is achieved. A histogram quantization of 256 per channel is used. Illuminants were constrained to lie on the curve of the black-body radiator. (CalibImage={})

Color cluster rotation	Described in Section 6.6. The cloud of points is aligned with the gray vector, positioned at the center of the RGB cube, and rescaled to fill most of the cube using the inverse of the square root of the largest eigenvalue as a scaling factor.
Comprehensive normalization	Described in Section 6.7. The lightness of the colors in the original image is used to scale the output colors to add back information about shading.
Risson (2003)	The segmentation threshold was set to 0.05. A minimum of five pixels were required for each segmented region. Regions with a saturation less than 0.12 were excluded. The curve of the black-body radiator was not used. The intersections were computed in RGB chromaticity space. All intersections between dichromatic lines were computed and the median, separately for the x- and y-direction, was taken as the position of the illuminant. (threshold=0.05, nLimit=5, minSaturation=0.12)
Horn (1974) / Blake (1985)	Described in Section 7.2. Zeros were removed from the input by transforming each channel with data in the range $[0, 1]$ according to $y = (255x + 1)/256$. The extensions of Blake (1985) are used, i.e. the threshold is applied after computing the first derivative. The threshold is set to 3. The inverse of the Laplacian is computed using an SOR iterative solver. (threshold=3)
Moore et al. (1991) retinex	Described in Section 7.3. Zeros were removed from the input by transforming each channel with data in the range $[0, 1]$ according to $y = (255x + 1)/256$. The logarithm is applied to the input signal. Local space average color is computed using a convolution with $e^{\frac{r^2}{2\sigma^2}}$ and $\sigma = 0.0932 n_{max}$ where $n_{max} = \max\{n_x, n_y\}$ and n_x and n_y is the width and height of the input image. The extensions to reduce color inductions across edges are not used. (sigma=σ)
Moore et al. (1991) extended	Described in Section 7.3. Zeros were removed from the input by transforming each channel with data in the range $[0, 1]$ according to $y = (255x + 1)/256$. Same as Moore et al. (1991) Retinex except that the extensions to reduce color inductions across edges are used.
Rahman et al. (1999)	Described in Section 7.4. Zeros were removed from the input by transforming each channel with data in the range $[0, 1]$ according to $y = (255x + 1)/256$. Three Gaussians are applied with $\sigma_j \in [0.4345 n_{max}, 0.0869 n_{max}, 0.0130 n_{max}]$, where $n_{max} = \max\{n_x, n_y\}$ and n_x and n_y are the width and height of the input image. The additional filter stage is not used. (sigma[]={ $\sigma_1, \sigma_2, \sigma_3$ })

PARAMETER SETTINGS

Homomorphic filtering	Described in Section 7.5. Zeros were removed from the input by transforming each channel with data in the range [0, 1] according to $y = (255x + 1)/256$. We have used the homomorphic filter shown in Figure 7.23(b). Rescaling is done at the third percentile. (pWhite=0.03)
Homomorphic filtering (HVS)	Described in Section 7.5. Zeros were removed from the input by transforming each channel with data in the range [0, 1] according to $y = (255x + 1)/256$. The filtering is done in a perceptually uniform coordinate system. The filter of Figure 7.27 was used to filter two chroma channels. Rescaling is done at the third percentile. (pWhite=0.03)
Intrinsic (minimum entropy)	Described in Section 7.6. Zeros were removed from the input by transforming each channel with data in the range [0, 1] according to $y = (255x + 1)/256$. The intrinsic image is computed by looping over all angles from 0° to 180° in 1° steps. Illumination information has been added such that the extra light shift is in the direction of the majority of the colors. We take the median of either the positive or the negative offset of the colors along the projection direction as the size of the extra light shift. The lightness of pixels from the input image were used to compute a shaded image. Shadows were not removed.
Local space average color (1)	Zeros were removed from the input by transforming each channel with data in the range [0, 1] according to $y = (255x + 1)/256$. Local space average color is computed using a convolution with $e^{\frac{r^2}{2\sigma^2}}$ and $\sigma = 0.0932 n_{max}$ where $n_{max} = \max\{n_x, n_y\}$ and n_x and n_y is the width and height of the input image. Output color is computed as described in Section 11.1. The parameter f is set to 2.0. (sigma=σ)
Local space average color (2)	Same as above except that output color is computed as described in Section 11.2.
Local space average color (3)	Same as above except that output color is computed as described in Section 11.3.
Local space average color (4)	Same as above except that output color is computed as described in Section 11.2 and saturation is increased by 30%.
Local space average color (5)	Same as above except that output color is computed as described in Section 11.3 and saturation is increased by 30%.
Local space average color (6)	Same as above except that output color is computed as described in Section 11.2 and saturation nonuniformly adjusted as described in Section 11.4.
Local space average color (7)	Same as above except that output color is computed as described in Section 11.3 and saturation nonuniformly adjusted as described in Section 11.4.

Anisotropic local space average color (1)		Zeros were removed from the input by transforming each channel with data in the range $[0, 1]$ according to $y = (255x + 1)/256$. Local space average color is computed using a convolution with $e^{\frac{r^2}{2\sigma^2}}$ and $\sigma = 0.14 n_{max}$, where $n_{max} = \max\{n_x, n_y\}$ and n_x and n_y are the width and height of the input image. The direction of the line of constant illumination is computed from the intensity of local space average color. Local space average color is computed along the line of constant illumination with $\sigma = 0.0869 n_{max}$ and $p = 1/(4\sigma^2 + 1)$. The curvature of the line of constant illumination is computed and data is extracted by linear interpolation using an 8-neighborhood. Output color is computed as described in Section 11.1. The parameter f is set to 2.0.
Anisotropic local space average color (2)		Same as above except that output color is computed as described in Section 11.2.
Anisotropic local space average color (3)		Same as above except that output color is computed as described in Section 11.3.
Anisotropic local space average color (4)		Same as above except that output color is computed as described in Section 11.2 and saturation is increased by 30%.
Anisotropic local space average color (5)		Same as above except that output color is computed as described in Section 11.3 and saturation is increased by 30%.
Anisotropic local space average color (6)		Same as above except that output color is computed as described in Section 11.2 and saturation nonuniformly adjusted as described in Section 11.4.
Anisotropic local space average color (7)		Same as above except that output color is computed as described in Section 11.3 and saturation nonuniformly adjusted as described in Section 11.4.
Combined white patch retinex and local space average color		Described in Section 11.5. Zeros were removed from the input by transforming each channel with data in the range $[0, 1]$ according to $y = (255x + 1)/256$. Local space average color is computed using a convolution with $e^{\frac{r^2}{2\sigma^2}}$ and $\sigma = 0.0932 n_{max}$ where $n_{max} = \max\{n_x, n_y\}$ and n_x and n_y is the width and height of the input image. The parameter p_d is set to $p_d = 10/(4\sigma^2 + 1)$. Output is computed by applying a sigmoidal function with $\sigma = 0.2$.

PARAMETER SETTINGS

Comprehensive local space average color	Described in Section 6.5. Zeros were removed from the input by transforming each channel with data in the range $[0, 1]$ according to $y = (255x + 1)/256$. Local space average color is computed using a convolution with $e^{\frac{r^2}{2\sigma^2}}$ and $\sigma = 0.0932 n_{\max}$ where $n_{\max} = \max\{n_x, n_y\}$ and n_x and n_y is the width and height of the input image. Comprehensive normalization was done using five levels. The intensity of the colors in the original image is used to scale the output colors to add back information about shading. (sigma=σ)

Bibliography

Akenine-Möller T and Haines E 2002 *Real-Time Rendering*. A K Peters, Natick, Massachusetts.
Almasi GS and Gottlieb A 1994 *Highly Parallel Computing*. The Benjamin/Cummings Publishing Company, Redwood City, California.
Banzhaf W, Nordin P, Keller RE and Francone FD 1998 *Genetic Programming – An Introduction: On The Automatic Evolution of Computer Programs and Its Applications*. Morgan Kaufmann Publishers, San Francisco, California.
Barber CB, Dobkin DP and Huhdanpaa H 1996 The quickhull algorithm for convex hulls. *ACM Transactions on Mathematical Software* **22**(4), 469–483.
Barnard K, Cardei V and Funt B 2002a A comparison of computational color constancy algorithms – part I: methodology and experiments with synthesized data. *IEEE Transactions on Image Processing* **11**(9), 972–984.
Barnard K, Martin L, Coath A and Funt B 2002b A comparison of computational color constancy algorithms – part II: experiments with image data. *IEEE Transactions on Image Processing* **11**(9), 985–996.
Barnard K, Martin L, Funt B and Coath A 2002c A data set for color research. *Color Research and Application* **27**(3), 147–151.
Barnard K, Ciurea F and Funt B 2001 Sensor sharpening for computational color constancy. *Journal of the Optical Society of America A* **18**(11), 2728–2743.
Barnard K, Finlayson G and Funt B 1997 Color constancy for scenes with varying illumination. *Computer Vision and Image Understanding* **65**(2), 311–321.
Barnard K, Martin L and Funt B 2000 Colour by correlation in a three dimensional colour space In *Proceedings of the 6th European Conference on Computer Vision, Dublin, Ireland* (ed. Vernon D), pp. 375–389. Springer-Verlag, Berlin.
Bayer BE. Color imaging array. United States Patent No. 3,971,065, July 1976.
Bentley PJ 1996 *Generic Evolutionary Design of Solid Objects using a Genetic Algorithm*. PhD thesis, Division of Computing and Control Systems, School of Engineering, The University of Huddersfield.
(ed. Bentley PJ) 1999 *Evolutionary Design by Computers*. Morgan Kaufmann Publishers.
Berwick D and Lee SW 1998 A chromaticity space for specularity, illumination color- and illumination pose-invariant 3-d object recognition *Sixth International Conference on Computer Vision*. Narosa Publishing, pp. 165–170.
Blake A 1985 Boundary conditions for lightness computation in mondrian world. *Computer Vision, Graphics, and Image Processing* **32**, 314–327.
Bockaert V 2005 Color filter array Digital Photography Review. http://www.dpreview.com.
Brainard DH and Freeman WT 1997 Bayesian color constancy. *Journal of the Optical Society of America A* **14**(7), 1393–1411.

Brainard DH and Wandell BA 1992 Analysis of the retinex theory of color vision In *Color* (eds. Healey GE, Shafer SA and Wolff LB), pp. 208–218. Jones and Bartlett Publishers, Boston.

Bräunl T, Feyrer S, Rapf W and Reinhardt M 1995 *Parallel Bildverarbeitung*. Addison-Wesley, Bonn.

Brill MH 1978 A device performing illuminant-invariant assessment of chromatic relations. *Journal of Theoretical Biology* **71**, 473–478.

Brill M and West G 1981 Contributions to the theory of invariance of color under the condition of varying illumination. *Journal of Mathematical Biology* **11**, 337–350.

Bronstein IN, Semendjajew KA, Musiol G and Mühling H 2001 *Taschenbuch der Mathematik*, fifth edn. Verlag Harri Deutsch, Thun und Frankfurt/Main.

Brown PK and Wald G 1964 Visual pigments in single rods and cones of the human retina. *Science* **144**, 45–52.

Buchsbaum G 1980 A spatial processor model for object colour perception. *Journal of the Franklin Institute* **310**(1), 337–350.

Cardei VC and Funt B 1999 Committee-based color constancy *Proceedings of the IS&T/SID Seventh Color Imaging Conference: Color Science, Systems and Applications*. Scottsdale, Arizona, pp. 311–313.

Castleman WL 2004 Resolution and 50% MTF with Canon CMOS and film cameras. comparison among the canon EOS-1Ds, EOS-1D Mark II, 20D, D60 and EOS-1V/Astia F100 film http://www.wlcastleman.com/equip/reviews/index.htm.

Ciurea F and Funt B 2004 Tuning retinex parameters. *Journal of Electronic Imaging* **13**(1), 58–64.

Coffin D 2004 *dcraw.c – Dave Coffin's raw photo decoder. Rev: 1.210* http://cynercom.net/~dcoffin/dcraw.

Colourware Ltd 2001 Frequently Asked Questions about Colour Physics. Colourware Ltd, Staffordshire, UK.

Cormen TH, Leiserson CE and Rivest RL 1990 *Introduction to Algorithms*. The MIT Press, Cambridge, Massachusetts.

Courtney SM, Finkel LH and Buchsbaum G 1995 A multistage neural network for color constancy and color induction. *IEEE Transactions on Neural Networks* **6**(4), 972–985.

Dartnall HJA, Bowmaker JK and Mollon JD 1983 Human visual pigments: microspectrophotometric results from the eyes of seven persons. *Proceedings of the Royal Society of London B* **220**, 115–130.

Darwin C 1996 *The Origin of Species*. Oxford University Press, Oxford, England.

De Valois RL and Pease PL 1971 Contours and contrast: responses of monkey lateral geniculate nucleus cells to luminance and color figures. *Science* **171**, 694–696.

Demmel J 1996 Cs267: Lectures 15 and 16. Solving the Discrete Poisson Equation using Jacobi, SOR, Conjugate Gradients, and the FFT. Computer Science Division, University of California at Berkeley.

Dennett DC 1995 *Darwin's Dangerous Idea: Evolution and the Meanings of Life*. Allen Lane, The Penguin Press, Harmondsworth, Middlesex, England.

Didas S, Weickert J and Burgeth B 2005 Stability and local feature enhancement of higher order nonlinear diffusion filtering In *Pattern Recognition, Proceedings of the 27th DAGM Symposium, Vienna, Austria* (eds. Wropatsch W, Sablatnig R and Hanbury A), pp. 451–458. Springer-Verlag, Berlin.

Dowling JE 1987 *The Retina: An Approachable Part of the Brain*. The Belknap Press of Harvard University Press, Cambridge, Massachusetts.

Drew MS, Finlayson GD and Hordley SD 2003 Recovery of chromaticity image free from shadows via illumination invariance *ICCV'03 Workshop on Color and Photometric Methods in Computer Vision*. Nice, France, pp. 32–39.

Dudek P and Hicks PJ 2001 A general-purpose cmos vision chip with a processor-per-pixel simd array *Proceedings of the 27th European Solid-State Circuits Conference*. IEEE, pp. 213–216.

Dufort PA and Lumsden CJ 1991 Color categorization and color constancy in a neural network model of V4. *Biological Cybernetics* **65**, 293–303.

D'Zmura M and Lennie P 1986 Mechanisms of color constancy. *Journal of the Optical Society of America A* **3**(10), 1662–1672.

Eastman Kodak Company, Austin Development Center 2004 DIGITAL SHO professional plug-in, Eastman Kodak Company.

Ebner M 2001 Evolving color constancy for an artificial retina In *Genetic Programming: Proceedings of the Fourth European Conference, EuroGP 2001, Lake Como, Italy, April 18–20* (eds. Miller J, Tomassini M, Lanzi PL, Ryan C, Tettamanzi AGB and Langdon WB), pp. 11–22. Springer-Verlag, Berlin.

Ebner M 2002 A Parallel Algorithm for Color Constancy. Technical Report 296, Universität Würzburg, Lehrstuhl für Informatik II, Am Hubland, 97074 Würzburg, Germany.

Ebner M 2003a Combining white-patch retinex and the gray world assumption to achieve color constancy for multiple illuminants In *Pattern Recognition, Proceedings of the 25th DAGM Symposium, Magdeburg, Germany* (eds. Michaelis B and Krell G), pp. 60–67. Springer-Verlag, Berlin.

Ebner M 2003b Evolutionary design of objects using scene graphs *Genetic Programming: Proceedings of the Sixth European Conference, EuroGP 2003*, Essex, UK, April 14–16, Springer-Verlag, Berlin, pp. 47–58.

Ebner M 2003c Verfahren und Vorrichtung zur Farbkorrektur von Bildern. *Deutsche Patentanmeldung, 18 Seiten, 6*. Oktober, DE 10346348 A1.

Ebner M 2004a Aktuelles Schlagwort: Evolvable hardware. *Künstliche Intelligenz* Böttcher IT Verlag, Volume 2, pp. 53–54, May.

Ebner M 2004b Color constancy using local color shifts In *Proceedings of the 8th European Conference on Computer Vision, Part III, Prague, Czech Republic, May, 2004* (eds. Pajdla T and Matas J), pp. 276–287. Springer-Verlag, Berlin.

Ebner M 2004c A parallel algorithm for color constancy. *Journal of Parallel and Distributed Computing* **64**(1), 79–88.

Ebner M 2004d Verfahren und Vorrichtung zur Farbkorrektur von Bildern mit nicht-linearen Beleuchtungsänderungen. Deutsche Patentanmeldung, 28 Seiten, 4. Juni, DE 102004027471 A1.

Ebner M 2006 Evolving color constancy. *Special Issue on Evolutionary Computer Vision and Image Understanding of Pattern Recognition Letters* **27**(11), 1220–1229.

Ebner M and Herrmann C 2005 On determining the color of the illuminant using the dichromatic reflection model In *Pattern Recognition, Proceedings of the 27th DAGM Symposium, Vienna, Austria* (eds. Kropatsch W, Sablatnig R and Hanbury A), pp. 1–8. Springer-Verlag, Berlin.

van Essen DC and Deyoe EA 1995 Concurrent processing in the primate visual cortex In *The Cognitive Neurosciences* (ed. Gazzaniga MS), pp. 383–400. The MIT Press, Cambridge, Massachusetts.

Fain GL and Dowling JE 1973 Intracellular recordings from single rods and cones in the mudpuppy retina. *Science* **180**, 1178–1181.

Faugeras OD 1979 Digital color image processing within the framework of a human visual model. *IEEE Transactions on Acoustics, Speech, and Signal Processing* **ASSP-27**(4), 380–393.

Ffytche DH, Guy CN and Zeki S 1995 The parallel visual motion inputs into areas v1 and v5 of human cerebral cortex. *Brain* **118**, 1375–1394.

Finlayson GD 1996 Color in perspective. *IEEE Transactions on Pattern Analysis and Machine Intelligence* **18**(10), 1034–1038.

Finlayson GD and Drew MS 2001 4-sensor camera calibration for image representation invariant to shading, shadows, lighting, and specularities. *Proceedings of the 8th IEEE International Conference on Computer Vision*, Volume 2, Vancouver, Canada, July 9–12, 2001, pp. 473–480.

Finlayson GD and Funt BV 1996 Coefficient channels: derivation and relationship to other theoretical studies. *Color Research and Application* **21**(2), 87–96.

Finlayson G and Hordley S 2001a Colour signal processing which removes illuminant colour temperature dependency. UK Patent Application GB 2360660A.

Finlayson GD and Hordley SD 2001b Color constancy at a pixel. *Journal of the Optical Society of America A* **18**(2), 253–264.

Finlayson GD, Chatterjee SS and Funt BV 1995 Color angle invariants for object recognition *Proceedings of the Third IS&T/SID Color Imaging Conference: Color Science, Systems and Applications, Nov. 7–10*. The Radisson Resort, Scottsdale, Arizona, pp. 44–47.

Finlayson GD, Drew MS and Funt BV 1994a Color constancy: generalized diagonal transforms suffice. *Journal of the Optical Society of America A* **11**(11), 3011–3019.

Finlayson GD, Drew MS and Funt BV 1994b Spectral sharpening: sensor transformations for improved color constancy. *Journal of the Optical Society of America A* **11**(4), 1553–1563.

Finlayson GD, Drew MS and Lu C 2004 Intrinsic images by entropy minimization In *Proceedings of the 8th European Conference on Computer Vision, Part III,Prague, Czech Republic, May, 2004* (eds. Pajdla T and Matas J), pp. 582–595. Springer-Verlag, Berlin.

Finlayson GD, Hordley SD and Drew MS 2002 Removing shadows from images *Proceedings of the European Conference on Computer Vision*. Springer-Verlag, Berlin, pp. 823–836.

Finlayson GD, Hubel PM and Hordley S 1997 Color by correlation *Proceedings of IS&T/SID. The Fifth Color Imaging Conference: Color Science, Systems, and Applications, Nov 17–20*. The Radisson Resort, Scottsdale, AZ, pp. 6–11.

Finlayson GD and Schaefer G 2001 Solving for colour constancy using a constrained dichromatic reflection model. *International Journal of Computer Vision* **42**(3), 127–144.

Finlayson GD, Schiele B and Crowley JL 1998 Comprehensive colour image normalization In *Fifth European Conference on Computer Vision (ECCV '98), Freiburg, Germany* (eds. Burkhardt H and Neumann B), pp. 475–490. Springer-Verlag, Berlin.

Fleyeh H 2005 Traffic signs color detection and segmentation in poor light conditions *Proceedings of the IAPR Conference on Machine Vision Applications*, Tsukuba Science City, Japan, Mai 16–18, pp. 306–309.

Foley JD, van Dam A, Feiner SK and Hughes JF 1996 *Computer Graphics: Principles and Practice. Second Edition in C*. Addison-Wesley Publishing Company, Reading, Massachusetts.

Forsyth DA 1988 A novel approach to colour constancy *Second International Conference on Computer Vision (Tampa, FL, Dec. 5–8)*. IEEE Press, pp. 9–18.

Forsyth DA 1992 A novel algorithm for color constancy In *Color* (eds. Healey GE, Shafer SA and Wolff LB), pp. 241–271. Jones and Bartlett Publishers, Boston.

Foveon Inc. 2002 X3 technology. why X3 is better.

Frigo M and Johnson SG 2004 FFTW for version 3.1.

Fuji Photo Film Co., LTD. 2005a Fujifilm data sheet. FUJICHROME 64T Type II Professional Ref. No. AF3–024E.

Fuji Photo Film Co., LTD. 2005b Fujifilm product information bulletin. FUJICHROME Velvia 100F Professional Ref. No. AF3–148E.

Funt B, Barnard K and Martin L 1998 Is machine colour constancy good enough? In *Fifth European Conference on Computer Vision (ECCV '98), Freiburg, Germany* (eds. Burkhardt H and Neumann B), pp. 445–459. Springer-Verlag, Berlin.

Funt B, Cardei V and Barnard K 1996 Learning color constancy *Proceedings of the IS&T/SID Fourth Color Imaging Conference*. Scottsdale, pp. 58–60.

Funt B, Ciurea F and McCann J 2004 Retinex in MATLAB. *Journal of Electronic Imaging* **13**(1), 48–57.

Funt BV and Drew MS 1988 Color constancy computation in near-mondrian scenes using a finite dimensional linear model In *Proceedings of the Computer Society Conference on Computer Vision and Pattern Recognition, Ann Arbor, MI* (eds. Jain R and Davis L), pp. 544–549. Computer Society Press.

Funt BV, Drew MS and Ho J 1991 Color constancy from mutual reflection. *International Journal of Computer Vision* **6**(1), 5–24.

Funt BV, Drew MS and Ho J 1992 Color constancy from mutual reflection In *Color* (eds. Healey GE, Shafer SA and Wolff LB), pp. 365–384. Jones and Bartlett Publishers, Boston.

Funt BV and Finlayson GD 1995 Color constant color indexing. *IEEE Transactions on Pattern Analysis and Machine Intelligence* **17**(5), 522–529.

Funt BV and Lewis BC 2000 Diagonal versus affine transformations for color correction. *Journal of the Optical Society of America A* **17**(11), 2108–2112.

Gershon R, Jepson AD and Tsotsos JK 1987 From [R,G,B] to surface reflectance: Computing color constant descriptors in images In *Proceedings of the Tenth International Joint Conference on Artificial Intelligence, Milan, Italy* (ed. McDermott JP), Volume 2, Morgan Kaufmann, pp. 755–758.

Glasser LG, McKinney AH, Reilly CD and Schnelle PD 1958 Cube-root color coordinate system. *Journal of the Optical Society of America* **48**(10), 736–740.

Goldberg DE 1989 *Genetic Algorithms in Search, Optimization, and Machine Learning*. Addison-Wesley Publishing Company, Reading, Massachusetts.

Gonzalez RC and Woods RE 1992 *Digital Image Processing*. Addison-Wesley Publishing Company, Reading, Massachusetts.

Haken H and Wolf HC 1990 *Atom- und Quantenphysik: Einführung in die Experimentellen und Theoretischen Grundlagen*, vierte edn. Springer-Verlag, Berlin, Heidelberg.

Hanbury A and Serra J 2003 Colour image analysis in 3D-polar coordinates In *Pattern Recognition. Proceedings of the 25th DAGM Symposium, Magdeburg, Germany, September 10–12* (eds. Michaelis B and Krell G), pp. 124–131. Springer-Verlag, Berlin.

Harold RW 2001 An introduction to appearance analysis. SecondSight No. 84. A reprint from GATFWorld, the magazine of the Graphic Arts Technical Foundation.

He XD, Torrance KE, Sillion FX and Greenberg DP 1991 A comprehensive physical model for light reflection. *Computer Graphics* **25**(4), 175–186.

Healey G and Slater D 1994 Global color constancy: recognition of objects by use of illumination-invariant properties of color distributions. *Journal of the Optical Society of America* **11**(11), 3003–3010.

Hedgecoe J 2004 *Fotografieren: die neue große Fotoschule*. Dorling Kindersley Verlag GmbH, Starnberg.

Helson H 1938 Fundamental problems in color vision. I. the principle governing changes in hue, saturation, and lightness of non-selective samples in chromatic illumination. *Journal of Experimental Psychology* **23**(5), 439–476.

Henderson B 2003 *Netpbm, Version 10.17* http://netpbm.sourceforge.net/.

Herault J 1996 A model of colour processing in the retina of vertebrates: from photoreceptors to colour opposition and colour constancy phenomena. *Neurocomputing* **12**, 113–129.

Herrmann C 2004 *Schätzung der Farbe einer Lichtquelle anhand eines digitalen Bildes durch Segmentierung und Filterung*, Projektpraktikum, Universität Würzburg, Institut für Informatik, Lehrstuhl für Informatik II.

Higuchi T, Niwa T, Tanaka T, Iba H, de Garis H and Furuya T 1993 Evolving hardware with genetic learning: a first step towards building a Darwin machine In *From Animals to Animats II: Proceedings of the Second International Conference on Simulation of Adaptive Behavior* (eds. Meyer JA, Roitblat HL and Wilson SW), pp. 417–424. The MIT Press, Cambridge, Massachusetts.

Ho J, Funt BV and Drew MS 1992 Separating a color signal into illumination and surface reflectance components: theory and applications In *Color* (eds. Healey GE, Shafer SA and Wolff LB), pp. 272–283. Jones and Bartlett Publishers, Boston.

Holland JH 1992 *Adaptation in natural and artificial systems: an introductory analysis with applications to biology, control, and artificial intelligence*. The MIT Press, Cambridge, Massachusetts.

Horn BKP 1974 Determining lightness from an image. *Computer Graphics and Image Processing* **3**, 277–299.

Horn BKP 1986 *Robot Vision*. The MIT Press, Cambridge, Massachusetts.

Hubel DH and Wiesel TN 1962 Receptive fields, binocular interaction and functional architecture in the cat's visual cortex. *Journal of Physiology* **160**, 106–154.

Hubel DH and Wiesel TN 1977 Functional architecture of macaque monkey visual cortex. *Proceedings of the Royal Society of London B* **198**, 1–59.

Hunt RWG 1957 Light energy and brightness sensation. *Nature* **179**, 1026–1027.

Hurlbert A 1986 Formal connections between lightness algorithms. *Journal of the Optical Society of America A* **3**(10), 1684–1693.

Hurlbert AC and Poggio TA 1987 Learning a Color Algorithm from Examples. Technical Report A.I. Memo No. 909, Massachusetts Institute of Technology, Artificial Intelligence Laboratory.

Hurlbert AC and Poggio TA 1988 Synthesizing a color algorithm from examples. *Science* **239**, 482–483.

International Color Consortium 2003 File format for color profiles (version 4.1.0) Specification ICC.1:2003–09.

International Commission on Illumination 1983 The Basis of Physical Photometry, second edition. Technical Report 18.2, International Commission on Illumination.

International Commission on Illumination 1988 CIE Photometric and Colorimetric Data. Technical Report D001, International Commission on Illumination.

International Commission on Illumination 1990 CIE 1988 2° Spectral luminous Efficiency Function for Photopic Vision. Technical Report 86, International Commission on Illumination.

International Commission on Illumination 1996 Colorimetry, second edition, corrected reprint. Technical Report 15.2, International Commission on Illumination.

Jacobsen RE, Ray SF, Attridge GG and Axford NR 2000 *The Manual of Photography. Photographic and Digital Imaging*. Focal Press, Oxford.

Jähne B 2002 *Digitale Bildverarbeitung*, fifth edn. Springer-Verlag, Berlin.

Jain R, Kasturi R and Schunck BG 1995 *Machine Vision*. McGraw-Hill, New York.

Johnson RA and Bhattacharyya GK 2001 *Statistics: Principles and Methods*, fourth edn. John Wiley & Sons, New York.

Judd DB 1940 Hue saturation and lightness of surface colors with chromatic illumination. *Journal of the Optical Society of America* **30**, 2–32.

Judd DB 1960 Appraisal of Land's work on two-primary color projections. *Journal of the Optical Society of America* **50**(3), 254–268.

Judd DB, MacAdam DL and Wyszecki G 1964 Spectral distribution of typical daylight as a function of correlated color temperature. *Journal of the Optical Society of America* **54**(8), 1031–1040.

Kalawsky RS 1993 *The Science of Virtual Reality and Virtual Environments*. Addison-Wesley Publishing Company, Workingham, England.

Kennard C, Lawden M, Morland AB and Ruddock KH 1995 Colour identification and colour constancy are impaired in a patient with incomplete achromatopsia associated with prestriate cortical lesions. *Proceedings of the Royal Society of London B* **260**, 169–175.

Koosh VF 2001 *Analog Computation and Learning in VLSI*. PhD thesis, California Institute of Technology Pasadena, California.

Kosko B 1992 *Neural Networks and Fuzzy Systems: A Dynamical Systems Approach to Machine Intelligence*. Prentice-Hall, A Simon & Schuster Company, Englewood Cliffs, New Jersey.

Koza JR 1992 *Genetic Programming. On the Programming of Computers by Means of Natural Selection*. The MIT Press, Cambridge, Massachusetts.

Koza JR 1994 *Genetic Programming II. Automatic Discovery of Reusable Programs*. The MIT Press, Cambridge, Massachusetts.

Koza JR, Bennett III FH, Andre D and Keane MA 1999 *Genetic Programming III. Darwinian Invention and Problem Solving*. Morgan Kaufmann Publishers.

Kuffler SW 1952 Neurons in the retina: organization, inhibition and excitation problems *Cold Spring Harbor Symposia on Quantitative Biology, Volume XVII, The Neuron*, pp. 281–292. The Biology Laboratory, Cold Spring Harbor, LI, New York.

Kuffler SW 1953 Discharge patterns and functional organisation of mammalian retina. *Journal of Neurophysiology* **16**(1), 37–68.

Kuypers F 1990 *Klassische Mechanik*, dritte edn. VCH Verlagsgesellschaft mbH, Weinheim.

Kyuma K, Nitta Y and Miyake Y 1997 Artificial retina chips for image processing. *Artificial Life and Robotics* **1**(1), 79–87.

Land EH 1959a Color vision and the natural image. Part I. *Proceedings of the National Academy of Sciences of the United States of America* **45**, 115–129.

Land EH 1959b Color vision and the natural image. Part II. *Proceedings of the National Academy of Sciences of the United States of America* **45**, 636–644.

Land EH 1959c Experiments in color vision. *Scientific American* **45**, 84–99.

Land EH 1960 Some comments on Dr. Judd's paper. *Journal of the Optical Society of America* **50**(3), 268.

Land EH 1962 Colour in the natural image. *Proceedings of the Royal Institution of Great Britain* **39**(176), 1–15.

Land EH 1964 The retinex. *American Scientist* **52**, 247–264.

Land EH 1974 The retinex theory of colour vision. *Proceedings of the Royal Institution of Great Britain* **47**, 23–58.

Land EH 1983 Recent advances in retinex theory and some implications for cortical computations: color vision and the natural image. *Proceedings of the National Academy of Sciences of the United States of America* **80**, 5163–5169.

Land EH 1986a An alternative technique for the computation of the designator in the retinex theory of color vision. *Proceedings of the National Academy of Sciences of the United States of America* **83**, 3078–3080.

Land EH 1986b Recent advances in retinex theory. *Vision Research* **26**(1), 7–21.

Land EH, Hubel DH, Livingstone MS, Perry SH and Burns MM 1983 Colour-generating interactions across the corpus callosum. *Nature* **303**, 616–618.

Land EH and McCann JJ 1971 Lightness and retinex theory. *Journal of the Optical Society of America* **61**(1), 1–11.

Lengyel E 2002 *Mathematics for 3D Game Programming and Computer Graphics*. Charles River Media, Hingham, Massachusetts.

Lennie P and D'Zmura M 1988 Mechanisms of color vision. *CRC Critical Reviews in Neurobiology* **3**(4), 333–400.

Levi P 1993 Radiometrische und fotometrische Grundlagen der Bildentstehung In *Verarbeiten und Verstehen von Bildern* (ed. Radig B), pp. 67–105. R. Oldenbourg Verlag, München.

Linnell KJ and Foster DH 1997 Space-average scene colour used to extract illuminant information In *John Dalton's Colour Vision Legacy. Selected Proceedings of the International Conference* (eds. Dickinson C, Murray I and Carden D), pp. 501–509. Taylor & Francis, London.

Livingstone MS and Hubel DH 1984 Anatomy and physiology of a color system in the primate visual cortex. *The Journal of Neuroscience* **4**(1), 309–356.

Mallot HA 1998 *Sehen und die Verarbeitung visueller Information, Eine Einführung*. Vieweg, Braunschweig.

Maloney LT 1992 Evaluation of linear models of surface spectral reflectance with small numbers of parameters In *Color* (eds. Healey GE, Shafer SA and Wolff LB), pp. 87–97. Jones and Bartlett Publishers, Boston.

Maloney LT and Wandell BA 1992 Color constancy: a method for recovering surface spectral reflectance In *Color* (eds. Healey GE, Shafer SA and Wolff LB), pp. 219–223. Jones and Bartlett Publishers, Boston.

Maloney LT and Wandell BA 1986 Color constancy: a method for recovering surface spectral reflectance. *Journal of the Optical Society of America A* **3**(1), 29–33.

Marks WB, Dobelle WH and MacNichol EF Jr 1964 Visual pigments of single primate cones. *Science* **143**, 1181–1183.

Marr D 1974 The computation of lightness by the primate retina. *Vision Research* **14**, 1377–1388.

Marr D 1982 *Vision*. W. H. Freeman & Company, New York.

Matsumoto M and Nishimura T 1998 Mersenne twister: A 623-dimensionally equidistributed uniform pseudo-random number generator. *ACM Transactions on Modeling and Computer Simulation* **8**(1), 3–30.

Maynard Smith J 1993 *The Theory of Evolution*. Cambridge University Press, Cambridge, England.

McCann JJ 1997 Adaptation or contrast: the controlling mechanism for colour constancy In *John Dalton's Colour Vision Legacy. Selected Proceedings of the International Conference* (eds. Dickinson C, Murray I and Carden D), pp. 469–473. Taylor & Francis, London.

McCann JJ, McKee SP and Taylor TH 1976 Quantitative studies in retinex theory. *Vision Research* **16**, 445–458.

McClelland JL and Rumelhart DE, The PDP Research Group 1986 *Parallel Distributed Processing, Explorations in the Microstructure of Cognition, Volume 2: Psychological and Biological Models*. The MIT Press, Cambridge, Massachusetts.

McKeefry DJ, Watson JDG, Frackowiak RSJ, Fong K and Zeki S 1997 The activity in human areas V1/V2, V3, and V5 during the perception of coherent and incoherent motion. *Neuroimage* **5**, 1–12.

Miller JF, Job D and Vassilev VK 2000 Principles in the evolutionary design of digital circuits – part I. *Genetic Programming and Evolvable Machines* **1**(1–2), 7–35.

Mitchell M 1996 *An Introduction to Genetic Algorithms*. The MIT Press, Cambridge, Massachusetts.

Möller T and Haines E 1999 *Real-Time Rendering*. A K Peters, Natick, Massachusetts.

Monteil J and Beghdadi A 1999 A new interpretation and improvement of the nonlinear anisotropic diffusion for image enhancement. *IEEE Transactions on Pattern Analysis and Machine Intelligence* **21**(9), 940–946.

Moore AJ and Allman J 1991 Saturable smoothing grid for image processing. United States Patent No. 5,294,989.

Moore A, Allman J and Goodman RM 1991 A real-time neural system for color constancy. *IEEE Transactions on Neural Networks* **2**(2), 237–247.

Moroney N and Sobel I 2001 Method and apparatus for performing local color correction. European Patent Application EP 1 139 284 A2.

Moutoussis K and Zeki S 1997 A direct demonstration of perceptual asynchrony in vision. *Proceedings of the Royal Society of London B* **264**, 393–399.

Movshon JA, Adelson EH, Gizzi MS and Newsome WT 1985 The analysis of moving visual patterns In *Pattern Recognition Mechanisms. Proceedings of a Study Week Organized by the Pontifical Academy of Sciences, Vatican City* (eds. Chagas C, Gattass R and Gross C), pp. 117–151. Springer-Verlag, Berlin.

Murray JD and van Ryper W 1994 *Encyclopedia of Graphics File Formats*. O'Reilly & Associates, Sebastopol, California.

Nascimento SMC and Foster DH 1997 Dependence of colour constancy on the time-course of illuminant changes In *John Dalton's Colour Vision Legacy. Selected Proceedings of the International Conference* (eds. Dickinson C, Murray I and Carden D), pp. 491–499. Taylor & Francis, London.

Nolting W 1992 *Grundkurs: Theoretische Physik, Band 3 Elektrodynamik*, zweite edn. Verlag Zimmermann-Neufang, Ulmen.

Normann RA and Perlman I 1979 The effects of background illumination on the photoresponses of red and green cones. *Journal of Physiology* **286**, 491–507.

Novak CL and Shafer SA 1992 Supervised color constancy for machine vision In *Color* (eds. Healey GE, Shafer SA and Wolff LB), pp. 284–299. Jones and Bartlett Publishers, Boston.

Oja E 1982 A simplified neuron model as a principal component analyzer. *Journal of Mathematical Biology* **15**, 267–273.

Orear J 1982 *Physik*. Carl Hanser Verlag, München, Wien.

Østerberg G 1935 *Topography of the Layer of Rods and Cones in the Human Retina*. Acta Ophthalmologica, Supplement 6 Copenhagen.

Parker JR 1997 *Algorithms for Image Processing and Computer Vision*. John Wiley & Sons, New York.

Paulus D, Csink L and Niemann H 1998 Color cluster rotation *Proceedings of the International Conference on Image Processing (ICIP)*, pp. 161–165. IEEE Computer Society Press.

Pomierski T and Groß HM 1995 Verfahren zur empfindungsgemäßen Farbumstimmung In *Mustererkennung 1995, Verstehen Akustischer und Visueller Informationen, 17. DAGM-Symposium, Bielefeld, 13.-15. September 1995* (eds. Sagerer G, Posch S and Kummert F), pp. 473–480. Springer-Verlag, Berlin.

Poynton C 1997 Frequently asked questions about color http://www.poynton.com/PDFs/ColorFAQ.pdf.

Poynton C 1998 Frequently asked questions about gamma http://www.poynton.com.

Poynton C 2003 *Digital Video and HDTV. Algorithms and Interfaces*. Morgan Kaufmann Publishers, San Francisco, California.

Press WH, Teukolsky SA, Vetterling WT and Flannery BP 1992 *Numerical Recipes in C: The Art of Scientific Computing*, second edn. Cambridge University Press.

Radig B 1993 *Verarbeiten und Verstehen von Bildern*. R. Oldenburg Verlag, München, Wien.

Rahman Z, Jobson DJ and Woodell GA 1999 Method of improving a digital image. United States Patent No. 5,991,456.

Ramachandran VS and Gregory RL 1978 Does colour provide an input to human motion perception? *Nature* **275**, 55–56.

Rechenberg I 1994 *Evolutionsstrategie '94*. Frommann-holzboog, Stuttgart.

Richards W and Parks EA 1971 Model for color conversion. *Journal of the Optical Society of America* **61**(7), 971–976.

Risson VJ 2003 Determination of an illuminant of digital color image by segmentation and filtering. United States Patent Application, Pub. No. US 2003/0095704 A1.

Rumelhart DE and McClelland JL, The PDP Research Group 1986 *Parallel Distributed Processing, Explorations in the Microstructure of Cognition*, Volume 1: Foundations. The MIT Press, Cambridge, Massachusetts.

Schiele B and Crowley JL 1996 Object recognition using multidimensional receptive field histograms In *Fourth European Conference On Computer Vision, Cambridge, UK, April 14–18* (eds. Buxton B and Cipolla R), pp. 610–619. Springer-Verlag, Berlin.

Schiele B and Crowley JL 2000 Recognition without correspondence using multidimensional receptive field histograms. *International Journal of Computer Vision* **36**(1), 31–52.

Schöneburg E, Heinzmann F and Feddersen S 1994 *Genetische Algorithmen und Evolutionsstrategien*. Addison-Wesley (Deutschland) GmbH, Bonn.

Schwefel HP 1995 *Evolution and Optimum Seeking*. John Wiley & Sons, New York.

Sedgewick R 1992 *Algorithmen*. Addison-Wesley (Deutschland) GmbH, Bonn.

Shapiro LG and Stockman GC 2001 *Computer Vision*. Prentice Hall, New Jersey.

Siggelkow S, Schael M and Burkhardt H 2001 SIMBA – Search IMages By Appearance In *Pattern Recognition. Proceedings of the 23th DAGM Symposium, München, Germany, September 12–14* (eds. Radig B and Florczyk S), pp. 9–16. Springer-Verlag, Berlin.

Smith AR 1978 Color gamut transform pairs. *SIGGRAPH '78 Conference Proceedings, Computer Graphics* **12**(3), 12–19.

Starkweather G 1998 Colorspace interchange using sRGB.

Stockham TG Jr 1972 Image processing in the context of a visual model. *Proceedings of the IEEE* **60**(7), 828–842.

Stokes M, Anderson M, Chandrasekar S and Motta R 1996 A Standard Default Color Space for the Internet – sRGB. Technical report, Version 1.10.

Swain MJ and Ballard DH 1991 Color indexing. *International Journal of Computer Vision* **7**, 11–32.

Tappen MF, Freeman WT and Adelson EH 2002 Recovering Intrinsic Images from a Single Image. Technical Report AI Memo 2002–015, Massachusetts Institute of Technology, Artificial Intelligence Laboratory.

Terzopoulos D 1986 Image analysis using multigrid relaxation methods. *IEEE Transactions on Pattern Analysis and Machine Intelligence* **PAMI-8**(2), 129–139.

Thompson A 1996 Silicon evolution In *Genetic Programming 1996, Proceedings of the First Annual Conference, July 28–31, 1996, Stanford University* (eds. Koza JR, Goldberg DE, Fogel DB and Riolo RL), pp. 444–452. The MIT Press, Cambridge, Massachusetts.

Tominaga S 1991 Surface identification using the dichromatic reflection model. *IEEE Transactions on Pattern Analysis and Machine Intelligence* **13**(7), 658–670.

Tominaga S and Wandell BA 1992 Component estimation of surface spectral reflectance In *Color* (eds. Healey GE, Shafer SA and Wolff LB), pp. 87–97. Jones and Bartlett Publishers, Boston.

Tovée MJ 1996 *An introduction to the visual system*. Cambridge University Press, Cambridge, Massachusetts.

Uschold A 2002 Scientific Test Report: Current Capabilities of Digital Cameras and the Comparison of the Classical Architecture of Digital Cameras based on 35mm SLR-Systems and a Digital Optimized Architecture. Scientific Test Report, Anders Uschold Digitaltechnik, Munich, Germany.

Usui S and Nakauchi S 1997 A neurocomputational model for colour constancy In *John Dalton's Colour Vision Legacy. Selected Proceedings of the International Conference* (eds. Dickinson C, Murray I and Carden D), pp. 475–482. Taylor & Francis, London.

Wandell BA 1987 The synthesis and analysis of color images. *IEEE Transactions on Pattern Analysis and Machine Intelligence* **PAMI-9**(1), 2–13.

Warnock J 1999 Coming out of the shadows Adobe Systems Incorporated.

Watt A 2000 *3D Computer Graphics*. Addison-Wesley, Harlow, England.

Weickert J 1997 A review of nonlinear diffusion filtering In *Scale-Space Theory in Computer Vision* (eds. ter Haar Romeny B, Florack L, Koenderink J and Viergever M), pp. 3–28. Springer-Verlag, Berlin.

Weickert J, ter Haar Romeny B and Viergever MA 1998 Efficient and reliable schemes for nonlinear diffusion filtering. *IEEE Transactions on Image Processing* **7**(3), 398–410.

Weiss Y 2001 Deriving intrinsic images from image sequences *Proceedings of the International Conference on Computer Vision, Vancouver, Canada, July 9–12, 2001*. IEEE.

Weisstein EW 1999a e. From *MathWorld* – A Wolfram Web Resource. http://mathworld.wolfram.com/e.html.

Weisstein EW 1999b Hankel transform From *MathWorld* – A Wolfram Web Resource. http://mathworld.wolfram.com/HankelTransform.html.

West G and Brill MH 1982 Necessary and sufficient conditions for von Kries chromatic adaptation to give color constancy. *Journal of Mathematical Biology* **15**, 249–258.

Wilson FAW, Scalaidhe SPO and Goldman-Rakic PS 1993 Dissociation of object and spatial processing domains in primate prefrontal cortex. *Science* **260**, 1955–1958.

Wyszecki G and Stiles WS 2000 *Color Science. Concepts and Methods, Quantitative Data and Formulae*, second edn. John Wiley & Sons, New York.

Young RA 1987 Color vision and the retinex theory. *Science* **238**, 1731–1732.

Zaidi Q 1998 Identification of illuminant and object colors: heuristic-based algorithms. *Journal of the Optical Society of America A* **15**(7), 1767–1776.

Zeki SM 1978 Review article: Functional specialisation in the visual cortex of the rhesus monkey. *Nature* **274**, 423–428.

Zeki S 1993 *A Vision of the Brain*. Blackwell Science, Oxford.

Zeki S 1999 *Inner Vision. An Exploration of Art and the Brain*. Oxford University Press, Oxford.

Zeki S and Bartels A 1999 The clinical and functional measurement of cortical (in)activity in the visual brain, with special reference to the two subdivisions (V4 and V4α) of the human colour centre. *Proceedings of the Royal Society of London B* **354**, 1371–1382.

Zeki S and Marini L 1998 Three cortical stages of colour processing in the human brain. *Brain* **121**, 1669–1685.

Zeki S, Watson JDG, Lueck CJ, Friston KJ, Kennard C and Frackowiak RSJ 1991 A direct demonstration of functional specialization in human visual cortex. *The Journal of Neuroscience* **11**(3), 641–649.

Zell A 1994 *Simulation Neuronaler Netze*. Addison-Wesley (Deutschland) GmbH, Bonn.

Zenger C and Bader M 2004 Algorithmen des wissenschaftlichen Rechnens. Schnelle Poisson-Löser – Lösungsvorschlag Technische Universität München, Lehrstuhl für Informatik V.

List of Symbols

The table below lists the symbols that were used in this book for quick reference. In some cases, the use of the same symbol to refer to two different things was inevitable. In such cases, the meaning should be clear from the context.

α	angle
a^*	color component, used with the $L^*a^*b^*$ color space
$\bar{a}$	average of color channels $\bar{a} = a_r + a_r + a_r$
a_i	component of space average color, activation of neuron i
$\mathbf{a}$	space average color $\mathbf{a} = [a_r, a_g, a_b]^T$
$\hat{\mathbf{a}}$	chromaticity coordinates of $\mathbf{a}$
$\check{\mathbf{a}}$	anisotropic space average color $\check{\mathbf{a}} = [\check{a}_r, \check{a}_g, \check{a}_b]^T$
A	area of a patch, maximum intensity given of by monitor
$\mathbf{A}$	lighting matrix
$\mathcal{A}$	algorithm
b^*	color component, used with the $L^*a^*b^*$ color space
$\bar{b}$	blue weight in color-matching experiments
$\mathbf{b}$	blue vector $\mathbf{b} = [0, 0, 1]^T$
B	intensity of the blue color channel
χ	chi-chromaticities, divergence measure
c_1, c_2	constants
$\bar{c}$	average of color channels $\bar{c} = c_r + c_r + c_r$
$\mathbf{c}$	color of pixel as stored in an image format $\mathbf{c} = [c_r, c_g, c_b]^T$. The individual components may or may not represent linear intensity information, and the details depend on the image format. Here, it is assumed that $c_i = I_i$.
$\hat{\mathbf{c}}$	chromaticity coordinates of $\mathbf{c}$
C	capacity or cyan
C_B	chroma component, used in HDTV
C_R	chroma component, used in HDTV
C^*_{uv}	chroma, used with the $L^*u^*v^*$ color space
$\mathbf{C}$	covariance matrix
δ	Dirac delta function
Δ	Laplace operator, small difference
∇^2	Laplace operator
d	diameter of lens, maximum deviation between color and space average color

Color Constancy M. Ebner
© 2007 John Wiley & Sons, Ltd

382　　　　　　　　　　　　　　　　　　　　　　　　　　　　　　　　　LIST OF SYMBOLS

D	descriptor, diffusion coefficient
$\mathbf{D}$	diagonal matrix
ϵ	small value larger than zero
e	base of the natural logarithm, Euler's number
$\mathbf{e}$	eigenvector
$E[x]$	expected value of x
Q	radiant energy, entropy, potential, error
Q_v	luminous energy
E_i	constants that denote $E_i = -\frac{c_2}{\lambda_i}$ with $i \in \{r, g, b\}$
E	irradiance
E_v	illuminance
Φ	radiant flux
Φ_v	luminous flux
f	focal length, scaling factor, bidirectional reflectance distribution function (BRDF)
γ	gamma factor
g	Green's function, gray value
$\bar{g}$	green weight in color-matching experiments
$\mathbf{g}$	green vector $\mathbf{g} = [0, 1, 0]^T$
gamma	gamma transfer function
G	geometry factor, intensity of the green color channel, Gaussian
Gate	gate function
η	refractive index
h_{uv}	hue angle, used with the $L^*u^*v^*$ color space
H	histogram or hue
$\mathcal{H}$	convex hull of a set of points
i	index used to reference one of the color bands red, green or blue, $i \in \{r, g, b\}$
I	intensity, radiant intensity, image sensor, current
$\mathbf{I}$	measured intensity vector
$\mathcal{I}$	entire image
I_v	luminous intensity
$\mathbf{j}$	flow
k	constant
K	curvature, von Kries coefficients
λ	wavelength
L	light source, illuminant
L	radiance
L_v	luminance
$\mathcal{L}$	dichromatic line
m	color constancy measure
M	radiant exitance or magenta
M_v	luminous exitance
$\mathbf{m}$	two or three-dimensional map, mean vector
$\mathbf{M}$	matrix
$\mathcal{M}$	set of maps
∇	Nabla operator

LIST OF SYMBOLS

N	neighborhood
$\mathbf{n}$	normalized vector
$\mathbf{N}$	normal vector of a surface
n	number of pixels, number of light sources, iteration counter, number of reflectances, number of scales, number of patches, general counter
ω	frequency
Ω	solid angle
$\mathbf{o}$	output color $\mathbf{o} = [o_r, o_g, o_b]^T$
p	probability, percentage
ϕ	azimuth
$\mathbf{p}$	point of intersection
P	power
P_B	chroma component, used in SDTV
P_R	chroma component, used in SDTV
$\mathcal{P}$	path
q	parameter used to adjust between isotropic and anisotropic averaging
ρ_{ij}	log color difference between color channels i and j
$\tilde{\rho}_{ij}$	log color difference between color channels i and j with mean subtracted
r	radius
$\bar{r}$	red weight in color-matching experiments
$\mathbf{r}$	red vector $\mathbf{r} = [1, 0, 0]^T$
$\mathbf{r}_{Obj}$	vector from lens to object patch
$\mathbf{r}_S$	vector from lens to sensor
R	reflectance, resistance, region, intensity of the red color channel
$\hat{R}$	reflectance basis function
$\mathbf{R}$	vector with the reflectance coefficients, rotation matrix
σ	scale factor, standard deviation
s	scale factor
s_{uv}	saturation, used with the $L^*u^*v^*$ color space
S	sensor response function, saturation
$\mathbf{S}$	vector of sensor's response functions, two- or three-dimensional scaling matrix
τ	time constant
θ	angle, polar angle, threshold
Θ	step or threshold function
t	time
T	temperature
$\mathbf{T}$	color transformation
u	value for bilinear interpolation
u^*	color component, used with the $L^*u^*v^*$ color space
$\mathbf{u}$	unit vector $\mathbf{u} = \frac{1}{3}[1, 1, 1]^T$
U	transformation to geometric mean chromaticity space
v	value for bilinear interpolation
v^*	color component, used with the $L^*u^*v^*$ color space
V	photopic function, voltage, value
V_M	modified photopic function

V'	scotopic function
$\mathbf{v}$	vertex, vector
w	factor for successive over-relaxation
w_i	weights
$\mathbf{w}$	white $\mathbf{w} = [1, 1, 1]^T$
$\bar{x}$	weight for a standard observer (used to compute X)
$\hat{x}$	chromaticity coordinate
X	one of the tristimulus values X, Y, and Z
x	generic coordinate, x-coordinate of image position
$\mathbf{x}$	position
y	y coordinate of image position
$\bar{y}$	weight for a standard observer (used to compute Y)
$\hat{y}$	chromaticity coordinate
Y	lightness, one of the tristimulus values X, Y, and Z, or yellow
L^*	lightness, used with the $L^*u^*v^*$ and $L^*a^*b^*$ color space
$\bar{z}$	weight for a standard observer (used to compute Z)
$\hat{z}$	chromaticity
Z	one of the tristimulus values X, Y, and Z

Index

Symbols
χ-chromaticities, 183
χ^2-divergence measure, 278, 287

A
absorbance characteristic, 10
absorbance curves, 71
absorbance of light, 39
achromatopsia, 26
adaptation, 84, 205
additive color generation, 68
additive color model, 68
adjusting saturation, 248
akinetopsia, 26
amacrine cell, 14
analog computation, 240
analog photography, 41
anisotropic diffusion, 257
anisotropic local space average color, 255
anti curl backing, 41, 43
anti halation layer, 41, 43
area of support, 229, 258
artificial retina, 241
artificial evolution, 198
artificial neuron, 194
artificially illuminated, 202

B
back-propagation, 195
backlight, 69
Bayer pattern, 46
bias, 240, 252

bidirectional reflectance distribution function (BRDF), 52
bilinear interpolation, 260
binarized shadow image, 215
biological neural network, 194
bipolar cell, 9, 14
bit string, 199
black and white film, 41
black-body radiator, 45, 56, 58, 126, 136, 140, 175
blind spot, 16
blob, 20
border of color space, 248
border of image, 157, 262
border pixels, 262
brain, 9

C
calibrated color space, 87
canonical gamut, 116, 117
capacitor, 234
capacity, 234
cathode-ray tube, 69
cell potential, 208
center-surround characteristic, 14
centered moment, 280
changing illumination, 279
chroma, 91, 93, 99, 101
chromaticities, 75
chromatopsia, 24
CIE, 69
classification, 205
cloudy day illumination, 56
CMY color space, 93
color blindness, 10, 13

color by correlation, 191
color classification, 209
color cluster rotation, 128, 195
color constancy, 1
color constancy measure, 295
color constant descriptors, 290, 324
color correction using filters, 45
color distribution, 276
color film, 42
color gamut, 69, 77
color histogram, 111, 275, 276, 279, 280, 282
color image enhancement, 170
color in perspective, 121
color interpolation, 46
color opponent cell, 30, 208, 326
color opponent mechanism, 16
color perception, 32
color pigment, 93
color print, 44
color ratios, 280
color reproduction, 67
color reversal film, 43
color sensation, 70
color shift, 241
color space of analog and digital video, 99
color spaces, 87
color temperature, 45
color transform, 79
committee-based color constancy, 197
component motion, 26
comprehensive color normalization, 129
computational theory of color constancy, 327
computational theory of color vision, 324
concentration gradient, 235
cone, 9
cone damage, 11
consumer photography, 3, 66, 67, 275, 327
continuity equation, 235
convergence, 132, 196, 208, 227–230, 236, 261, 267
convex hull, 116

convolution, 153, 155, 166, 169, 194, 220, 233, 236
corpus callosum, 26, 35
coupler compounds, 43
covariance matrix, 129, 180, 280
crossover, 202
cube root relationship, 32
current, 230
curvature, 264
curved line, 264

D

developer, 43
diagonal color transform, 84
diagonal map, 116, 126
dichromatic line, 136
dichromatic reflection model, 134
diffusion coefficient, 235
diffusion process, 235
digital camera, 46
digital photography, 46
Dirac delta function, 54, 62, 64, 83, 103, 104, 106, 116, 241, 322, 329
display device, 67, 68, 77, 79, 81–83
double opponent cell, 22, 30, 206, 208, 209
dye, 43
dynamic range compression, 169
dyschromatopsia, 26

E

eigenvectors for object recognition, 280
elementary function, 200
emulsion, 43
emulsion layer, 41
energy minimization, 210
enlarger light, 44, 215
estimating the illuminant locally, 219, 239
evaluation of color constancy algorithms, 275, 282
evolution, 198, 205
evolution strategy, 199
evolutionary algorithm, 198
evolving color constancy, 198
evolving computer programs, 198

executable program, 200
expected value, 109
experimental psychology, 303
exponential kernel, 221, 230
extended light source, 53
eye, 9
eye movements, 205
eyeball, 9

F
filling in derivatives, 215
film, 41
film development, 41
film exposure, 41
finite set of basis functions, 63, 84, 210
firing rate, 208
fitness function, 205
fixation, 42
flat panel display, 69
fluorescent illuminant, 176
flux, 47
Fourier transform, 170, 232, 233, 282
fovea, 13
frequency domain, 233
frequency space, 170

G
gamma correction, 79, 81, 240
gamut of display device, 68
gamut-constraint method, 115
ganglion cell, 9
gap junction, 14
Gaussian, 5, 137, 169, 170, 210, 215, 220, 236, 241, 324
gelatin, 41
genetic material, 198
genetic operator, 198, 202
genetic programming, 198
geometry factor, 61
global motion, 26
gray vector, 30, 94, 116, 128, 129, 196, 205, 241, 242, 246, 248, 250
gray world assumption, 106, 108, 109, 114, 132, 133, 211, 220, 221, 229, 239, 241, 251, 257

Green's function, 155
grid of processing elements, 221, 233

H
heuristics based algorithm, 191
hidden layer, 195
hippocampus, 29
histogram, 105
histogram-based object recognition, 275
histogram intersection, 276
homogeneous color space, 92
homogeneous diffusion, 236
homomorphic filtering, 170
horizontal cell, 14
HSI color space, 93
HSL color space, 96
HSV color space, 96
hue, 75, 91, 93, 94, 96, 205
human color constancy, 324, 327
human eye, 9
human visual system, 9, 174

I
illuminance, 331, 332
illuminants, 56
illumination gradient, 194, 258
image boundary, 229
image sequence, 188
imaging chip, 217, 241, 244
impulse, 154
inferior temporal cortex, 27, 29
inheritable trait, 198
inner nodes, 200
input layer, 195
integrated reflectance, 2, 32
intensity, 75, 87, 91, 93–95
interblob region, 20
International Commission on Illumination (CIE), 59, 70
interpolation methods, 259
intrinsic image, 175, 185, 213
invariant direction, 180
iris, 9
irradiance, 47, 332
isotropic local space average color, 258

J
Jacobi's method, 160

K
kernel, 220
Kirchhoff's law, 163, 230

L
Lab color space, 92
Lambertian radiator, 58
Lambertian surface, 54, 106
Laplace operator, Laplacian, 113, 154, 157, 190, 213, 281
latent image, 41, 43
lateral geniculate nucleus, 18, 21, 22
learning, 196
learning a linear filter, 193
learning algorithm, 195
learning color constancy, 193
lens, 9, 45
light adaptation, 16
light sensitive emulsion, 42
lighting matrix, 64
lightness, 31, 93, 154
lightness constancy, 16
line of constant illumination, 257
linear filter, 193
liquid crystal display (LCD), 69
local space average color, 219
locally estimating the illuminant, 219
log-chromaticity difference space, 180
logarithm, 113, 115, 153, 154, 157, 166, 170, 172, 174, 175, 177, 179, 189, 214, 281
logarithmic receptors, 148
logarithmic relationship, 32
luma, 99
luminance, 74, 80, 90, 99, 331, 332
luminous energy, 332
luminous exitance, 331, 332
luminous flux, 72, 332
luminous intensity, 331, 332
Luv color space, 89

M
Macbeth color checker, 237
Mach band, 153
magnocellular (M) cell, 18
matte object, 39
maximum deviation, 253
mesh of processing elements, 221
metamers, 35, 78
model image, 288
monitor, 69
monochromatic colors, 75, 77
monochromatic primaries, 69, 70
morphological operator, 214
motor neuron, 194
multi-dimensional histogram, 278
multiple illuminants, 109
multiple light sources, 53
multiple local light sources, 219
multiple scales, 169
Munsell book of colors, 33, 151, 303
mutation, 202

N
narrow band filter, 33, 34, 78
narrow band sensor, 64, 84, 103, 172, 175, 280, 282
nature, 205
negative image, 43
neighborhood, 220
neighbors, 224
network of resistors, 163
neural architecture, 206, 210
neural network, 194
neural network model, 208
neurons, 9
neutral-density filter, 37, 151
node, 194, 230
nonlinear change of illuminant, 255
nonlinear signal, 99
nontrivial output, 275
nonuniform illuminant, 219, 240
nonuniform illumination, 143
normalized color, 75
normalized color shift, 246
normalized histogram, 276
normalized space average color, 246
number of iterations, 229

INDEX

O
object recognition, 1, 3, 26, 29, 65, 185, 205, 275, 279, 282, 327
observed gamut, 116
ocular muscles, 9
ocular dominance column, 19
off-center cell, 14
on-center cell, 14
on–off ganglion cell, 14
on–off response, 14
opsin, 13
optical axis, 52
oscillatory behavior, 225
outer nodes, 200
output layer, 195

P
parallel architecture, 200
parallelization, 221
parameter optimization, 199
parietal cortex, 26
parse tree, 200
parvocellular (P) cell, 18
parvocellular-blob pathway, 27
parvocellular-interblob pathway, 27
perpendicular component, 242
phenomenological descriptions, 324
phosphor, 69, 79
photographic paper, 44
photometric units, 331
photometry, 331
photon, 10
photopic luminous efficiency function, 73, 74
photopic vision, 10, 70
photopigment, 10
photoreceptor, 13
plastic film base, 41, 43
point light source, 53
point of equal energy, 75
Poisson equation, 155, 161, 215
population, 198
positive image, 43
posterior parietal cortex, 27
power, 332
primaries, 69
primary color, 68
primary intensities, 69
principal axis, 196
principal component, 128, 129
principal component analysis, 63
printing, 68
processing element, 200, 222, 255
professional photographer, 45
projected coordinate, 179
pupil, 9
purple line, 75

R
radiance, 48, 332
radiant energy, 332
radiant exitance, 332
radiant flux, 72, 332
radiant intensity, 332
radiometry, 47, 331
radiosity, 48
radius of curve, 265
random path, 149
ratio, 144
receptive field, 14
receptor, 9
recognition rate, 288
reflectance, 2, 3, 13, 31
reflectance models, 52
reflected light, 39
refraction, 39
reproduction, 198, 202
resistance, 230, 231
resistive grid, 166, 167, 217, 230, 231, 244
resistor, 230
response characteristic, 62
response of photoreceptors, 326
resting potential, 208
retina, 2, 9
retinex theory, 143
RGB color space, 87
rod, 9
root mean squared error, 32, 197, 263, 282, 284, 299
rotated coordinates, 260

S

saturation, 75, 91, 93, 94, 96, 137, 170, 216, 248
scaling input values, 239
scaling of color channels, 64, 205
scanner, 46
scotomas, 23
scotopic vision, 10, 70
scratch resistive coating, 41
secondary color, 68
selection, 198
self-luminous colors, 90
sensor array, 46, 51
sensor response, 60
sensor sharpening, 62, 83
sequential product, 146
sequential sum, 148
shadow, 184, 190, 213
shadow brightening, 213, 215
shadow edge, 190, 214
shadow removal, 213
sharpening, 175
shifted white point, 248
shiny object, 39
sigmoidal function, 195, 227, 253, 284
silver halide crystals, 41
single instruction multiple data (SIMD), 201
single-lens reflex camera, 46
sky, 53
smoothly changing illuminant, 255
smoothly varying illuminant, 219
Snell's law, 40
solid angle, 48, 53
spatial domain, 233
spatially varying illuminants, 219
spotlight, 258, 263, 284
square root relationship, 32
sRGB color space, 87, 240
standard observer, 70
standardized color space, 67
sub-tree, 202
subtract space average color, 251
subtractive color generation, 68
subtractive color model, 68

successive over-relaxation (SOR), 160, 229, 261
surface colors, 92
symbolic expression, 200

T

temporal visual cortex, 26
terminal symbols, 200
test image, 287
theoretical models, 320
theory of evolution, 198
threshold, 149, 152, 158, 159, 195, 240
threshold operation, 113, 150, 159, 213
threshold operator, 154
training vectors, 193
tree structure, 200
tree-based genetic programming, 200

U

uniform coordinate system, 92
uniform illumination, 103

V

V1, 19, 21, 23–27, 29
V2, 23–25, 27, 36
V3, 23–26
V3A, 25
V4, 23–26, 29, 36, 324
V5, 23, 26, 27
variant of Horn's algorithm, 113
variation, 198
video color space, 99
visual cortex, 3, 6, 16, 35
visual system, 9
voltage, 231
von Kries coefficients, 83, 84, 322

W

weight, 194
white balance, 2, 105, 110
white patch retinex algorithm, 104, 165, 251

X

XYZ color space, 70

Permissions

Figures	Permission
1.2, 7.4, 7.5	Reproduced from Land EH and McCann JJ 1971 Lightness and retinex theory. Journal of the Optical Society of America 61(1), 1–11, by permission from The Optical Society of America.
2.1, 2.2, 2.5, 2.7, 8.2	Contain illustrations from LifeArt Collection Images © 1989–2001 by Lippincott Williams & Wilkins used by permission from SmartDraw.com.
2.3	Reprinted with permission from Gordon L. Fain and John E. Dowling. Intracellular recordings from single rods and cones in the mudpuppy retina. Science, Vol. 180, pp. 1178–1181, June, Copyright 1973 AAAS
2.5	Frontal view of retina from LifeART Collection Images © 1989–2001 by Lippincott Williams & Wilkins used by permission from SmartDraw.com.
2.8	Reproduced from R. A. Normann and I. Perlman. The effects of background illumination on the photoresponses of red and green cones. Journal of Physiology, Vol. 286, pp. 491–507, 1979, by permission of Blackwell Publishing, UK.
2.9	Reproduced from Semir Zeki. A Vision of the Brain. Blackwell Science, Oxford, 1993, by permission of Blackwell Science, UK.
2.10	Copyright 1984 by the Society for Neuroscience. Reproduced by permission of Society for Neuroscience. M. S. Livingstone and David H. Hubel. Anatomy and physiology of a color system in the primate visual cortex. The Journal of Neuroscience, Vol. 4, No. 1, pp. 309–356, Jan., 1984.
2.11, 2.12	Reproduced from D. H. Hubel and T. N. Wiesel. Receptive fields, binocular interaction and functional architecture in the cat's visual cortex. Journal of Physiology, Vol. 160, pp. 106–154, 1962, by permission of Blackwell Publishing, UK.
2.14, 2.16	Reproduced from S. Zeki. An Exploration of Art and the Brain. Oxford University Press, Oxford, 1999 by permission of S. Zeki, University College London, UK.

Figures	Permission
2.15	Redrawn from Figure 1 (page 1373) S. Zeki and A. Bartels. The clinical and functional measurement of cortical (in)activity in the visual brain, with special reference to the two subdivisions (V4 and V4α) of the human colour centre. Proceedings of the Royal Society of London. Series B, 354, pp. 1371–1382, The Royal Society, 1999, used by permission.
2.17	Car image from SmartDraw used by permission from SmartDraw.com.
2.18	Reproduced by permission from Massachusetts Institute of Technology. D. C. Van Essen and E. A Deyoe. Concurrent Processing in the Primate Visual Cortex. M. S. Grazzanida (ed.), The Cognitive Neurosciences, The MIT Press, pp. 383–400, 1995.
2.19	Reprinted with permission from F. A. W. Wilson, S. P. Ó Scalaidhe, and P. S. Goldman-Rakic. Dissociation of object and spatial processing domains in primate prefrontal cortex. Science, Vol. 260, pp. 1955–1958, June, Copyright 1993 AAAS.
2.20	Reprinted from Vision Res., Vol. 26, No. 1, Edwin H. Land, Recent advances in retinex theory. pp. 7–21, Copyright 1986, with permission from Elsevier
2.23, 2.24	Adapted by permission from Macmillan Publishers Ltd: Nature, E. H. Land, D. H. Hubel, M. S. Livingstone, S. Hollis Perry, M. M. Burns, Nature, Vol. 303, No. 5918, pp. 616–618, Copyright 1983.
3.3	Illustration on page 41 based on Dorling Kindersley (©) 2003 illustration used with permission
3.4	Illustration on page 42 based on Dorling Kindersley (©) 2003 illustration used with permission
3.5	Reproduced by permission of FUJIFILM. Ref. No. AF3-148E (EIGI-03.4-HB.10-1) No. 21 Ref. No. AF3-024E (EIGI-01.11-HB.3-2) No. 19
3.6	Reproduced by permission of FUJIFILM. Ref. No. AF3-024E (EIGI-01.11-HB.3-2) No. 15
3.7	Reproduced by permission of FUJIFILM. Ref. No. AF3-148E (EIGI-03.4-HB.10-1) No. 23 Ref. No. AF3-024E (EIGI-01.11-HB.3-2) No. 21
4.9, 6.29	Reproduced by permission of Pearson Education from 3D Computer Graphics Third Edition, Alan Watt, Pearson Education Limited, © Pearson Education Limited 2000
5.10, 5.12	Reprinted from Digital Video and HDTV. Algorithms and Interfaces, Charles Poynton, Morgan Kaufmann Publishers, San Francisco, CA, Copyright 2003, with permission from Elsevier
7.8	Reproduced by permission of The MIT Press. Horn BKP 1986 Robot Vision, The MIT Press, Cambridge, MA
7.11	Reprinted from Computer Graphics and Image Processing, Vol. 3, Berthold K. P. Horn, Determining Lightness from an Image, pp. 277–299. Copyright 1974 with permission from Elsevier

PERMISSIONS

Figures	Permission
7.17, 7.19	Reproduced by permission of IEEE. Moore A and Allman J and Goodman RM 1991 A real-time neural system for color constancy. IEEE Transactions on Neural Networks, IEEE, 2(2), 237–247, March
7.25	Reproduced by permission of IEEE. Faugeras OD 1979 Digital color image processing within the framework of a human visual model. IEEE Transactions on Acoustics, Speech and Signal Processing, ASSP-27(4) 380–393
8.1	Data kindly supplied by Anya Hurlbert, University of Newcastle, UK
8.9	Reproduced from Ebner M. 2006 Evolving color constancy. Special issue on evolutionary computer vision and image understanding of pattern recognition letters, Elsevier, 27(11), 1220–1229, by permission from Elsevier
8.10, 8.11	Reproduced from D'Zmura M and Lennie P 1986. Mechanisms of color constancy. Journal of the Optical Society of America A, 3(10), 1662–1672, by permission from The Optical Society of America
8.12	Redrawn from Figure 3 (page 298) from Dufort PA and Lumsden CJ 1991 Color categorization and color constancy in a neural network model of V4. Biological Cybernetics, Springer-Verlag, Vol. 65, pp. 293–303, Copyright Springer-Verlag 1991, with kind permission from Springer Science and Business Media
8.13	Redrawn from Figure 6ab (page 299) from Dufort PA and Lumsden CJ 1991 Color categorization and color constancy in a neural network model of V4. Biological Cybernetics, Springer-Verlag, Vol. 65, pp. 293–303, Copyright Springer-Verlag 1991, with kind permission from Springer Science and Business Media
8.14	Redrawn from Figure 8.4.1 (page 477) Usui S and Nakauchi S 1997 A neurocomputational model for colour constancy. In (eds. Dickinson C, Murray I and Carded D), John Dalton's Colour Vision Legacy. Selected Proceedings of the International Conference, Taylor & Francis, London, pp. 475–482, by permission from Taylor & Francis Books, UK
10.17, 12.14, 13.6, D.1, D.2, D.3, D.4, D.5	Original image data from "Data for Computer Vision and Computational Colour Science" made available through http://www.cs.sfu.ca/~colour/data/index.html. See Barnard K, Martin L, Funt B and Coath A 2002 A data set for color research, Color Research and Application, Wiley Periodicals, 27(3), 147–151. Reproduced by permission of Kobus Barnard